VOLUME NINETY NINE

ADVANCES IN VIRUS RESEARCH

In Loeffler's Footsteps – Viral Genomics in the Era of High-Throughput Sequencing

VOLUME NINETY NINE

ADVANCES IN VIRUS RESEARCH

In Loeffler's Footsteps – Viral Genomics in the Era of High-Throughput Sequencing

Edited by

MARTIN BEER

Institute of Diagnostic Virology,
Friedrich-Loeffler-Institut,
Greifswald-Insel Riems, Germany

DIRK HÖPER

Institute of Diagnostic Virology,
Friedrich-Loeffler-Institut,
Greifswald-Insel Riems, Germany

ELSEVIER

ACADEMIC PRESS
An imprint of Elsevier

Academic Press is an imprint of Elsevier
50 Hampshire Street, 5th Floor, Cambridge, MA 02139, United States
525 B Street, Suite 1800, San Diego, CA 92101-4495, United States
The Boulevard, Langford Lane, Kidlington, Oxford OX5 1GB, United Kingdom
125 London Wall, London, EC2Y 5AS, United Kingdom

First edition 2017

ISBN: 978-0-12-812598-4
ISSN: 0065-3527

Publisher: Zoe Kruze
Acquisition Editor: Ashley M. Jackman
Senior Editorial Project Manager: Helene Kabes
Production Project Manager: James Selvam
Senior Cover Designer: Greg Harris

Typeset by SPi Global, India

CONTENTS

CONTRIBUTORS

Houssam Attoui
UMR1161 Virologie, ANSES, INRA, Ecole Nationale Vétérinaire d'Alfort, Université Paris-Est, Maisons-Alfort, France

Martin Beer
Institute of Diagnostic Virology, Friedrich-Loeffler-Institut, Greifswald-Insel Riems, Germany

Guillaume Belbis
UMR1161 Virologie, ANSES, INRA, Ecole Nationale Vétérinaire d'Alfort, Université Paris-Est, Maisons-Alfort, France

Sandra Blome
Institute of Diagnostic Virology, Friedrich-Loeffler-Institut, Greifswald-Insel Riems, Germany

Christiane Bouchier
Institut Pasteur, Genomics Platform, Paris, France

Céline L. Boujon
University of Bern, Bern, Switzerland

Hervé Bourhy
Institut Pasteur, Unit Lyssavirus Dynamics and Host Adaptation, WHO Collaborating Centre for Reference and Research on Rabies, Paris, France

Emmanuel Bréard
UMR1161 Virologie, ANSES, INRA, Ecole Nationale Vétérinaire d'Alfort, Université Paris-Est, Maisons-Alfort, France

Grégory Caignard
UMR1161 Virologie, ANSES, INRA, Ecole Nationale Vétérinaire d'Alfort, Université Paris-Est, Maisons-Alfort, France

Laurent Dacheux
Institut Pasteur, Unit Lyssavirus Dynamics and Host Adaptation, WHO Collaborating Centre for Reference and Research on Rabies, Paris, France

Stephan Drewes
Institute of Novel and Emerging Infectious Diseases, Friedrich-Loeffler-Institut, Greifswald-Insel Riems, Germany

Jan F. Drexler
Charité—Universitätsmedizin Berlin, Corporate Member of Freie Universität Berlin, Humboldt-Universität zu Berlin, and Berlin Institute of Health, Institute of Virology, Berlin; German Center for Infection Research (DZIF), Germany

Elisa Eggerbauer
Institute of Molecular Virology and Cell Biology, WHO Collaborating Centre for Rabies Surveillance and Research, Friedrich-Loeffler-Institut, Greifswald-Insel Riems, Germany

Conrad M. Freuling
Institute of Molecular Virology and Cell Biology, WHO Collaborating Centre for Rabies Surveillance and Research, Friedrich-Loeffler-Institut, Greifswald-Insel Riems, Germany

Stephanie Haberl
Bavarian Health and Food Safety Authority, Oberschleißheim, Germany

Martin Hölzer
RNA Bioinformatics and High Throughput Analysis, Faculty of Mathematics and Computer Science, Friedrich Schiller University Jena; European Virus Bioinformatics Center (EVBC), Jena, Germany

Dirk Höper
Institute of Diagnostic Virology, Friedrich-Loeffler-Institut, Greifswald-Insel Riems, Germany

Jens Jacob
Julius Kühn-Institute, Federal Research Centre for Cultivated Plants, Institute for Plant Protection in Horticulture and Forests, Vertebrate Research, Münster, Germany

Michel C. Koch
University of Bern, Bern, Switzerland

Manja Marz
RNA Bioinformatics and High Throughput Analysis, Faculty of Mathematics and Computer Science, Friedrich Schiller University Jena; European Virus Bioinformatics Center (EVBC); FLI Leibniz Institute for Age Research, Jena, Germany

Thomas C. Mettenleiter
Institute of Molecular Virology and Cell Biology, WHO Collaborating Centre for Rabies Surveillance and Research, Friedrich-Loeffler-Institut, Greifswald-Insel Riems, Germany

Thomas Müller
Institute of Molecular Virology and Cell Biology, WHO Collaborating Centre for Rabies Surveillance and Research, Friedrich-Loeffler-Institut, Greifswald-Insel Riems, Germany

Antonie Neubauer-Juric
Bavarian Health and Food Safety Authority, Oberschleißheim, Germany

Karsten Passior
Local Bat Conservation, Lower Saxony, Nordstemmen, Germany

Florian Pfaff
Institute of Diagnostic Virology, Friedrich-Loeffler-Institut, Greifswald-Insel Riems, Germany

Corinne Sailleau
UMR1161 Virologie, ANSES, INRA, Ecole Nationale Vétérinaire d'Alfort, Université Paris-Est, Maisons-Alfort, France

Torsten Seuberlich
University of Bern, Bern, Switzerland

Petra Straková
Institute of Vertebrate Biology v.v.i., Academy of Sciences, Brno, Czech Republic

Cécile Troupin
Institut Pasteur, Unit Lyssavirus Dynamics and Host Adaptation, WHO Collaborating Centre for Reference and Research on Rabies, Paris, France

Rainer G. Ulrich
Institute of Novel and Emerging Infectious Diseases, Friedrich-Loeffler-Institut; German Center for Infection Research (DZIF), Partner site Hamburg-Luebeck-Borstel-Insel Riems, Greifswald-Insel Riems, Germany

Damien Vitour
UMR1161 Virologie, ANSES, INRA, Ecole Nationale Vétérinaire d'Alfort, Université Paris-Est, Maisons-Alfort, France

Kerstin Wernike
Institute of Diagnostic Virology, Friedrich-Loeffler-Institut, Greifswald-Insel Riems, Germany

Claudia Wylezich
Institute of Diagnostic Virology, Friedrich-Loeffler-Institut, Greifswald-Insel Riems, Germany

Stéphan Zientara
UMR1161 Virologie, ANSES, INRA, Ecole Nationale Vétérinaire d'Alfort, Université Paris-Est, Maisons-Alfort, France

CHAPTER ONE

The First "Virus Hunters"

Thomas C. Mettenleiter[1]
Institute of Molecular Virology and Cell Biology, WHO Collaborating Centre for Rabies Surveillance and Research, Friedrich-Loeffler-Institut, Greifswald-Insel Riems, Germany
[1]Corresponding author: e-mail address: thomas.mettenleiter@fli.de

Contents

Abstract

The history of virology is a history of conceptual and technological inventions and breakthroughs. The development of filters made of porcelain or kieselgur by the end of the 19th century which withheld bacteria allowed the identification of infectious agents smaller than bacteria and noncultivable on the media known at that time and used to grow bacteria. Even finer-grain filters resulted in the observation that the ultravisible novel infectious agents are in fact of particulate nature. Infections of plants and animals were the first to be attributed to these tiny entities. Proof resulted from experimental infection of the natural hosts (including humans). Thus, of the first 30 viruses identified, 20 are veterinary viruses, i.e. infectious agents of poultry and livestock. The discovery that bacteria also have viruses in the 1910s expanded the viral universe which continues today. Filterability and ultravisibility remained a hallmark for the identification of viruses until the advent of the electron microscope in the late 1930s marking another technological breakthrough in virology. Cell culture techniques allowed virus propagation outside the infected organism. In the past decades, the advent and development of molecular biology has brought more innovations culminating in the rapid and accurate determination of genomic material of a variety of living beings including viruses in a hitherto unknown speed and depth using next-generation sequencing and metagenomic analyses. Thus, it is no surprise that new viruses are detected constantly including specimens of unprecedented size and shape. Virologists agree that the viral universe is immense, and only a small fraction has been explored yet.

Advances in Virus Research, Volume 99
ISSN 0065-3527
https://doi.org/10.1016/bs.aivir.2017.07.005

1. INTRODUCTION

Infections now known to be caused by viruses have been recorded since the beginning of historical documentation. Polio and rabies are among the earliest descriptions of these infections demonstrating their association with humans and livestock for millennia. The devastating epidemics of variola from the middle ages into modern times sparked Edward Jenner's attempts to prevent lethal disease by inoculation of humans with "benevolent" infectious material from cows (lat: *vacca*), thereby establishing a common preventive principle, vaccination. This was followed up by Louis Pasteur who obtained rabies virus vaccine after repeated passaging in rabbits (DiMaio, 2016).

In a previous issue of Advances in Virus Research, former series editor Fred A. Murphy reviewed in a "Historical Perspective" what discovery of a new virus actually constitutes (Murphy, 2016). When looking into textbooks the early times of virus discovery usually only get cursory attention. However, since this essentially set the stage for all subsequent virological research, culminating in the current explosion in the identification of novel viruses and viral genomic sequences by new technologies, e.g., next-generation sequencing and metagenomic analysis, it is worth of (re-) evaluation how the principle of viruses as infectious entities separate from bacteria and with specific properties developed.

2. THE FIRST "VIRUS HUNTERS"

Virology started with a technological breakthrough, the development of porcelain and kieselgur (diatomacious earth) filters for sterilization of bacteria-containing material (reviewed in Mudd, 1928). Charles Edouard Chamberland (1851–1908) introduced his porcelain filters in 1884 working together with Louis Pasteur (Chamberland, 1884). These filters were produced in graded porosities from L1 (most coarse) to L13 (finest). Kieselgur filters which use a similar principle were invented and marketed by Wilhelm Berkefeld (1836–1897). A particularly fine porcelain filter had been developed by Shibasaburo Kitasato (1853–1931) for his studies, together with Robert Koch, on the tetanus toxin (Kitasato, 1891).

It was the Russian scientist Dimitri Iwanowski (1864–1920) who first analyzed the causative agent of tobacco mosaic disease by using these newly developed filters. He demonstrated that the agent causing this disease which had been shown to be infectious by Adolf Mayer (1843–1942), a German chemist working in Wageningen in the Netherlands (Mayer, 1886), passed through the bacteria-tight filters (Iwanowski, 1892). Thus, he first described "filterability" as a hallmark of the agent (Lvov, 1993). However, until his death he did not realize that he actually discovered a new world of pathogens but rather believed conservatively that filterability does not reflect a novel class of pathogens but simply considered the "filterable" agent of tobacco mosaic disease as a different, smaller form of bacteria. Thus, he was unable to grasp the fundamental change in concept his filtration experiments were indicative for.

The Dutch scientist Martinus Wilhelmus Beijerinck (1851–1931) also studied tobacco mosaic disease and was interested in identification of the causative agent. Apparently unaware of the previous work by Iwanowski, he essentially repeated his experiments with the same results, i.e., unhindered passage of the agent through Chamberland filters. In contrast to Iwanowski, however, Beijerinck understood that he discovered something novel, not resembling any of the known biological entities. He coined the term "contagium vivum fluidum" to describe this novel biological principle, which can be translated into "infectious liquid" (Beijerinck, 1898). He used this new term prominently in the title of his seminal publication highlighting the importance he attributed to this finding. *Contagium vivum fluidum* was used as opposed to *contagium vivum fixum*, which would indicate a particulate, corpuscular agent.

However, viruses are clearly no "infectious liquids" and that they are in fact ultrafilterable, but particulate was correctly observed for the first time by Friedrich Loeffler (1852–1915) (Fig. 1) working with Paul Frosch (1860–1928) on the agent causing foot-and-mouth disease in cloven-hoofed animals using a combination of Chamberland and the newly developed fine-grain Kitasato filters (Horzinek, 1995). The latter technological breakthrough was available to them since Shibasaburo Kitasato who worked in the laboratory of Robert Koch was familiar with the research of Loeffler and Frosch (both also former and present coworkers of Koch). Thus, technology was instrumental in defining this new class of infectious agents and Loeffler and Frosch proved its filterability, in vivo replication and corpuscular nature in their seminal 1898 "Third Report to the Prussian Minister of Culture," dated January 8 (published on February 10 in the *Deutsche Medizinische Wochenschrift* and on March 10 in the *Zentralblatt für Bakteriologie*). It is interesting that Loeffler and Frosch also went further and hypothesized that this new concept might also apply to other infectious diseases caused

Fig. 1 Friedrich Loeffler (*left*) and Robert Koch, ca. 1880.

by hitherto unknown microbes. They listed variola, cowpox, scarlet fever, measles, spotted typhus, and rinderpest as presumably being caused by filterable viruses (Loeffler and Frosch, 1898). Retrospectively, they were correct in four of the six examples but, more importantly, this testifies to Loeffler and Frosch's recognition that a new class of infectious agents had indeed been found.

In summary, Beijerinck, in contrast to Iwanowski, was clearly aware that he had come across a change in concept, i.e., filterable novel infectious agents (in this context he first used the latin term "*virus*"), and has to be credited by the notion that "the contagion, to reproduce itself, must be incorporated into the living cytoplasm of the cell, into whose multiplication, it is, as it were, passively drawn" (Horzinek, 1995). However, his *contagium vivum fluidum* did not reflect the nature of the novel "virus." Thus, Witz (1998) and Van Regenmortel (2010) summarize the available evidence to conclude that Loeffler and Frosch's interpretation of their filtration experiments "came much closer to the modern concept of a virus than anybody else at the time" (Witz, 1998) and "so they should be acknowledged as the founders of virology" (Van Regenmortel, 2010) which has found support more recently also by Murphy (2016). Interestingly, neither Loeffler nor Frosch followed this up by more detailed studies on the nature of this agent. They were only interested in finding a prevention or cure for foot-and-mouth disease (Loeffler, 1898) with only limited success. Their serum therapy was costly and unreliable and was actually discouraged from being used (Fröhner, 1912).

Unfortunately and incorrectly, this conceptual leap which actually led to the foundation of a new discipline, virology, is still not universely attributed to Loeffler and Frosch but rather to Beijerinck or even Iwanowski (see Bos, 1995; Enquist and Racaniello, 2013; Lvov, 1993; Wilkinson, 2001).

3. THE HENLE–LOEFFLER–KOCH POSTULATES

Loeffler and Frosch as well as their contemporaries were unable to fulfill the recently established Henle–Loeffler–Koch postulates. In fact, while Jacob Henle (1809–1885) has given the conceptual background (and coined the term "contagium vivum" for potentially infectious microorganisms), this was eventually taken up by Koch and Loeffler to establish principles which are still valid today. Loeffler, in his work *On the importance of microorganisms for the development of diphtheria in humans, dove, and calves*, formulated and numbered them for the first time. He writes that *if also diphtheria is a disease caused by microorganisms, those three postulates have to be fulfilled whose fulfilment is absolutely required for the strict proof of the parasitic nature of any such disease.* He phrased the postulates as: (1) *There have to be constantly detected in the local diseased areas organisms in typical arrangement.* (2) *The organisms, to which by their association to the diseased parts an importance for the development of these alterations has to be assigned, must be isolated and cultivated in pure.* (3) *With the pure culture the disease must be reproduced.*

However, Loeffler and Frosch were aware that they could not actually prove their concept along these postulates since a pure, extracorporal culture of the new filterable agents was not possible. Subsequently, the classical postulates have since been revised repeatedly to accommodate the specific nature of viruses including recognition and incorporation of molecular data which are relevant for today's "virus hunting" (Fredericks and Relman, 1996; Mokili et al., 2012).

Whereas Iwanowski published his studies in 1892, both Beijerinck's and Loeffler and Frosch's results appeared in print in 1898. So, which year would mark the "beginning of virology" (Horzinek, 1995)? Based on the above concept, I am convinced that 1898 should be labeled the year when virology started and, consequently, symposia commemorating 100 years of virology were organized in 1998 in Greifswald as well as in Amsterdam (Calisher and Horzinek, 1999). However, this has been disputed and the start of "virology" as a separate scientific discipline has been suggested to be placed into the 1950s based on the unifying concept combining "filterability" with the "eclipse phase" viruses show after infection of their host (Van Helvoort, 1996).

4. THE WORLD'S FIRST VIRUS RESEARCH INSTITUTE: GREIFSWALD/INSEL RIEMS

Working with highly contagious infectious agents was hazardous (and still is). Since Loeffler and Frosch were totally dependent on animal experimentation to demonstrate infectivity in their experimental assays, they used pigs and cattle as the natural hosts of foot-and-mouth disease virus. First studies were carried out in Berlin in sheds built into two railway arches of the Berlin local train network (Hinz-Wessels and Thiel, 2010). Subsequently, when Loeffler had returned to his chair of hygiene at Greifswald university which he occupied since 1888, he continued his experiments in the urban center of the city. However, since the smell, in particular of the porcine experimental animals, disturbed the neighbors, he was forced to move to the rural, agricultural outskirts of the city. Here, his experiments, which were performed in a rented barn (Fig. 2) under less than ideal conditions, resulted in repeated foot-and-mouth disease outbreaks in the region culminating in the phrase *Germany could be free of Foot-and-Mouth Disease if there weren't Professor Loeffler's experiments* written by local farmers to the department of agriculture in Berlin. In 1907, the ministerial office eventually forbade Loeffler all further experimentation unless he could find a site where biosecurity could be significantly enhanced. This did not come unexpected for Loeffler, who had written already in 1906 that *the location of a future institute would have to be such that it could by itself completely exclude any spread of the infective matter. Undoubtedly, an island would be best suited for this purpose.* Thus, the island concept of a biosecure site was initiated, later followed by virus research institutes on Lindholm island in Denmark and Plum Island in the United States of America. Loeffler settled on the nearby island of Riems (Insel Riems) and informed the Prussian Department (Ministry) of Spiritual,

Fig. 2 Loeffler's experimental animal facility in Greifswald.

Educational, and Medical Affairs on the start of his experimental studies by writing *Your Excellency, I obediently report that, immediately after the arrival of a fresh lymph sample from Witkowo on Monday, 10 October, I commenced work on the Isle of Riems*. This was 1910, so 10/10/10 is the birthday of the world's first and oldest virus research institute (Hinz-Wessels and Thiel, 2010) which was named after his founder in 1952. Interestingly, one of the boats shuttling staff between mainland and island carried the name "Virusjäger" (virus hunter) (Fig. 3) and Loeffler, who never obtained the Nobel prize for his groundbreaking discovery (nor did any of the first virus hunters; Norrby, 2008), was commemorated with his portrait on a stamp (his first ever recognition on a stamp compared to the plethora of stamps recognizing Pasteur or Koch) (Fig. 4) issued by the German Federal ministry of finance in 2010 celebrating the centenary of the institute he founded (Fig. 5).

5. EARLY VIRUS DISCOVERY

Filterability remained the differentiating feature for the new class of infectious agents, setting them apart from bacteria. Since these novel agents were "uncultivable" extracorporally, proof of infectivity relied on infection experiments in the relevant hosts. Thus, the filterability of yellow fever virus was actually proven by infection of healthy human volunteers using *bacteria-free serum filtrate obtained by passing the diluted serum through a Berkefeld laboratory filter* (Reed, 1902). Interestingly, Reed (1902) also comments that his experiments *indicate that yellow fever, like the foot-and-mouth disease of cattle, is caused*

Fig. 3 "Virusjäger"—Virus hunter. A motor launch in use after WW II for transport of staff to and from the institute on Insel Riems.

Fig. 4 Stamp issued in Oct 2010 by the German Federal ministry of finance commemorating the centenary of the foundation of the Friedrich-Loeffler-Institut.

Fig. 5 The Friedrich-Loeffler-Institut on Insel Riems today.

by a micro-organism so minute in size that it might be designated as ultra-microscopic demonstrating that he was aware of Loeffler and Frosch's studies. Not surprisingly, thus, 20 of the first 31 described "filterable viruses" were animal pathogens (Table 1).

Table 1 The First "Filterable Viruses"

1	1892/1898	D. Ivanovsky, M. Beijerinck	Tobacco mosaic virus (the first virus?)
2	1898	F. Loeffler, P. Frosch	Foot-and-mouth disease virus (the first virus?, the first animal virus)
3	1898	G. Sanarelli	Myxoma virus (the first poxvirus)
4	1900	W. Reed, J. Carroll, A. Agramonte, J. Lazear	Yellow fever virus (the first human virus, the first arbovirus, the first flavivirus)
5	1900	J. M'Fadyean, T. Edgington, A. Theiler	African horse sickness virus (the first orbivirus)
6	1901	E. Centanni, E. Savonuzzi, A. Lode, J. Gruber	Fowl plague virus (avian influenza virus, the first orthomyxovirus)
7	1902	A. Aujeszky	Pseudorabies virus (the first herpesvirus)
8	1902	A. Borrel	Sheeppox virus
9	1902	M. Nicolle, M. Adil-Bey	Rinderpest virus (the first morbillivirus)
10	1902	J. Spruell, A. Theiler	Bluetongue viruses
11	1903	M. Dorset, E. De Schweinitz	Classical swine fever virus (hog cholera virus, the first pestivirus)
12	1903	M. Remlinger, Riffat-Bay, A. di Vestea	Rabies virus (the first rhabdovirus)
13	1904	H. Vallée, H. Carré	Equine infectious anemia virus (the first retrovirus)
14	1905	H. Carré, P. Laidlaw, G. Dunkin	Canine distemper virus
15	1907	P. Ashburn, C. Craig	Dengue viruses
16	1908	V. Ellermann, O. Bang	Avian leukemia virus (the first leukemia virus)
17	1909	K. Landsteiner, E. Popper	Poliovirus (the first enterovirus)
18	1909	R. Doerr, K. Franz, S. Taussig	Sandfly fever viruses (the first phleboviruses)

Continued

Table 1 The First "Filterable Viruses"—cont'd

19	1911	J. Goldberger, J. Anderson	Measles virus
20	1911	P. Rous, J. Beard	Rous sarcoma virus (the first solid tumor virus)
21	1914	A. Hess, Y. Hiro, S. Tasaka	Rubella virus (the only rubivirus)
22	1917	F. Twort, F. d'Herelle	Bacteriophages
23	1917	R. Montgomery	Nairobi sheep disease virus (the first nairovirus, the first tick-borne virus)
24	1918	A. Breinl, J. Cleland, E. French	Murray Valley encephalitis virus
25	1919	A. Löwenstein	Herpes simplex virus
26	1921	R. Montgomery	African swine fever virus (the only asfarvirus)
27	1925	K. Kundratitz, H. Ruska, T. Weller, M. Stoddard	Varicella-zoster-virus—chickenpox and zoster
28	1926	W. Cotton, P. Olitsky, J. Traum, H. Schoening	Vesicular stomatitis viruses
29	1926	F. Kraneveld, T. Doyle	Newcastle disease virus (the first paramyxovirus)
30	1928	J. Verge, N. Christoforoni	Feline panleukopenia virus (the first parvovirus)

Light gray, plant viruses; *medium gray*, animal viruses; *dark gray*, human viruses; *darkest gray*, bacteriophages. Murphy, F., 2014. The Foundations of Medical and Veterinary Virology: Discoverers and Discoveries, Inventors and Inventions, Developers and Technologies. http://www.utmb.edu/virusimages/.

In 1928, Thomas M. Rivers edited a series of reviews in the book "Filterable Viruses" (Fig. 6) after writing an extensive review on the same topic 1 year earlier (Rivers, 1927). Dedicated *To the memory of CHARLES E. SIMON in evidence of appreciation of his leadership and foresight in the development in the United States of the first Department of Filterable Viruses*, herein he summarized in a table the knowledge on filterable viruses that had accumulated in the intervening 30 years since Beijerinck's and Loeffler and Frosch's groundbreaking experiments (Fig. 7). In his first chapter on general aspects of filterable viruses, he references 539 publications! Although there are, of course, knowledge gaps and misinterpretations (e.g., mycoplasms had not

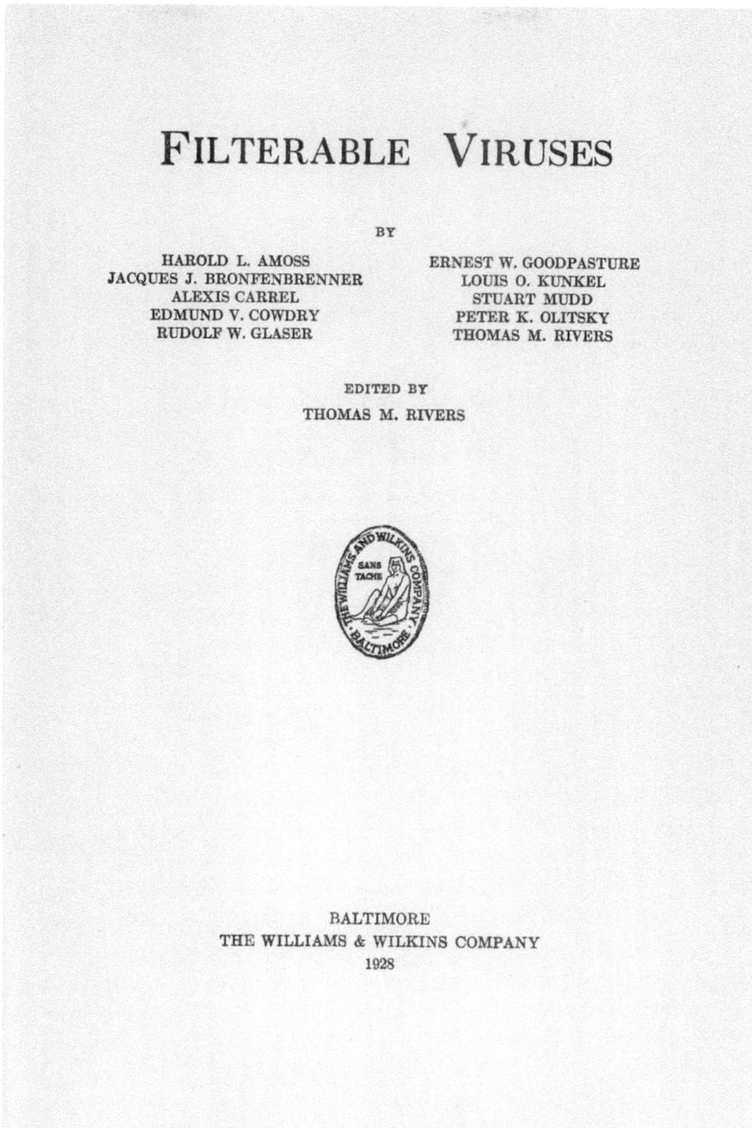

FILTERABLE VIRUSES

BY

HAROLD L. AMOSS
JACQUES J. BRONFENBRENNER
ALEXIS CARREL
EDMUND V. COWDRY
RUDOLF W. GLASER

ERNEST W. GOODPASTURE
LOUIS O. KUNKEL
STUART MUDD
PETER K. OLITSKY
THOMAS M. RIVERS

EDITED BY
THOMAS M. RIVERS

BALTIMORE
THE WILLIAMS & WILKINS COMPANY
1928

Fig. 6 "Title page" of: Rivers, T.M. (Ed.), 1928. Filterable Viruses. The Williams & Wilkins Co., Baltimore.

yet been recognized), the plethora of information is remarkable. The book also contains chapters on filters and filtration and on the use of tissue cultures in the study of viruses and summarizes the information on specific virus infections such as poliomyelitis, foot-and-mouth disease, fowlpox, insect

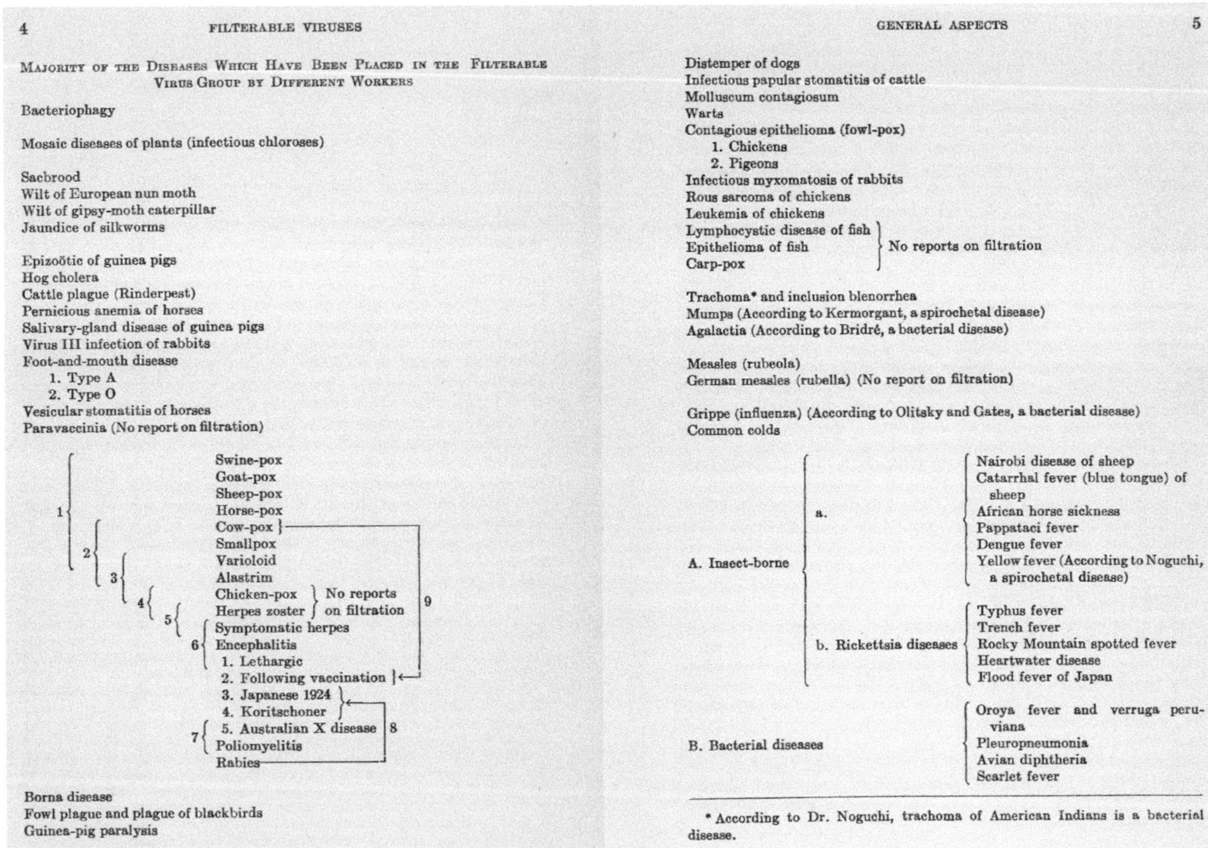
4 FILTERABLE VIRUSES

MAJORITY OF THE DISEASES WHICH HAVE BEEN PLACED IN THE FILTERABLE VIRUS GROUP BY DIFFERENT WORKERS

Bacteriophagy

Mosaic diseases of plants (infectious chloroses)

Sacbrood
Wilt of European nun moth
Wilt of gipsy-moth caterpillar
Jaundice of silkworms

Epizoötic of guinea pigs
Hog cholera
Cattle plague (Rinderpest)
Pernicious anemia of horses
Salivary-gland disease of guinea pigs
Virus III infection of rabbits
Foot-and-mouth disease
1. Type A
2. Type O
Vesicular stomatitis of horses
Paravaccinia (No report on filtration)

1 { Swine-pox, Goat-pox, Sheep-pox, Horse-pox, 2 { Cow-pox, 3 { Smallpox, Varioloid, Alastrim, 4 { Chicken-pox, Herpes zoster } No reports on filtration
5 { Symptomatic herpes, 6 { Encephalitis
1. Lethargic
2. Following vaccination
3. Japanese 1924
4. Koritschoner
7 { 5. Australian X disease, Poliomyelitis
Rabies
9 (Cow-pox → 2. Following vaccination)
8 (Rabies → 3. Japanese 1924, 4. Koritschoner)

Borna disease
Fowl plague and plague of blackbirds
Guinea-pig paralysis

GENERAL ASPECTS 5

Distemper of dogs
Infectious papular stomatitis of cattle
Molluscum contagiosum
Warts
Contagious epithelioma (fowl-pox)
1. Chickens
2. Pigeons
Infectious myxomatosis of rabbits
Rous sarcoma of chickens
Leukemia of chickens
Lymphocystic disease of fish, Epithelioma of fish, Carp-pox } No reports on filtration

Trachoma* and inclusion blenorrhea
Mumps (According to Kermorgant, a spirochetal disease)
Agalactia (According to Bridré, a bacterial disease)

Measles (rubeola)
German measles (rubella) (No report on filtration)

Grippe (influenza) (According to Olitsky and Gates, a bacterial disease)
Common colds

A. Insect-borne
a. { Nairobi disease of sheep; Catarrhal fever (blue tongue) of sheep; African horse sickness; Pappataci fever; Dengue fever; Yellow fever (According to Noguchi, a spirochetal disease) }
b. Rickettsia diseases { Typhus fever; Trench fever; Rocky Mountain spotted fever; Heartwater disease; Flood fever of Japan }

B. Bacterial diseases { Oroya fever and verruga peruviana; Pleuropneumonia; Avian diphtheria; Scarlet fever }

* According to Dr. Noguchi, trachoma of American Indians is a bacterial disease.

Fig. 7 Table on members of the "filterable virus group" in: Rivers, T.M. (Ed.), 1928. Filterable Viruses. The Williams & Wilkins Co., Baltimore, pp. 4–5.

virus diseases, and virus diseases of plants. A chapter on bacteriophages, which had been detected as filterable pathogens of bacteria by Frederick Twort (1877–1950; Twort, 1915) and Felix d'Herelle (1873–1949; d'Herelle, 1917), is also included (Fig. 8). This pioneering compilation is a treasure trove of information on the early days of virology.

CONTENTS

CHAPTER PAGE

PREFACE ix

I. SOME GENERAL ASPECTS OF FILTERABLE VIRUSES 3
Thomas M. Rivers, M. D., Member of The Rockefeller Institute for Medical Research, New York.

II. FILTERS AND FILTRATION. (1 text-figure.) 55
Stuart Mudd, M. D., Assistant Professor of Experimental Pathology, School of Medicine, Associate in Pathology, Henry Phipps Institute, University of Pennsylvania, Philadelphia.

III. TISSUE CULTURES IN THE STUDY OF VIRUSES 97
Alexis Carrel, M. D., Member of The Rockefeller Institute for Medical Research, New York.

IV. INTRACELLULAR PATHOLOGY IN VIRUS DISEASES. (1 plate and 25 text-figures.) 113
Edmund V. Cowdry, Ph. D., Associate Member of The Rockefeller Institute for Medical Research, New York.

V. VIRUS DISEASES OF MAN AS EXEMPLIFIED BY POLIOMYELITIS 159
Harold L. Amoss, M. D., Associate Professor of Medicine, Johns Hopkins Medical School, Baltimore.

VI. VIRUS DISEASES OF MAMMALS AS EXEMPLIFIED BY FOOT-AND-MOUTH DISEASE AND VESICULAR STOMATITIS 205
Peter K. Olitsky, M. D., Associate Member of The Rockefeller Institute for Medical Research, New York.

VII. VIRUS DISEASES OF FOWLS AS EXEMPLIFIED BY CONTAGIOUS EPITHELIOMA (FOWL-POX) OF CHICKENS AND PIGEONS. (3 plates.) .. 235
Ernest W. Goodpasture, M. D., Professor of Pathology, Vanderbilt School of Medicine, Nashville.

VIII. VIRUS DISEASES OF INSECTS.
Sacbrood of honey bees. (3 plates and 1 text-figure.) 279
The polyhedral diseases. (3 plates.) 301
Rudolf W. Glaser, Sc. D., Associate Member of The Rockefeller Institute for Medical Research, Princeton.

IX. VIRUS DISEASES OF PLANTS. (3 plates.) 335
Louis O. Kunkel, Ph. D., Boyce Thompson Institute for Plant Research, Yonkers.

X. VIRUS DISEASES OF BACTERIA—BACTERIOPHAGY. (2 plates.) 373
Jacques J. Bronfenbrenner, Ph. D., Associate Member of The Rockefeller Institute for Medical Research, New York.

INDEX 419

vii

Fig. 8 "Table of Contents" of: Rivers, T.M. (Ed.), 1928. Filterable Viruses. The Williams & Wilkins Co., Baltimore, p. VII.

6. THE NATURE OF VIRUSES

Interestingly, Rivers (1927) in his review already included a paragraph on size calculations indicating ranges from 20 to 200 nm which is surprisingly accurate. More detailed studies were possible after Theodore (The) Svedberg (1884–1971) developed in the 1920s the first ultracentrifuges (Svedberg, 1927) allowing for the first time accurate size determinations of the "filterable viruses" as exemplified by Schlesinger's work on bacteriophages (1932).

Although the dispute continued whether viruses would indeed be considered "living matter" as suggested by Beijerinck and Loeffler and Frosch, in 1935 Stanley (1887–1955) was able to crystallize tobacco mosaic virus and show that the crystals retained infectivity. His analyses indicated that the crystals only contained protein prompting the notion the "filterable viruses" actually represent something like self-amplifying enzymes. However, in 1936 it was demonstrated by Frederic Charles Bawden (1908–1972) that tobacco mosaic virus was actually a combination of nucleic acid and protein (*infective nucleoprotein*; Bawden et al., 1936) and, thus, compatible with the definition of "life."

Finally, in the 1930s, Ernst Ruska (1906–1988) and colleagues developed the first electron microscopes which were introduced into virus research by his brother, Helmut (1908–1973; reviewed in Gelderblom and Krüger, 2014; Krüger et al., 2000). In 1938, the first electron micrograph of a virus, ectromelia virus, an orthopoxvirus, was published (Von Borries et al., 1938). In 1939, they reported on the first visualization of tobacco mosaic virus (Kausche et al., 1939). With these extraordinary technological developments the basis was laid for virology as we know it today (Méthot, 2016).

ACKNOWLEDGMENTS

I am deeply grateful and indebted to Frederick A. Murphy, my predecessor as one of the series editors of Advances in Virus Research, for his wonderful powerpoint compilation 'Foundations of Virology' (http://www.utmb.edu/virusimages/). This is a must for all virus affectionados interested in the history of this field. I also thank my secretary Viola Damrau for securing copies of mostly 'historical' literature which, I admit, I read for the first time in full in preparation of this short chapter. I am aware that the dispute on who actually 'founded' Virology will likely continue but I may be forgiven for championing Friedrich Loeffler (and I stated my reasons), the founder of the institute which bears his name and which I have the honor to lead for more than 20 years now. I am also grateful to all staff at the FLI for their support in pursuing a career in virology since my first encounter with a picture of a T4 bacteriophage at the age of 15.

REFERENCES

Bawden, F.C., Pirie, N.W., Bernal, J.D., Fankuchen, I., 1936. Liquid crystalline substances from virus infected plants. Nature 138, 1051–1052.

Beijerinck, M.W., 1898. Über ein Contagium vivum fluidum als Ursache der Fleckenkrankheit der Tabaksblätter. Verhandelingen Koninklijke Akademie van Wetenschappen te Amsterdam (Tweede Sectie) 6, 3–21.

Bos, L., 1995. The embryonic beginnings of virology: unbiased thinking and dogmatic stagnation. Arch. Virol. 140, 613–619.

Calisher, C., Horzinek, M., 1999. 100 Years of virology—the birth and growth of a discipline. Arch. Virol. Suppl. 15.

Chamberland, C., 1884. Sur une filter donnant de l'eau physiologiquement pure. C. R. Acad. Sci. 99, 247–248.

d'Herelle, F.H., 1917. Sur une microbe invisible antagoniste des bacilles dysentériques. C. R. Acad. Sci. 165, 373–375.

DiMaio, D., 2016. Thank you Edward. Merci, Louis. PLoS Pathog. 12, e1005320.

Enquist, L.W., Racaniello, V.R., 2013. Virology: from contagium fluidum to virome. In: Knipe, D., Howley, P.M. (Eds.), Fields Virology, sixth ed. Wolters Kluwer, Lippincott Williams & Wilkins, Philadelphia, pp. 1–20.

Fredericks, D.N., Relman, D.A., 1996. Sequence-based identification of microbial pathogens: a reconsideration of Koch's postulates. Clin. Microbiol. Rev. 9, 18–33.

Fröhner, E., 1912. Kompendium der speziellen Pathologie und Therapie für Tierärzte. Enke, Stuttgart.

Gelderblom, H., Krüger, D.H., 2014. Helmut Ruska (1908–1973): his role in the evolution of electron microscopy in the life sciences, and especially virology. Adv. Imag. Elect. Phys. 182, 1–94.

Hinz-Wessels, A., Thiel, J., 2010. The Friedrich-Loeffler-Institute 1910–2010: 100 Years of Research for Animal Health. Bebra Wissenschaft, Berlin.

Horzinek, M., 1995. The beginnings of animal virology in Germany. Arch. Virol. 140, 1157–1162.

Iwanowski, D., 1892. Über die Mosaikkrankheit der Tabakspflanze. Bull. Sci. Acad. Imp. Sci. St. Petersb. 35, 67–70.

Kausche, G.A., Pfannkuch, P.F., Ruska, H., 1939. Die Sichtbarmachung von pflanzlichem Virus im Übermikroskop. Naturwissenschaften 27, 292–299.

Kitasato, S., 1891. Experimentelle Untersuchungen über das Tetanusgift. Z. Hyg. 10, 267–305.

Krüger, D.H., Schneck, P., Gelderblom, H.R., 2000. Helmut Ruska and the visualization of viruses. Lancet 355, 1713–1717.

Loeffler, F., 1898. IV. Bericht der Kommission zur Erforschung der Maul- und Klauenseuche bei dem Institut für Infektionskrankheiten in Berlin. Centrabl. Bakteriol. 24, 569–574. 1. Abt.

Loeffler, F., Frosch, P., 1898. Berichte der Kommission zur Erforschung der Maul- und Klauenseuche bei dem Institut für Infektionskrankheiten in Berlin. Centralbl. Bakteriol. 23, 371–391. 1. Abt.

Lvov, D.K., 1993. Centenary of virology. In: Mahy, B.W.J., Lvov, D.K. (Eds.), Concepts in Virology: From Ivanovsky to the Present. Harwood Academic Publishers, Chur, pp. 3–14.

Mayer, A., 1886. Über die Mosaikkrankheit des Tabaks. Die landwirtschaftlichen Versuchsstationen 32, 451–467.

Méthot, P.O., 2016. Writing the history of virology in the twentieth century: discovery, disciplines, and conceptual change. Stud. Hist. Philos. Biol. Biomed. Sci. 59, 145–153.

Mokili, J.L., Rohwer, F., Dutilh, B.E., 2012. Metagenomics and future perspectives in virus discovery. Curr. Opin. Virol. 2, 63–77.
Mudd, S., 1928. Filters and filtration. In: Rivers, T.M. (Ed.), Filterable Viruses. The Williams and Wilkins Co., Baltimore, pp. 55–96.
Murphy, F.A., 2016. Historical perspective: what constitutes the discovery (of a new virus)? Adv. Virus Res. 95, 197–220.
Norrby, E., 2008. Nobel prizes and the emerging virus concept. Arch. Virol. 153, 1109–1123.
Reed, W., 1902. Recent researches concerning the etiology, propagation, and prevention of yellow fever, by the United States Army Commission. J. Hyg. 11, 101–119.
Rivers, T.M., 1927. Filterable viruses—a critical review. J. Bacteriol. 14, 217–258.
Schlesinger, M., 1932. Die Bestimmung von Teilchengröße und spezifischem Gewicht des Bakteriophagen durch Zentrifugierversuche. Z. Hyg. Infektionskrankh. 114, 161–176.
Stanley, W.M., 1935. Isolation of a crystalline protein possessing the properties of tobacco mosaic virus. Science 81, 644–645.
Svedberg, T., 1927. The Ultracentrifuge. Nobel Lecture. May 19, 1927.
Twort, F., 1915. An investigation on the nature of ultra-microscopic viruses. Lancet 186, 1241–1243.
Van Helvoort, T., 1996. When did virology start? ASM News 62, 142–145.
Van Regenmortel, M.H.V., 2010. The nature of viruses. In: Mahy, B.W.J., van Regenmortel, M.H.V. (Eds.), Desk Encyclopedia of General Virology. Academic Press, Oxford, pp. 19–23.
Von Borries, B., Ruska, E., Ruska, H., 1938. Bakterien und Virus in übermikroskopischer Aufnahme. Klin. Wochenschr. 17, 921–925.
Wilkinson, L., 2001. History of virology. In: Encyclopedia of Life Sciences. John Wiley & Sons, pp. 1–22.
Witz, J., 1998. A reappraisal of the contribution of Friedrich Loeffler to the development of the modern concept of virus. Arch. Virol. 143, 2261–2263.

FURTHER READING

Loeffler, F., 1884. Untersuchungen über die Bedeutung der Mikroorganismen für die Entstehung der Diphtherie beim Menschen: bei der Taube und beim Kalbe. Mittheilungen aus dem kaiserlichen Gesundheitsamte 2, 421–499.
Rivers, T.M., 1928a. Some general aspects of filterable viruses. In: Rivers, T.M. (Ed.), Filterable Viruses. The Williams and Wilkins Co., Baltimore, pp. 3–54.
Rivers, T.M. (Ed.), 1928b. Filterable viruses. The Williams & Wilkins Co., Baltimore.

CHAPTER TWO

Loeffler 4.0: Diagnostic Metagenomics

Dirk Höper[1], Claudia Wylezich, Martin Beer
Institute of Diagnostic Virology, Friedrich-Loeffler-Institut, Greifswald-Insel Riems, Germany
[1]Corresponding author: e-mail address: dirk.hoeper@fli.de

Contents

Abstract

A new world of possibilities for "virus discovery" was opened up with high-throughput sequencing becoming available in the last decade. While scientifically metagenomic analysis was established before the start of the era of high-throughput sequencing, the availability of the first second-generation sequencers was the kick-off for diagnosticians to use sequencing for the detection of novel pathogens. Today, diagnostic metagenomics is becoming the standard procedure for the detection and genetic characterization of new viruses or novel virus variants. Here, we provide an overview about technical considerations of high-throughput sequencing-based diagnostic metagenomics together with selected examples of "virus discovery" for animal diseases or zoonoses and metagenomics for food safety or basic veterinary research.

Advances in Virus Research, Volume 99
ISSN 0065-3527
https://doi.org/10.1016/bs.aivir.2017.08.001

1. INTRODUCTION

Since nucleic acids are the general genetic material of all organisms and viruses, shotgun sequencing is the most generic approach for the detection of any pathogen. Hence, it is logical to apply the different high-throughput sequencing technologies for the detection of novel or unexpected viruses. Therefore, already with the advent of the first commercially available second-generation sequencing platforms, diagnosticians and researchers started evaluating and optimizing the detection of pathogens by sequencing. Likewise, researchers interested in exploring the full diversity of the virome started applying shotgun high-throughput sequencing for the analysis of diverse samples to identify novel viruses and virus variants.

Expanding our knowledge of the virosphere is not only scientifically interesting but also important for disease detection and control. As shown later in selected examples, various diseases of unknown etiology are caused by pathogens that were recently detected by novel technologies, namely, by microarrays or high-throughput sequencing. Regardless of the platform or the specific sample selection and preprocessing, an unambiguous association of the detected pathogen with the observed disease requires the fulfillment of Koch's postulates as he formulated them in a talk (Redactions Comité, 1891). Due to the problems that may arise in fulfilling these, a number of amendments of Koch's postulates have been suggested (Mokili et al., 2012). Nevertheless, still the stringent criteria formulated in the late 19th century by Koch and coworkers are the golden standard. In some of the examples provided in a later section of the article and discussed in detail in the following chapters, Koch's postulates were fulfilled, showing that in veterinary science this important proof can be achieved.

Another advantage of metagenomics for diagnostics is the possibility to detect coinfections. This may be of rising importance as we progress with detecting new agents that may infect the host without causing clinical signs of illness in a certain host without a coinfecting agent but cause serious or varying symptoms when coming together with different other facultative pathogens, as for instance analyzed by Blomström et al. (2016), Zhang et al. (2014), or Hanke et al. (2017). In case of coinfections, Koch's postulates might be hard to fulfill if not even impossible. Nevertheless, selecting individuals with the same symptoms and from epidemiologically related vs unrelated cases may help gather evidence of the coinfecting agents to be the cause of the observed disease.

2. TECHNICAL CONSIDERATIONS OF DIAGNOSTIC METAGENOMICS

2.1 The Right Sample

The choice of the samples is of utmost importance for successful virus detection. Even though a systematic assessment of the impact of the sample so far has not been conducted, published data imply this (Hoffmann et al., 2015; Pfaff et al., 2017a). Moreover, it is logical that due to differing organ tropism of different pathogens different organs will be positive for pathogen nucleic acids. Noteworthy to mention that also the time point of sampling is crucial. Looking at the studies of Wernike and colleagues (Bilk et al., 2012; Hoffmann et al., 2012; Wernike et al., 2012, 2013) dealing with Schmallenberg virus (SBV), it is evident that owing to the short viremia sampling will yield genome copies only in a relatively narrow period and hence allow virus detection.

Fig. 1 displays the impact of the pathogen content on detection probability. The plots on the one hand illustrate that at a certain pathogen load it is nearly impossible to fail with detection. On the other hand, Fig. 1 also shows that if the pathogen content is too low, astronomic datasets may be necessary for the detection. However, it is logical that the larger and more complex the dataset gets, the longer the data analysis to find hints at the potential causative agent takes. Although substantial efforts have been put into optimizing sample preprocessing for pathogen detection, the intention of all these is to bias the sequencing (Briese et al., 2015; Conceicao-Neto et al., 2015; Kohl et al., 2015). However, this bias naturally comes together with information loss due to the selection of only a fraction of the nucleic acids, and this loss can render the analysis useless if the necessary information is lost.

As a consequence, the applied preprocessing procedures often sacrifice the generic character of the shotgun sequencing approach. Instead, selection of a more suitable sample can solve the problem of the detection limits. As shown in Fig. 2 for the variegated squirrel bornavirus 1 (VSBV-1), the probability of detection of a certain number of viral reads in different samples from the same individual greatly varies. This variation clearly is a result of the organ tropism of the virus in quest. Therefore, analyzing a variety of sample materials might be necessary for success. As also shown in Fig. 2, samples from different animals infected with the same viral species may have a greatly varying pathogen content, as depicted for a pair of rabies virus (RABV) infected animals (data taken from Hanke et al., 2016). This is to

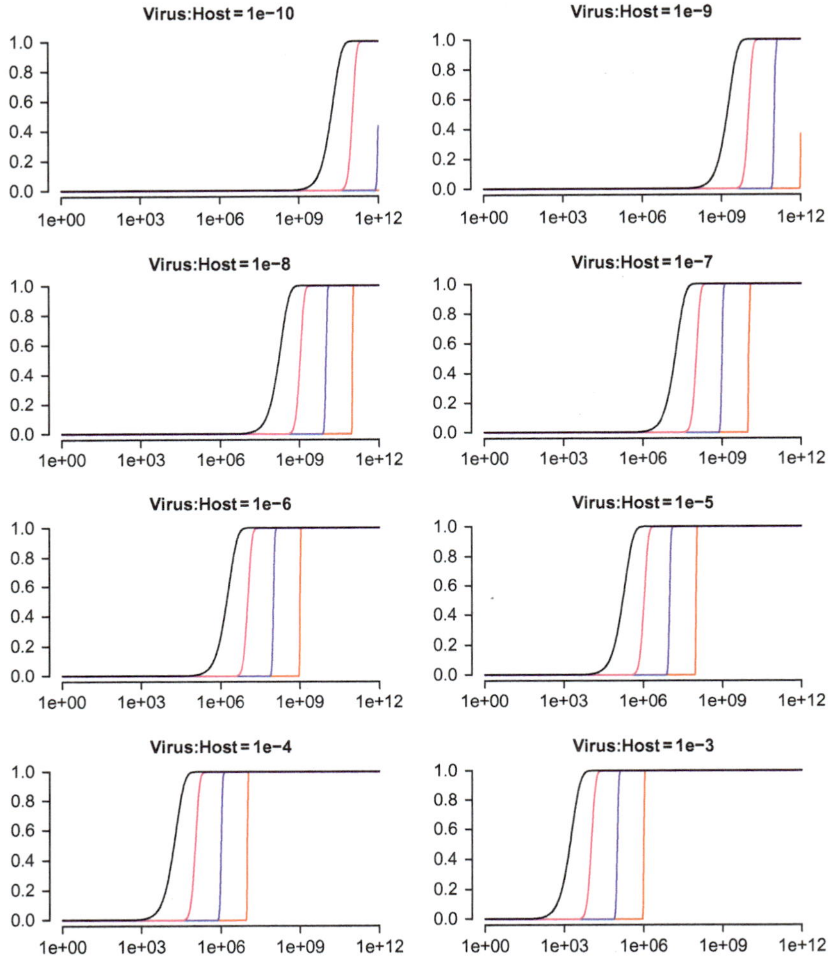

Fig. 1 Theoretical probabilities for pathogen detection. The graphs display the probabilities (plotted on the ordinate) to detect at least one read (plotted *black*), at least 10 reads (*magenta*), at least 100 reads (*blue*), or at least 1000 reads (*red*) representing a pathogen present in a sample with a certain pathogen: host ratio (plotted as the title of each graph) in a dataset with a certain size (plotted on the abscissa).

a lesser extent also true for a pair of samples from cases of ovine astrovirus (OvAstV) infected animals (data taken from Pfaff et al., 2017a).

2.2 Pooling

In some cases, pooling of samples might appear useful to reduce the expenses per sample. However, calculating the probabilities for detection of VSBV-1 in pooled samples shows (Fig. 2) that the sequencing effort increases, at least

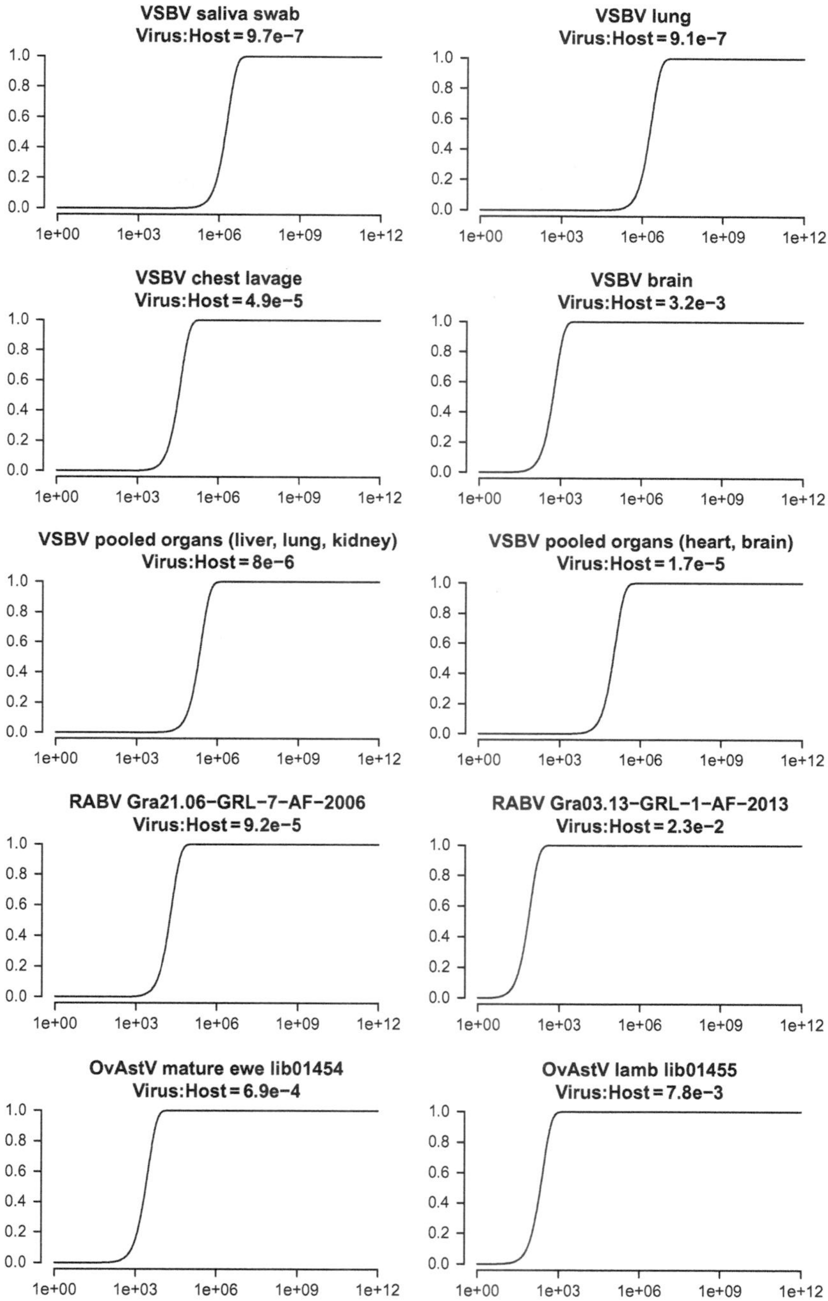

Fig. 2 Example probabilities for pathogen detection. Observed virus: host ratios taken from published studies in which pathogens were detected by shotgun metagenomics. The graphs display the probabilities (plotted on the ordinate) to detect at least one read representing a pathogen present in a sample with a certain pathogen: host ratio (plotted as the title of each graph) in a dataset with a certain size (plotted on the abscissa). *VSBV*, variegated squirrel bornavirus, data taken from Hoffmann et al. (2015); *RABV*, Rabies virus, data taken from Hanke et al. (2016); *OvAstV*, ovine astrovirus, data taken from Pfaff et al. (2017a).

when negative samples or samples with a low virus content are part of the pool. As shown, eventually the expenses can be reduced without significantly reducing the probability of detection if a sample with sufficiently high pathogen content is part of the sample pool. However, when working with pooled samples and aiming for complete genomes, individual positive samples have to be identified after detection of the virus, and individual libraries and sequencing have to be conducted afterward.

2.3 DNA or RNA?

Another issue that has to be considered carefully is the choice of the nucleic acids to sequence, DNA or RNA. Possible criteria to take into account for this decision are at first of course the expected type of virus, i.e., DNA or RNA virus. Next, if dealing with samples where virus replication can be expected, preferably RNA should be chosen since this will enable the detection of both DNA and RNA viruses. On the contrary, choosing DNA will make the detection of RNA viruses fundamentally impossible. In case of samples in which no viral replication can be expected, for instance feces or environmental samples, DNA and RNA must be analyzed and should preferably be processed separately (also see Section 2.2).

2.4 Considerations for Sequencing Platform Selection

Despite the fact that some HTS platforms are no longer available, there is still a considerable range of technologies and instruments from which researchers and diagnosticians can select. All of these have their specific pros and cons and all proved to be more or less suitable for pathogen detection. We will not evaluate the different platforms here nor will we argue for the use of a specific platform, rather we will only briefly discuss a number of characteristics to consider for platform selection. Inter alia, characteristics influencing the choice of the most suitable sequencing platform are the read length, which is a major determinant of the reliable taxonomic classification, the runtime to complete an analysis, the possible size of the datasets, or the overall cost for an analysis.

Fig. 3 and Table 1 show an example for the impact of read length on the sensitivity of detection. The original dataset (from Hoffmann et al., 2012) and the same dataset with shortened reads were analyzed using the software pipeline RIEMS (Scheuch et al., 2015) with identical settings and databases. It is clearly visible that longer reads are more readily classified than shorter ones. On the contrary, to a certain extent errors can be tolerated and do not

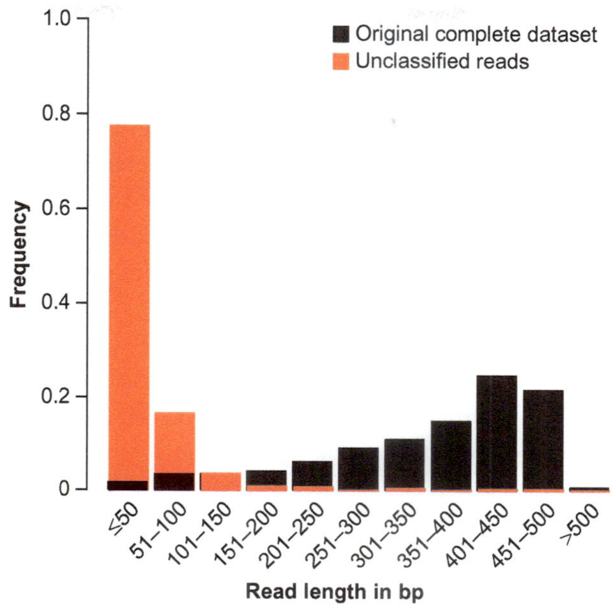

Fig. 3 Read length dependency of sensitive classification. The figure displays two overlaid histograms of the length distributions of the complete Schmallenberg virus dataset (black; Hoffmann et al., 2012) and those reads for which no sequence with significant identity could be detected (red) using an early version of the RIEMS analysis pipeline (Scheuch et al., 2015).

Table 1 Main Figures of the Classification Efficiency Depending on Read Length Using the Dataset From the Schmallenberg Virus Identification (Hoffmann et al., 2012)

	Original Read Length	Shortened Reads
Mean read length	315	96
Total no. reads	27,420	27,420
Classified reads	26,128	25,310
Orthobunyavirus hits	7	1
Identity of the orthobunyavirus reads with subject sequence	68.4%–95.6%	98.7%
Unclassified reads	1292	2110

Data were analyzed using an early version of the RIEMS analysis pipeline (Scheuch et al., 2015) with identical settings and databases.

influence a sensitive reliable classification (Scheuch et al., 2015) since in case of diagnostic metagenomics not exact genotyping but classification at higher taxonomic ranks is the aim. This tolerance depends on the software workflow in use (Scheuch et al., 2015). Of course, if the error rate rises too high, no sufficient sequence identity will be detectable and hence no classification possible. Finally, it has to be concluded that there is a trade-off between read length and error rates that can vary depending on the availability of reference sequences with sufficient similarity enabling recognition of the pathogen.

Of utmost importance are the potential inherent sources of contamination like carryover between runs or missorting of/mislabeling with molecular barcodes due to impurity of the barcoded adapters (Sigma Aldrich, 2017) used for deconvolution of the datasets. Such missorting is the likely reason for errors seen in the INSDC databases. For instance, Mukherjee et al. (2015) reported on integration of the PhiX phage into bacterial genomes of diverse families, even those that had never before been associated with PhiX. The authors report that 10% of the contaminated genomes had even been published in literature. By thorough analysis, they found out that the integration of the phage genome into the bacterial genomes was by missorting of the raw data, which included the used sequencing control sample into all datasets. Fig. 4 depicts another example of probable missorting and its result in the generated sequence. In this case, the Influenza A neuraminidase gene was assembled into a genome sequence of a bacterium taxonomically classified as *Bacillus* sp. In the respective genome assembly, also the Influenza A virus Hemagglutinin gene was included (not shown). These two examples of the incorporation of foreign sequences are most likely also a result of improper data deconvolution. Moreover, these examples show that care must be taken when analyzing datasets, since such errors will always occur and need manual inspection and correction of the database content by trained data curators.

2.5 Read Classification

After sample selection, processing, and sequencing, with data analysis another important part of the detection process starts. Since the huge datasets that are generated by shotgun metagenomics may comprise mostly host sequences masking the relevant information, sensitive and specific workflows for classification of the obtained reads are urgently needed.

```
Query= AR_01-17-I_lib02115_NA_1 [6 - 1415] 1..1425 length=1425
numreads=9389

Length=470
                                                                      Score     E
Sequences producing significant alignments:                          (Bits)  Value

gb|KQL43666.1|  hypothetical protein AN962_05950, partial [Bacill...  965     0.0
gb|AJS16200.1|  neuraminidase [Influenza A virus (A/duck/Eastern ...  964     0.0
gb|ALP30086.1|  neuraminidase [Influenza A virus (A/goose/Zhejian...  962     0.0

>gb|KQL43666.1| hypothetical protein AN962_05950, partial [Bacillus sp. FJAT-21955]
Length=480

 Score = 965 bits (2494),  Expect = 0.0, Method: Compositional matrix adjust.
 Identities = 464/470 (99%), Positives = 466/470 (99%), Gaps = 0/470 (0%)

Query  1    MNPNQKIATIGSISLGLAVFNVLLHAVSIILMVLALGKSENNGICKGTIVREYNETVRIE  60
            MNPNQKI TIGSISLGL VFNVLLHAVSIIL VLALGKSENNGICKGT+VREYNETVRIE
Sbjct  11   MNPNQKIVTIGSISLGLVVFNVLLHAVSIILTVLALGKSENNGICKGTVVREYNETVRIE  70

Query  61   KVTQWYNTSVVEYVPHWNEGTYINNTEPICDVKGFAPFSKDNGIRVGSRGHIFVIREPFV  120
            KVTQWYNTSVVEYVPHWNEGTYINNTEPICDVKGFAPFSKDNGIRVGSRGHIFVIREPFV
Sbjct  71   KVTQWYNTSVVEYVPHWNEGTYINNTEPICDVKGFAPFSKDNGIRVGSRGHIFVIREPFV  130

Query  121  SCSPVECRTFFLTQGALLNDKHSNGTVKDRSPFRTLMSVEVGQSPNVYQARFEAVAWSAT  180
            SCSPVECRTFFLTQGALLNDKHSNGTVKDRSPFRTLMSVEVGQSPNVYQARFEAVAWSAT
Sbjct  131  SCSPVECRTFFLTQGALLNDKHSNGTVKDRSPFRTLMSVEVGQSPNVYQARFEAVAWSAT  190
```

Fig. 4 Example for a result of improper data deconvolution and review. An Influenza A virus neuraminidase gene included in the whole-genome shotgun assembly of a *Bacillus* sp. An Influenza A H5N8 neuraminidase amino acid sequence was searched using blastp (Camacho et al., 2009) with default settings in the NCBI nr database.

A major pitfall of data analysis is the fact that if dealing with a novel pathogen, no suitable reference sequence allowing its easy recognition may be available. Hence, although generally seen as a method that works without a priori knowledge because the starting point of the complete procedure can be true shotgun sequencing, regardless of the chosen type of nucleic acid, DNA or RNA, metagenomics heavily relies on a priori knowledge. Until we overcome the necessity that initial identification of the pathogen is based on similarity of the sequencing reads with known sequences available in the INSDC databases, we will not have a system that is independent of a priori knowledge and hence truly unbiased. This stresses the need for algorithms that can determine the source of a sequence with sufficient reliability solely based on low-level characteristics without calculating sequence similarities. On the one hand, novel optimized procedures like for instance Kraken (Wood and Salzberg, 2014) or Diamond (Buchfink et al., 2015) have the potential to speed up the classification. On the other hand, if it is possible to determine characteristics for classes of genomes (classified by higher orders like for instance a superkingdom), e.g., composition rather than sequence-based characteristics, this may help increase the sensitivity of detection.

Approaches for classification of sequences that do not rely on calculation of alignments but are based on oligonucleotide frequencies have been proposed and also shown to be suitable for classification (A. Belka et al., unpublished; Diaz et al., 2009; Gregor et al., 2016; Leung et al., 2011; McHardy et al., 2007). The mentioned approaches let it seem likely that it will in the near future be possible to improve the detection of novel viruses without a suitable known reference. Furthermore, the increasing number of sequencing projects results in a fast growing number of sequences from novel virus species, genera, or even families. This new data sets continuously improve the analysis and detection rate for new pathogens.

2.6 Data Quality

Another important issue that is frequently stressed (Byrd et al., 2014; Wood and Salzberg, 2014) is the quality of the sequences used for building the reference datasets for classification. As pointed out in Section 2.4, this is at the time of writing a problem. This problem may even be increasing due to the increasing democratization of sequencing and the consequent increase of sequencing efforts. Therefore, it is urgently necessary to have (more) trained personnel for database curation to ensure the quality and reliability of the nucleotide sequence database content.

2.7 Sensitivity and Specificity of Diagnostic Metagenomics

Even after some years of continuous use, it is still not reliably possible to establish the sensitivity and specificity of metagenomic pathogen detection. With regard to the specificity, it is noteworthy to mention that with a truly unbiased sample preparation and sequencing protocol, only the final step of the identification procedure, i.e., the algorithms and databases used for taxonomic sequence classification, determines the specificity. On the contrary, all protocols including sample preprocessing will strongly influence the specificity of the overall process. All these procedures in the worst case will render detection of pathogens that are not in the focus of the preprocessing procedure nearly impossible. Nevertheless, due to common and divergent characteristics of groups of pathogens, the specificity of the preprocessing protocols cannot be assessed.

At the same time, sample processing and data analysis have a strong impact on the sensitivity. As an example, searching for an unknown pathogen but only using a viral sequence database for data analysis will cause the sensitivity for the detection of bacterial pathogens drop to zero, and vice versa. The same applies for novel pathogens for which no suitable reference sequence is available. In these cases, the sensitivity may also drop to zero when no suitable alternative criteria can be applied to determine the pathogen. Likewise, even in case of the combination of unbiased sequencing with unbiased data analysis, the algorithm at least in part determines the sensitivity (see for instance (Scheuch et al., 2015) for a comparison of the sensitivities of different software). In summary, a general assessment of the sensitivity and specificity of high-throughput sequencing-based pathogen detection is not possible.

Taken together, in approximately 10 years of high-throughput sequencing-based diagnostic metagenomics, mostly driven by second-generation sequencing, a substantial number of viruses has been detected and genomically characterized, thereby proving the suitability of sequencing approaches for diagnostics.

3. RECENT EXAMPLES IN "VIRUS DISCOVERY" FOR ANIMAL DISEASES AND ZOONOSES

In former times, the discovery of new viruses or novel virus variants happened mainly as the consequence of isolation on cell culture, embryonated chicken eggs, or in animal models. Virus growth was detected due to disease development, cytopathic effects, and the use of broad diagnostics like electron microscopy or antisera for neutralization or staining. Detection of completely new viruses was often by accident only, depending, e.g., on the possibility to grow the virus. Today, "virus discovery" is driven by the new developments in molecular diagnostics, mainly broad PCRs, microarrays, and new sequencing technologies. Especially for infectious animal diseases or zoonoses, and in the field of virus reservoirs, like viruses transmitted from bats or voles, tremendous progress in the detection and characterization of new viruses was achieved.

A growing number of examples exists now where viruses were detected in samples from diseased animals or in animal reservoirs. Table 2 provides a short overview of some prominent examples. It is important to mention that in several of the listed cases, the experimental confirmation of the causation was possible, and the conclusions were not based exclusively on the sequence information. Various important examples are presented later in much more detail in this issue.

One of the most prominent examples is the detection of SBV in serum samples from cattle showing rather unspecific signs of illness in 2011 in

Table 2 Overview of Recent Examples in Virus Discovery for Animal and Zoonotic Diseases Detected Using High-Throughput Sequencing Technologies

Year	Newly Detected Virus	References
2009	Lujo virus/hemorrhagic fever-associated arenavirus	Briese et al. (2009)
2010	Shaking mink syndrome astovirus	Blomström et al. (2010)
2011	Hepacivirus from a dog	Kapoor et al. (2011)
2012	Middle East respiratory syndrome (MERS)	Zaki et al. (2012)
2012	Schmallenberg orthobunyavirus	Hoffmann et al. (2012)
2012	Bat Influenza A virus	Tong et al. (2012)
2013	Swine Influenza C virus	Hause et al. (2013)
2013	Rodent hepaciviruses	Drexler et al. (2013)
2013	Primate pegi- and hepaciviruses	Quan et al. (2013)
2013/2017	Ovine astroviruses	Li et al. (2013) and Pfaff et al. (2017a)
2014	Bokeloh bat lyssavirus	Nolden et al. (2014)
2014/2016	Bovine astroviruses	Bouzalas et al. (2014) and Schlottau et al. (2016)
2015	Bovine hepacivirus	Baechlein et al. (2015)
2015	Novel tick-borne flavivirus subtype (louping-ill-related) in goats	Mansfield et al. (2015)
2015	Bluetongue virus serotype 27	Jenckel et al. (2015)
2015	Variegated squirrel bornavirus 1	Hoffmann et al. (2015)
2015	Atypical porcine pestivirus	Hause et al. (2015)
2016	Porcine pegivirus	Baechlein et al. (2016)
2017	Penguin alphaherpesvirus	Pfaff et al. (2017b)
2017	Lateral-shaking inducing neuro degenerative agent (LINDA)	Lamp et al. (2017)

Germany. The impact of SBV was huge since it spread all over Europe within only a few years due to the naïve population of ruminants. Metagenomics allowed the detection and whole-genome characterization of virus from acutely infected animals before the further spread and also before malformed offspring was born in 2012. Furthermore, the sequence information from sequencing enabled the fast development of specific molecular diagnostics. SBV and the current situation is reviewed in a later chapter of this issue.

Further examples include also the detection of diverse astroviruses from various ruminant species, especially sheep and cattle. Most importantly, these cases were all associated with neurological disease. Before, astrovirus-induced encephalitis was only reported for very few human cases. This additional and new knowledge about astroviruses and their impact on differential diagnostics of encephalitis cases is also presented in detail in the chapter "The expanding field of mammalian astroviruses: opportunities and challenges in clinical virology" by Boujon et al.

Another field in veterinary virology where metagenomics provided completely new insights and expanded our knowledge on animal diseases was the first detection of the so-called atypical porcine pestiviruses (APPV). Although initially not associated with clear symptoms and identified during the NGS-based screening of porcine samples in the United States, several groups could link APPV infection with the congenital tremor of newborn piglets. More details on the novel pestiviruses can be found in the chapter "New leaves in the growing tree of pestiviruses" by Blome et al.

With the same NGS-based analyses, the *Orbivirus* family was substantially expanded by a number of new serotypes of bluetongue viruses. The metagenomics-driven growth of the species BTV is shown in a dedicated chapter summarizing not only the detection but also the further characterization of isolated BTV in vitro and in vivo. A very similar "growth" happened in the genus *Lyssavirus of the Rhabdoviridae* family (see chapter about "Novel lyssaviruses" by Eggerbauer et al.).

Examples which are not further elaborated in this issue but which are of importance comprise, e.g., a novel Herpesvirus from Penguins (SpAHV). This report is a good example for the use of novel sequences to complete the phylogeny of a virus family and to provide data for a better understanding of the relationship of similar viruses from different animal species (Pfaff et al., 2017b). The same is true for all the novel members of the genus Hepacivirus with new complete genomes of viruses detected in dogs, horses, voles, rats, and cattle (Baechlein et al., 2015; Drexler et al., 2013; Kapoor et al., 2011; Quan et al., 2013). These sequences changed the picture of Hepaciviruses as a virus restricted to only humans into that of a huge virus group present in many animal species including humans. This is one of the most impressive examples how NGS-based metagenomics can change our view on important viruses within years or even a few months. The same is true for pegiviruses (Baechlein et al., 2016; Quan et al., 2013) and influenza A viruses.

Especially the novel bat influenza viruses H17N10 and H18N11 expanded even the well-studied influenza A viruses (Tong et al., 2012) in a new reservoir host. Bat influenza virus is in addition a perfect example for the de novo generation of a replicating virus by using only the sequence information for reverse genetics and recovery of infectious virus particles without the need of virus isolation from positive sample material (Moreira et al., 2016). High-quality whole genomes of viruses are therefore in many cases the basis of further studies like gene expression, construction of chimeric viruses, or generation of recombinant clones.

4. METAGENOMICS FOR FOOD SAFETY

Beside veterinary virology, food safety, which is closely linked to veterinary virology by, for instance, zoonotic pathogens, is another field for which diagnostic metagenomics can improve pathogen detection (Stasiewicz et al., 2015). Foodborne pathogenic bacteria like *Salmonella, Listeria,* or toxin-producing *Escherichia coli* strains, but also norovirus and Hepatitis A virus and parasites (*Trichinella, Giardia*) can cause disease outbreaks accompanied by vomiting and more or less severe diarrhea. For instance, a severe outbreak of gastroenteritis and the hemolytic–uremic syndrome caused by Shiga toxin-producing *E. coli* in Germany in 2011 mostly affecting adult women was likely caused be the consumption of sprouts (Frank et al., 2011). In another case, frozen strawberries caused a norovirus gastroenteritis outbreak in Germany in 2012. Samples from this outbreak analyzed by real-time RT-PCR assay revealed a combination of three different genotypes that had not been reported in Germany so far resulting in the suggestion that the strawberries were polluted from sewage rather than from a single infected food handler (Mäde et al., 2013).

Genomes of foodborne pathogens have been often sequenced for genomic characterization of strains of special interest and for comparative genomic studies, but food products were rather seldom handled and evaluated for metagenomics. To systematically evaluate the suitability of shotgun metagenomics for the assessment of the quality of animal-derived foods and foods in general, we conducted a pilot study in which we investigated conventional food samples of animal or plant origin. We analyzed various untreated and highly processed food sample using our in-house metagenomics sample processing workflow. The investigated sample matrices included rocket, mushrooms, ham, salmon, meat loaf, cheese, oat flakes pizza, chocolate, oysters, strawberries, tap water, and parasite-containing wild boar meat. The used workflow starts with RNA instead of DNA extraction. In general, we could obtain RNA from all tested food products. The resulting concentration and especially the quality of the RNA obtained from processed food samples were clearly lower since their RNA content and integrity is inherently very low. Special matrices like fatty (e.g., cheese) and fiber-containing foods (e.g., oat flakes) were challenging and possibly need sample-dependent pretreatments, like defatting. Nucleic acids of samples with a low pH value (e.g., frozen berries suspicious to be contaminated by norovirus) are difficult to isolate and should be pretreated with a neutralization step.

Nevertheless, we were able to detect viruses in different data sets resulting from both untreated and processed foods. In a rocket sample, we found a novel yellows virus (0.3% of reads) that was only about 91% identical (at nucleotide level) to yellows viruses known from turnip and brassica. Meat loaf is an example of processed food in which we were able to detect a well-known pepper mild mottle virus (0.1%) that might cause different symptoms like fever or pruritus (Colson et al., 2010). Interestingly, we discovered a novel mycovirus in untreated (mushrooms) and processed samples (pizza with mushrooms). This virus is only 35% similar (at amino acid level) to known mycoviruses; however, mycovirus sequences found in both mushroom samples are nearly identical to each other. In the examples mentioned here, the detected viruses are RNA viruses, which prove our RNA-based strategy to be very successful in shedding light on the virosphere. In addition, parasite-specific sequences matching the expected pathogens could be detected in wild boar meat containing parasites (trichina 0.02% and liver fluke 0.000002%). Thus, the used protocols seem to be applicable over a wide range of foodborne pathogens and food categories. Routinely applied for foodborne pathogens, resulting metagenomics data can be helpful for surveillance as well as foodborne outbreak investigation and will improve the hazard identification by increased specificity and potentially by a fundamental change in the definition of the hazard being rather a specific virulent strain, subtype, or gene instead of a not well-specified species.

5. METAGENOMICS FOR BASIC RESEARCH IN (VETERINARY) VIROLOGY

While the aforementioned examples illustrate the impact of high-throughput sequencing on veterinary diagnostics and the closely connected field of food safety, there are of course diverse applications beyond. Whereas anthropologists already successfully analyzed ancient samples immediately after the early second-generation sequencers became available (Green et al., 2006, 2010; Maricic and Pääbo, 2009; Noonan et al., 2006), researchers only recently started using shotgun high-throughput sequencing to systematically assess the pathogen content of historic samples available in museums or other collections and ancient DNA from archaeologic specimens. Systematic investigation of historic samples will not only help discover the diversity of the virome but will essentially contribute to the elucidation of viral evolution and the potential impact of vaccination on the evolving virome, for instance. At the moment, however, no reports on the analysis of historic animal samples have been published, but the studies of historic DNA samples are focused to human samples. For instance, in a recently published paper (Feldman et al., 2016), the authors describe the analysis of an ancient *Yersinia pestis* genome. The DNA was recovered from a 6th-century skeleton, a putative victim of the Justinianic plague, found in a southern German grave. The authors were able to assemble a complete genome with high coverage. Their sequence analysis revealed a number of unique variants and some structural differences compared to the previously available *Y. pestis* sequences. In another report, Duggan et al. (2016) sequenced a complete variola virus genome from a 17th century child mummy from Lithuania. The authors were able to reconstruct the complete viral genome, which was found to be basal to all strains from the 20th century. The authors concluded that much of variola virus evolution and diversification occurred recently driven by the impact of vaccination. Another study published by Pajer et al. (2017) shed further light on the long-term evolution of variola virus. In their study, they report on the sequencing and analysis of two complete variola virus genomes from historic human specimens from a museum in Prague. Together these two studies provide a new level of insight into long-term virus evolution.

This impressively demonstrates the power of high-throughput sequencing for the investigation of the evolutionary history of viruses. Another example for the possibilities that high-throughput sequencing opens in this field is the work published by Toppinen et al. (2015). They analyzed DNA obtained from bones of putative Finnish casualties from World War II for parvovirus B19 DNA. Interestingly, they found only viral genotypes that disappeared from Northern Europe in the 1970s or had never previously been reported to be found there. Moreover, using molecular clock analysis the authors were able to date the most recent common ancestors for all sequenced viruses back to the early 19th century. Noteworthy to mention that in addition to virus analysis, the obtained sequences also enabled tracing the origin of the casualties. The authors were able to show that one of the casualties was most likely of Russian origin. This highlights an additional facet of genome analysis using shotgun high-throughput sequencing. A similar approach was used for in-depth spatio-temporal investigation of arctic rabies viruses and their reservoir hosts in Greenland (Hanke et al., 2016). The study explored virus dynamics over a period of roughly 10 years. The authors did not find a link between viral genotypes and the reservoir host population. All these studies together demonstrate the utility of shotgun high-throughput sequencing for investigations into virus dynamics and evolution.

6. CONCLUSIONS AND FUTURE DIRECTIONS

For sensitive, reliable, and in the optimal case comparable results, a certain degree of standardization is necessary. This encompasses the procedures from sample to sequence and the subsequent analyses of the generated datasets for the detection of the viruses. An important determinant of success is the selection of suitable samples and the sampling and sample handling and storage conditions. To this end, awareness must be raised as to the preservation of samples that are intended for metagenomic analysis. A common preservation is formalin fixation followed by paraffin embedding. However, this minimizes the prospects of success for sequencing-based pathogen detection. Moreover, since the complete characterization of viruses and their unambiguous connection with the disease imperatively requires virus isolates, native material must be available. Therefore, it is necessary to preserve samples by other means than fixation and embedding. Possible procedures have to be evaluated for their suitability and subsequently need to be validated on a range of samples. Finally, this will help improve the complete procedure of diagnostic metagenomics.

The constantly growing number of available viral genome sequences will improve the chances of recognizing new pathogens. Moreover, the ongoing efforts for the implementation of algorithms to identify sequences of viral origin without sequences comparison will enable the identification of viruses even stronger deviating from those known.

Despite all these successes, further improvements and standardization are urgently necessary for routine use of shotgun metagenomics in diagnostics. With the ongoing dissemination of high-throughput sequencing in diagnostic laboratories, the need for quality control will rise. Therefore, in the near future it will be of utmost importance to establish platform-independent quality measures and ring trials for diagnostic laboratories that use high-throughput sequencing for pathogen detection in general and for virus detection in particular. To make high-throughput sequencing-based diagnostics as generic as its foundation is, the necessary procedures must be designed in a way to abolish the necessity of differentiation between viral, bacterial, and parasitic pathogens for reliable detection.

REFERENCES

Baechlein, C., Fischer, N., Grundhoff, A., Alawi, M., Indenbirken, D., Postel, A., et al., 2015. Identification of a novel hepacivirus in domestic cattle from Germany. J. Virol. 89 (14), 7007–7015. https://doi.org/10.1128/JVI.00534-15.

Baechlein, C., Grundhoff, A., Fischer, N., Alawi, M., Hoeltig, D., Waldmann, K.-H., Becher, P., 2016. Pegivirus infection in domestic pigs, Germany. Emerg. Infect. Dis. 22 (7), 1312–1314. https://doi.org/10.3201/eid2207.160024.

Bilk, S., Schulze, C., Fischer, M., Beer, M., Hlinak, A., Hoffmann, B., 2012. Organ distribution of Schmallenberg virus RNA in malformed newborns. Vet. Microbiol. 159 (1–2), 236–238. https://doi.org/10.1016/j.vetmic.2012.03.035.

Blomström, A.L., Fossum, C., Wallgren, P., Berg, M., 2016. Viral metagenomic analysis displays the co-infection situation in healthy and pmws affected pigs. PLoS One 11 (12), e0166863. https://doi.org/10.1371/journal.pone.0166863.

Blomström, A.-L., Widén, F., Hammer, A.-S., Belák, S., Berg, M., 2010. Detection of a novel astrovirus in brain tissue of mink suffering from shaking mink syndrome by use of viral metagenomics. J. Clin. Microbiol. 48 (12), 4392–4396. https://doi.org/10.1128/JCM.01040-10.

Bouzalas, I.G., Wüthrich, D., Walland, J., Drögemüller, C., Zurbriggen, A., Vandevelde, M., et al., 2014. Neurotropic astrovirus in cattle with nonsuppurative encephalitis in Europe. J. Clin. Microbiol. 52 (9), 3318–3324. https://doi.org/10.1128/JCM.01195-14.

Briese, T., Paweska, J.T., McMullan, L.K., Hutchison, S.K., Street, C., Palacios, G., et al., 2009. Genetic detection and characterization of Lujo virus, a new hemorrhagic fever-associated arenavirus from southern Africa. PLoS Pathog. 5 (5), e1000455. https://doi.org/10.1371/journal.ppat.1000455.

Briese, T., Kapoor, A., Mishra, N., Jain, K., Kumar, A., Jabado, O.J., Lipkin, W.I., 2015. Virome capture sequencing enables sensitive viral diagnosis and comprehensive virome analysis. mBio 6 (5), e01491–15. https://doi.org/10.1128/mBio.01491-15.

Buchfink, B., Xie, C., Huson, D.H., 2015. Fast and sensitive protein alignment using DIAMOND. Nat. Methods 12 (1), 59–60. https://doi.org/10.1038/nmeth.3176.

Byrd, A.L., Perez-Rogers, J.F., Manimaran, S., Castro-Nallar, E., Toma, I., McCaffrey, T., et al., 2014. Clinical PathoScope. Rapid alignment and filtration for accurate pathogen identification in clinical samples using unassembled sequencing data. BMC Bioinformatics 15, 262. https://doi.org/10.1186/1471-2105-15-262.

Camacho, C., Coulouris, G., Avagyan, V., Ma, N., Papadopoulos, J., Bealer, K., Madden, T.L., 2009. BLAST+. Architecture and applications. BMC Bioinformatics 10, 421. https://doi.org/10.1186/1471-2105-10-421.

Colson, P., Richet, H., Desnues, C., Balique, F., Moal, V., Grob, J.-J., et al., 2010. Pepper mild mottle virus, a plant virus associated with specific immune responses, fever, abdominal pains, and pruritus in humans. PLoS One 5 (4), e10041. https://doi.org/10.1371/journal.pone.0010041.

Conceicao-Neto, N., Zeller, M., Lefrere, H., de Bruyn, P., Beller, L., Deboutte, W., et al., 2015. Modular approach to customise sample preparation procedures for viral metagenomics. A reproducible protocol for virome analysis. Sci. Rep. 5, 16532. https://doi.org/10.1038/srep16532.

Diaz, N.N., Krause, L., Goesmann, A., Niehaus, K., Nattkemper, T.W., 2009. TACOA. Taxonomic classification of environmental genomic fragments using a kernelized nearest neighbor approach. BMC Bioinformatics 10, 56. https://doi.org/10.1186/1471-2105-10-56.

Drexler, J.F., Corman, V.M., Müller, M.A., Lukashev, A.N., Gmyl, A., Coutard, B., et al., 2013. Evidence for novel hepaciviruses in rodents. PLoS Pathog. 9 (6), e1003438. https://doi.org/10.1371/journal.ppat.1003438.

Duggan, A.T., Perdomo, M.F., Piombino-Mascali, D., Marciniak, S., Poinar, D., Emery, M.V., et al., 2016. 17(th) century variola virus reveals the recent history of smallpox. Curr. Biol. 26 (24), 3407–3412. https://doi.org/10.1016/j.cub.2016.10.061.

Feldman, M., Harbeck, M., Keller, M., Spyrou, M.A., Rott, A., Trautmann, B., et al., 2016. A high-coverage Yersinia pestis genome from a sixth-century justinianic plague victim. Mol. Biol. Evol. 33 (11), 2911–2923. https://doi.org/10.1093/molbev/msw170.

Frank, C., Werber, D., Cramer, J.P., Askar, M., Faber, M., der Heiden, M.a., et al., 2011. Epidemic profile of Shiga-toxin-producing Escherichia coli O104:H4 outbreak in Germany. N. Engl. J. Med. 365 (19), 1771–1780. https://doi.org/10.1056/NEJMoa1106483.

Green, R.E., Krause, J., Ptak, S.E., Briggs, A.W., Ronan, M.T., Simons, J.F., et al., 2006. Analysis of one million base pairs of Neanderthal DNA. Nature 444 (7117), 330–336. https://doi.org/10.1038/nature05336.

Green, R.E., Krause, J., Briggs, A.W., Maricic, T., Stenzel, U., Kircher, M., et al., 2010. A draft sequence of the Neandertal genome. Science 328 (5979), 710–722. https://doi.org/10.1126/science.1188021.

Gregor, I., Dröge, J., Schirmer, M., Quince, C., McHardy, A.C., 2016. PhyloPythiaS+. A self-training method for the rapid reconstruction of low-ranking taxonomic bins from metagenomes. PeerJ 4. e1603. https://doi.org/10.7717/peerj.1603.

Hanke, D., Freuling, C.M., Fischer, S., Hueffer, K., Hundertmark, K., Nadin-Davis, S., et al., 2016. Spatio-temporal analysis of the genetic diversity of arctic rabies viruses and their reservoir hosts in Greenland. PLoS Negl. Trop. Dis. 10 (7). e0004779. https://doi.org/10.1371/journal.pntd.0004779.

Hanke, D., Pohlmann, A., Sauter-Louis, C., Höper, D., Stadler, J., Ritzmann, M., et al., 2017. Porcine epidemic diarrhea in Europe: In-detail analyses of disease dynamics and molecular epidemiology. Viruses 9 (7), 177. https://doi.org/10.3390/v9070177.

Hause, B.M., Ducatez, M., Collin, E.A., Ran, Z., Liu, R., Sheng, Z., et al., 2013. Isolation of a novel swine influenza virus from Oklahoma in 2011 which is distantly related to human influenza C viruses. PLoS Pathog. 9 (2), e1003176. https://doi.org/10.1371/journal.ppat.1003176.

Hause, B.M., Collin, E.A., Peddireddi, L., Yuan, F., Chen, Z., Hesse, R.A., et al., 2015. Discovery of a novel putative atypical porcine pestivirus in pigs in the USA. J. Gen. Virol. 96 (10), 2994–2998. https://doi.org/10.1099/jgv.0.000251.

Hoffmann, B., Scheuch, M., Höper, D., Jungblut, R., Holsteg, M., Schirrmeier, H., et al., 2012. Novel orthobunyavirus in Cattle, Europe, 2011. Emerg. Infect. Dis. 18 (3), 469–472. https://doi.org/10.3201/eid1803.111905.

Hoffmann, B., Tappe, D., Höper, D., Herden, C., Boldt, A., Mawrin, C., et al., 2015. A variegated squirrel bornavirus associated with fatal human encephalitis. N. Engl. J. Med. 373 (2), 154–162. https://doi.org/10.1056/NEJMoa1415627.

Jenckel, M., Breard, E., Schulz, C., Sailleau, C., Viarouge, C., Hoffmann, B., et al., 2015. Complete coding genome sequence of putative novel bluetongue virus serotype 27. Genome Announc. 3 (2), e00016-15. https://doi.org/10.1128/genomeA.00016-15.

Kapoor, A., Simmonds, P., Gerold, G., Qaisar, N., Jain, K., Henriquez, J.A., et al., 2011. Characterization of a canine homolog of hepatitis C virus. Proc. Natl. Acad. Sci. U. S. A. 108 (28), 11608–11613. https://doi.org/10.1073/pnas.1101794108.

Kohl, C., Brinkmann, A., Dabrowski, P.W., Radonić, A., Nitsche, A., Kurth, A., 2015. Protocol for metagenomic virus detection in clinical specimens. Emerg. Infect. Dis. 21 (1), 48–57. https://doi.org/10.3201/eid2101.140766.

Lamp, B., Schwarz, L., Högler, S., Riedel, C., Sinn, L., Rebel-Bauder, B., et al., 2017. Novel pestivirus species in pigs, Austria, 2015. Emerg. Infect. Dis. 23 (7), 1176–1179. https://doi.org/10.3201/eid2307.170163.

Leung, H.C.M., Yiu, S.M., Yang, B., Peng, Y., Wang, Y., Liu, Z., et al., 2011. A robust and accurate binning algorithm for metagenomic sequences with arbitrary species abundance ratio. Bioinformatics 27 (11), 1489–1495. https://doi.org/10.1093/bioinformatics/btr186.

Li, L., Diab, S., McGraw, S., Barr, B., Traslavina, R., Higgins, R., et al., 2013. Divergent astrovirus associated with neurologic disease in cattle. Emerg. Infect. Dis. 19 (9), 1385–1392. https://doi.org/10.3201/eid1909.130682.

Mäde, D., Trübner, K., Neubert, E., Höhne, M., Johne, R., 2013. Detection and typing of norovirus from frozen strawberries involved in a large-scale gastroenteritis outbreak in Germany. Food Environ. Virol. 5, 162–168. https://doi.org/10.1007/s12560-013-9118-0.

Mansfield, K.L., Morales, A.B., Johnson, N., Ayllón, N., Höfle, U., Alberdi, P., et al., 2015. Identification and characterization of a novel tick-borne flavivirus subtype in goats (Capra hircus) in Spain. J. Gen. Virol. 96 (Pt. 7), 1676–1681. https://doi.org/10.1099/vir.0.000096.

Maricic, T., Pääbo, S., 2009. Optimization of 454 sequencing library preparation from small amounts of DNA permits sequence determination of both DNA strands. Biotechniques 46 (1), 51–57. https://doi.org/10.2144/000113042.

McHardy, A.C., Martin, H.G., Tsirigos, A., Hugenholtz, P., Rigoutsos, I., 2007. Accurate phylogenetic classification of variable-length DNA fragments. Nat. Methods 4 (1), 63–72. https://doi.org/10.1038/nmeth976.

Mokili, J.L., Rohwer, F., Dutilh, B.E., 2012. Metagenomics and future perspectives in virus discovery. Curr. Opin. Virol. 2 (1), 63–77. https://doi.org/10.1016/j.coviro.2011.12.004.

Moreira, É.A., Locher, S., Kolesnikova, L., Bolte, H., Aydillo, T., García-Sastre, A., et al., 2016. Synthetically derived bat influenza A-like viruses reveal a cell type-but not species-specific tropism. Proc. Natl. Acad. Sci. U. S. A. 113 (45), 12797–12802. https://doi.org/10.1073/pnas.1608821113.

Mukherjee, S., Huntemann, M., Ivanova, N., Kyrpides, N.C., Pati, A., 2015. Large-scale contamination of microbial isolate genomes by Illumina PhiX control. Stand Genomic. Sci. 10, 18. https://doi.org/10.1186/1944-3277-10-18.

Nolden, T., Banyard, A.C., Finke, S., Fooks, A.R., Hanke, D., Hoper, D., et al., 2014. Comparative studies on the genetic, antigenic and pathogenic characteristics of Bokeloh bat lyssavirus. J. Gen. Virol. 95 (Pt. 8), 1647–1653. https://doi.org/10.1099/vir.0.065953-0.

Noonan, J.P., Coop, G., Kudaravalli, S., Smith, D., Krause, J., Alessi, J., et al., 2006. Sequencing and analysis of Neanderthal genomic DNA. Science 314 (5802), 1113–1118. https://doi.org/10.1126/science.1131412.

Pajer, P., Dresler, J., Kabíckova, H., Písa, L., Aganov, P., Fucik, K., et al., 2017. Characterization of two historic smallpox specimens from a Czech museum. Viruses 9 (8), 200. https://doi.org/10.3390/v9080200.

Pfaff, F., Schlottau, K., Scholes, S., Courtenay, A., Hoffmann, B., Höper, D., Beer, M., 2017a. A novel astrovirus associated with encephalitis and ganglionitis in domestic sheep. Transbound. Emerg. Dis. 64 (3), 677–682. https://doi.org/10.1111/tbed.12623.

Pfaff, F., Schulze, C., König, P., Franzke, K., Bock, S., Hlinak, A., et al., 2017b. A novel alphaherpesvirus associated with fatal diseases in banded penguins. J. Gen. Virol. 98, 89–95. https://doi.org/10.1099/jgv.0.000698.

Quan, P.-L., Firth, C., Conte, J.M., Williams, S.H., Zambrana-Torrelio, C.M., Anthony, S.J., et al., 2013. Bats are a major natural reservoir for hepaciviruses and pegiviruses. Proc. Natl. Acad. Sci. U. S. A. 110 (20), 8194–8199. https://doi.org/10.1073/pnas.1303037110.

Redactions Comité, 1891. In: Ueber bakteriologische Forschung. X. internationaler medicinischer Congress. Berlin, 1891. Verhandlungen des X. internationalen medicinischen Congresses, 1, vol. 5. Verlag von August Hirschwald, Berlin.

Scheuch, M., Höper, D., Beer, M., 2015. RIEMS. a software pipeline for sensitive and comprehensive taxonomic classification of reads from metagenomics datasets. BMC Bioinformatics 16 (1), 69. https://doi.org/10.1186/s12859-015-0503-6.

Schlottau, K., Schulze, C., Bilk, S., Hanke, D., Höper, D., Beer, M., Hoffmann, B., 2016. Detection of a novel bovine astrovirus in a cow with encephalitis. Transbound. Emerg. Dis. 63 (3), 253–259. https://doi.org/10.1111/tbed.12493.

Sigma Aldrich, 2017. Traditional HPLC is Incapable of Reducing Cross Contamination of Custom Next-Generation Sequencing Adapters to Acceptable Levels. Sigma Aldrich, Saint Louis, MO. Available online at http://www.sigmaaldrich.com/technical-documents/articles/biology/traditional-hplc-is-incapable.html. checked on 8/1/2017.

Stasiewicz, M.J., den Bakker, H.C., Wiedmann, M., 2015. Genomics tools in microbial food safety. Curr. Opin. Food Sci. 4, 105–110. https://doi.org/10.1016/j.cofs.2015.06.002.

Tong, S., Li, Y., Rivailler, P., Conrardy, C., Castillo, D.A.A., Chen, L.-M., et al., 2012. A distinct lineage of influenza A virus from bats. Proc. Natl. Acad. Sci. U. S. A. 109 (11), 4269–4274. https://doi.org/10.1073/pnas.1116200109.

Toppinen, M., Perdomo, M.F., Palo, J.U., Simmonds, P., Lycett, S.J., Söderlund-Venermo, M., et al., 2015. Bones hold the key to DNA virus history and epidemiology. Sci. Rep. 5, 17226. https://doi.org/10.1038/srep17226.

Wernike, K., Eschbaumer, M., Breithaupt, A., Hoffmann, B., Beer, M., 2012. Schmallenberg virus challenge models in cattle. Infectious serum or culture-grown virus? Vet. Res. 43, 84. https://doi.org/10.1186/1297-9716-43-84.

Wernike, K., Hoffmann, B., Bréard, E., Bøtner, A., Ponsart, C., Zientara, S., et al., 2013. Schmallenberg virus experimental infection of sheep. Vet. Microbiol. 166 (3–4), 461–466. https://doi.org/10.1016/j.vetmic.2013.06.030.

Wood, D.E., Salzberg, S.L., 2014. Kraken: ultrafast metagenomic sequence classification using exact alignments. Genome Biol. 15 (3), R46. https://doi.org/10.1186/gb-2014-15-3-r46.

Zaki, A.M., van Boheemen, S., Bestebroer, T.M., Osterhaus, A.D., Fouchier, R.A., 2012. Isolation of a novel coronavirus from a man with pneumonia in Saudi Arabia. N. Engl. J. Med. 367 (19), 1814–1820. https://doi.org/10.1056/NEJMoa1211721.

Zhang, B., Tang, C., Yue, H., Ren, Y., Song, Z., 2014. Viral metagenomics analysis demonstrates the diversity of viral flora in piglet diarrhoeic faeces in China. J. Gen. Virol. 95 (Pt 7), 1603–1611. https://doi.org/10.1099/vir.0.063743-0.

CHAPTER THREE

Schmallenberg Virus: A Novel Virus of Veterinary Importance

Kerstin Wernike[1], Martin Beer
Institute of Diagnostic Virology, Friedrich-Loeffler-Institut, Greifswald-Insel Riems, Germany
[1]Corresponding author: e-mail address: kerstin.wernike@fli.de

Contents

Abstract

In late 2011, unspecific clinical symptoms such as fever, diarrhea, and decreased milk production were observed in dairy cattle in the Dutch/German border region. After exclusion of classical endemic and emerging viruses by targeted diagnostic systems, blood samples from acutely diseased cows were subjected to metagenomics analysis. An insect-transmitted orthobunyavirus of the Simbu serogroup was identified as the causative agent and named Schmallenberg virus (SBV). It was one of the first detections of the introduction of a novel virus of veterinary importance to Europe using the new technology of next-generation sequencing. The virus was subsequently isolated from identical samples as used for metagenomics analysis in insect and mammalian cell lines and disease symptoms were reproduced in calves experimentally infected with both, this culture-grown virus and blood samples of diseased cattle.

Since its emergence, SBV spread very rapidly throughout the European ruminant population causing mild unspecific disease in adult animals, but also premature birth or stillbirth and severe fetal malformation when naive dams were infected during a critical phase of gestation. In the following years, SBV recirculated regularly to a larger extend; in the 2014 and 2016 vector seasons the virus was again repeatedly detected in the blood of adult ruminants, and in the following winter and spring months, a number of malformed calves and lambs was born.

Advances in Virus Research, Volume 99
ISSN 0065-3527
https://doi.org/10.1016/bs.aivir.2017.07.001

The genome of viruses present in viremic adult animals showed a very high sequence stability; in sequences generated between 2012 and 2016, only a few amino acid substitutions in comparison to the initial SBV isolate could be detected. In contrast, a high sequence variability was identified in the aminoterminal part of the glycoprotein Gc-encoding region of viruses present in the brain of malformed newborns. This mutation hotspot is independent of the region or host species from which the samples originated and is potentially involved in immune evasion mechanisms.

1. FIRST DETECTION OF SCHMALLENBERG VIRUS AND INITIAL SPREAD

In summer and autumn 2011, an unidentified disease in cattle characterized by fever, diarrhea, and decreased milk production was reported from farmers and veterinarians to the animal health services, local diagnostic laboratories, and national research institutes in Germany and the Netherlands. After the exclusion of classical endemic and emerging viruses such as bovine herpesvirus type 1, pestiviruses, bluetongue virus, epizootic hemorrhagic disease virus, foot-and-mouth disease virus, Rift Valley fever virus, or bovine ephemeral fever virus with targeted diagnostic systems, pools of representative blood samples from acutely diseased cows were subjected to next-generation sequencing-based metagenomics. By using this approach, the causative agent of the observed clinical signs, a novel orthobunyavirus (family *Bunyaviridae*), was identified (Hoffmann et al., 2012), and based on the origin of the samples named Schmallenberg virus (SBV). Phylogenetic analyses suggested that SBV belongs together with, e.g., Simbu virus, Akabane virus (AKAV), Aino virus (AINOV), Douglas virus, Shamonda virus, and Sathuperi virus to the Simbu serogroup of orthobunyaviruses (Goller et al., 2012; Yanase et al., 2012). Simbu viruses are widely distributed in Asia, Africa, and Oceania (Fig. 1), but were until the emergence of SBV not present in Europe (Saeed et al., 2001).

Based on viral sequences obtained by the initial metagenomic analysis, first diagnostic real-time RT-PCR assays were established and used to confirm the presence of the virus on further affected cattle farms, and the PCRs combined with newly developed serological test systems were applied in epidemiological investigations (Bilk et al., 2012; Hoffmann et al., 2012; Wernike et al., 2014a).

Blood samples taken before summer 2011 from wild and captive ruminants in Germany, the Netherlands, Belgium, or France were seronegative for SBV suggesting the absence of the virus from Central Europe before this

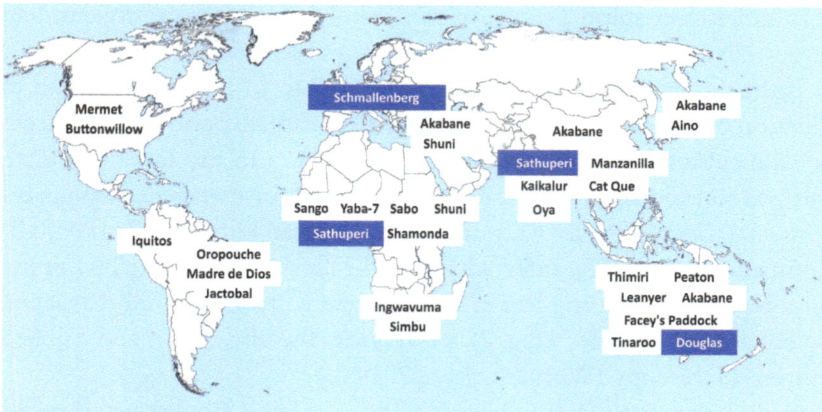

Fig. 1 Distribution of representative members of the Simbu serogroup. Viruses of the Sathuperi species are highlighted in *blue*.

time (Garigliany et al., 2012a; Mouchantat et al., 2015; Veldhuis et al., 2013; Wernike et al., 2014d; Zanella et al., 2013). Since its first appearance, however, SBV spread very rapidly throughout the European ruminant population reaching Belgium, France, Luxembourg, Italy, Spain, and Southern England during the first vector season, i.e., in summer and autumn 2011 (EFSA, 2012), thereby causing a very high seroprevalence of approximately 70% to nearly 100% in the initially most affected areas (Elbers et al., 2012; Garigliany et al., 2012a; Wernike et al., 2014a).

SBV reappeared in the following year, spread across the British Isles to Scotland (Anonymous, 2013a; Mason et al., 2013) and Ireland (Bradshaw et al., 2012), reached the Scandinavian countries (ProMED-mail, 2012a, b, 2013; Rasmussen et al., 2014; Wisloff et al., 2014), Eastern European countries such as Poland or Lithuania (Larska et al., 2013a; Lazutka et al., 2014), crossed the Alpine countries (Balmer et al., 2014; Steinrigl et al., 2014), and spread to Southern Europe as far as Spain and Greece (Chaintoutis et al., 2014; Fernandez-Aguilar et al., 2014).

2. IN LOEFFLER'S FOOTSTEPS: FROM FIRST DETECTION TO THE FULFILLMENT OF KOCH'S POSTULATES

In the 19th century, general guidelines for the identification of pathogens that could be isolated with the techniques of that time were described and summarized in the so-called Koch's postulates (Loeffler, 1884). They represent an important benchmark for microbiologists worldwide even in our days, more than a century after the era of Henle, Koch, and Loeffler.

In order to establish a causal relationship between a microorganism and the observed clinical symptoms, Koch's postulates require the isolation of the pathogen from the diseased host, the growth in pure culture, and the induction of clinical symptoms in healthy organisms experimentally infected with the cultured microorganism (Loeffler, 1884). In the case of SBV, the virus was isolated from identical samples as used for metagenomics analysis in insect (*Culicoides* cells KC) and baby hamster kidney cells (BHK-21) (Hoffmann et al., 2012); infected cultured cells could be stained by immunofluorescence tests using sera of convalescent animals or monoclonal antibodies raised against SBV (Fig. 2). In addition, the virus was made visible by electron microscopy (Wernike et al., 2014a).

As described for the diphtheria pathogen in the 19th century (Loeffler, 1884), microorganisms putatively causative for a specific disease should be able to reproduce the clinical symptoms when injected in pure culture into healthy hosts. Such an assumption could be proven also for SBV, the culture-grown virus was injected into healthy calves which subsequently showed the Simbu virus typical short-lived viremia and in some cases fever and diarrhea (Hoffmann et al., 2012; Wernike et al., 2012b).

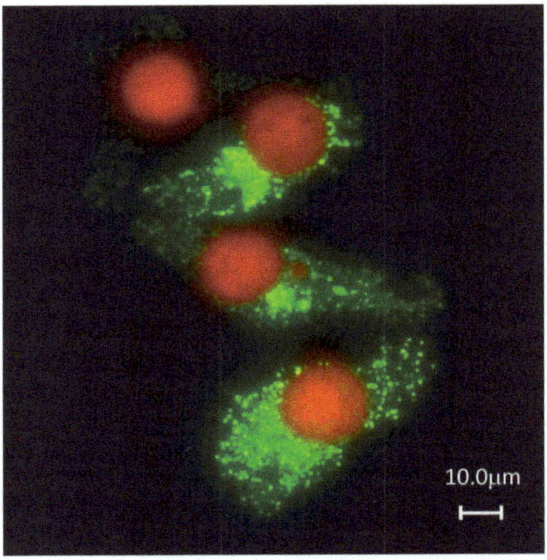

Fig. 2 SBV infected baby hamster kidney cells stained with an N-protein-specific monoclonal antibody (Wernike et al., 2015a) raised against the first SBV isolate in combination with a fluorescein isothiocyanate (FITC-) conjugated goat antimouse IgG (Sigma-Aldrich Co.); the nuclei were stained with propidium iodide.

3. KNOWLEDGE GAINED SINCE 2011 ABOUT THE VIRUS, ITS TRANSMISSION, AND CLINICAL MANIFESTATIONS

The members of the Simbu serogroup of orthobunyaviruses to which SBV belongs are generally transmitted by mosquitoes and/or *Culicoides* biting midges (Elliott and Blakqori, 2011; Saeed et al., 2001). The same holds true for SBV, since viral RNA has been, e.g., detected in field-collected *Culicoides* midges of various species (De Regge et al., 2012a; Elbers et al., 2013; Goffredo et al., 2013; Larska et al., 2013b; Rasmussen et al., 2012). However, mosquitoes do not play an important role, if any, in the transmission of SBV or its persistence in Europe (Balenghien et al., 2014; Scholte et al., 2014; Wernike et al., 2014c). The transmission via insect vectors, i.e., *Culicoides* biting midges, is most likely the only one, in any case the most important route of infection of ruminants. A direct horizontal transmission of SBV is highly unlikely as oronasal inoculation of calves with culture-grown SBV did not lead to infection (Wernike et al., 2013a), and in-contact sheep and cattle remained seronegative in numerous experimental infection studies (Wernike et al., 2012b, 2013a, b). In contrast, a vertical transmission of SBV from the pregnant dam to the developing fetus by transplacental infection occurs (Martinelle et al., 2015; Veldhuis et al., 2014; Wernike et al., 2014b). In addition, viral genome has been detected in bull semen (Hoffmann et al., 2013; Ponsart et al., 2014; van der Poel et al., 2014b), and it was shown by subcutaneous inoculation of interferon receptor-deficient mice or calves with PCR-positive semen straws that some samples contained infectious virus (Ponsart et al., 2014; Schulz et al., 2014). Whether an infection can also be induced by the insemination of cows and, if so, the epidemiological relevance remains to be examined.

When adult ruminants are infected, none or only mild, unspecific clinical signs such as fever, diarrhea, or reduced milk production for a few days are induced (Hoffmann et al., 2012; Wernike et al., 2013b, 2014d). As demonstrated for closely related viruses (St. George and Kirkland, 2004; Radostits et al., 2007) the viremic period of SBV is likewise short, viral RNAs are only detectable for a few days in the blood of infected animals (Hoffmann et al., 2012; Laloy et al., 2015; Schulz et al., 2015; Wernike et al., 2012b, 2013b).

The host species comprise beside the major target species cattle, sheep, and goats various wild and captive ruminants and some zoo animals (Chiari et al., 2014; Laloy et al., 2016; Larska et al., 2013a; Linden et al., 2012;

Molenaar et al., 2015; Mouchantat et al., 2015). Antibodies against SBV were also found in a single dog in Sweden and viral RNA was detected in the brain of a puppy (Sailleau et al., 2013; Wensman et al., 2013); however, large-scale serological surveys performed in dogs in Belgium and wild carnivores in Germany did not provide any further indication of SBV infections of carnivores, since anti-SBV antibodies were not detected in any sample (Garigliany et al., 2013; Mouchantat et al., 2015).

Although type I interferon-receptor knockout mice are susceptible to an experimental SBV infection (Kraatz et al., 2015; Wernike et al., 2012a, 2017), no SBV-specific antibodies have been detected in wild-type mice and shrews indicating that free-living shrews and rodents are most likely not a reservoir for SBV (Mouchantat et al., 2015). Members of the Suidae family, however, may be receptive. SBV-specific antibodies were found in free-ranging wild boars in Germany and Belgium, and a few domestic pigs transiently seroconverted after experimental infection; however, no SBV RNA was detectable and no clinical signs were observed (Desmecht et al., 2013; Mouchantat et al., 2015; Poskin et al., 2014).

As mentioned above, SBV causes only a mild, transient disease in ruminants of most age classes. However, much more important than this effect is certainly the induction of severe congenital malformations, premature birth or stillbirth, or the birth of mummified fetuses, when naïve dams are infected during a critical phase of pregnancy. The induced malformations are summarized as arthrogryposis–hydranencephaly syndrome. For further, well-known Simbu viruses such as AKAV, the critical timeframe in cattle is between the third and sixth months of pregnancy (Kirkland et al., 1988; Konno et al., 1982; Kurogi et al., 1977), and in sheep between days 30 and 50 of gestation (Hashiguchi et al., 1979; Parsonson et al., 1977, 1988). This critical period is most likely similar for SBV (Martinelle et al., 2015; Wernike et al., 2014b). Nevertheless, not every infection of pregnant dams with SBV or other Simbu serogroup viruses during that timeframe leads to an abnormal course of gestation or induces a fetal antibody response (Parsonson et al., 1975; Veldhuis et al., 2014; Wernike et al., 2014b). Even the birth of one malformed and one healthy lamb or kid within the same litter has been reported (Anonymous, 2012b; van den Brom et al., 2012). Lambs or calves infected in utero with SBV may show varying degrees of deformation especially in the central nervous (Fig. 3) and/or skeletal-muscular system which includes arthrogryposis, kyphosis, lordosis, torticollis, scoliosis, ankylosis, brachygnathia, mild-to-severe hypoplasia of the central nervous system, porencephaly, narrow spinal cords, or poliomyelitis (Anonymous,

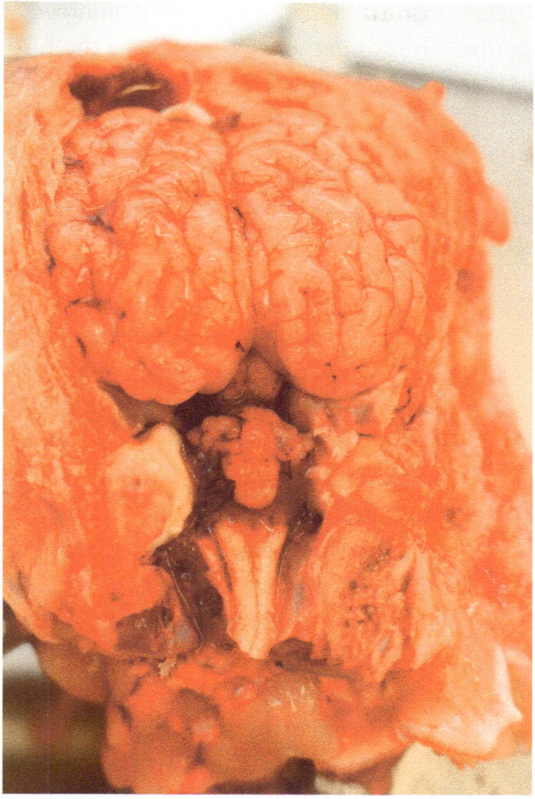

Fig. 3 Hypoplasia of the cerebellum caused by fetal infection with SBV.

2012a; Garigliany et al., 2012b; Herder et al., 2012; Peperkamp et al., 2012; van den Brom et al., 2012). In most cases, the calves/lambs are born at term leading to a time lag of several months between the infection of the dam and the visibility of the effects on the fetuses/newborns which is correlated to the length of pregnancy (approximately 9 months in cattle and 150 days in sheep and goats). Apart from this main effect, SBV might be further associated with embryonic or fetal death and early abortion. Increased numbers of repeated estrus suggesting early embryonic or fetal loss have been reported in SBV-affected sheep, goat, and cattle flocks (Dominguez et al., 2014; Helmer et al., 2013; Lievaart-Peterson et al., 2012; Luttikholt et al., 2014; Saegerman et al., 2014).

The transmission by insect vectors, whose peak season is during the summer and autumn months, and the delay until the most important effect of SBV infections, i.e., stillbirth or birth of severely malformed offspring, which is determined by the length of pregnancy, becomes visible, is responsible for the seasonal dynamics of SBV-induced disease: acutely infected adult animals in summer and autumn, and the birth of malformed lambs starting in December followed by an increasing number of malformed calves.

4. MOLECULAR CHARACTERIZATION AND GENETIC VARIABILITY

Like in typical orthobunyaviruses, Schmallenberg virions which are about 100 nm in diameter contain three segments of negative-stranded RNA genome: a large (L), a medium (M), and a small (S) segment. These three genome segments encode a total of six proteins. The L-segment encodes the RNA-dependent RNA polymerase (RdRp), while the M-segment encodes the viral glycoproteins Gn and Gc, as well as a nonstructural protein (NSm), and the S-segment encodes the nucleocapsid (N) protein and in an alternative overlapping reading frame the nonstructural protein NSs (Chowdhary et al., 2012; Elliott, 2014; Walter and Barr, 2011). The terminal nucleotides at the 3′ and 5′ ends of each orthobunyaviral genome segment are complementary enabling the formation of a panhandle structure (Elliott, 2014).

The L-segment-encoded viral protease is crucial for viral RNA synthesis since it catalyzes both, transcription and genome replication. In addition to its polymerase function, the RdRp contains an endonuclease domain at the aminoterminal end that cleaves capped oligonucleotides from host mRNAs. These capped oligomers are subsequently used to prime the viral mRNA synthesis (Elliott, 2014). The bunyaviral N-protein, which is the most abundant protein in the virion and in infected cells, forms ribonucleoprotein complexes with the three RNA segments, and these complexes are associated with the RdRp (Elliott, 2014; Plyusnin et al., 2012). Apart from its involvement in transcription and replication the major role of the N-protein is encapsidation of the genome (Dong et al., 2013). The remaining two structural proteins, the glycoproteins Gn and Gc, are integral transmembrane proteins which form a heterodimer in the endoplasmic reticulum of infected cells and are important for viral attachment and membrane fusion (Elliott, 2014).

The nonstructural protein NSs is not essential for in vitro growth (Kraatz et al., 2015; Varela et al., 2013), but it is a major virulence factor in the mammalian host as it counteracts the induction of interferon and the shutoff of host cell protein synthesis. In addition, it is involved in the regulation of translation, apoptosis, and viral polymerase activity (Barry et al., 2014; Gouzil et al., 2017). The function of NSm, the second nonstructural protein which is dispensable for viral replication in vitro and in vivo, is mostly unknown, but it may be involved in virus assembly and morphogenesis and could interact with the NSs protein since an SBV NSs/NSm double deletion mutant virus showed a growth defect in cell culture and higher levels of α-/β-interferon were induced compared to an NSs single mutant virus (Kraatz et al., 2015).

Of all three genomic segments, the M-segment has been shown to be the most variable one. This is also described for related viruses such as AKAV (Kobayashi et al., 2007). However, the sequence variability of SBVs M-segment heavily depends on the sample material. A geographically and host species independent highly variable region was found in the Gc-coding sequence of viruses present in malformed newborns (Fischer et al., 2013a). Even in samples collected in a confined area (Germany) and during a limited time period (winter 2011/12) a high sequence variability was found. Nucleotide sequence comparisons of the complete glycoprotein-coding open reading frame revealed a minimum of 5 nucleotide substitutions and a maximum of 77 nucleotide alterations which included up to 36 nucleotide deletions between individual samples. Amino acid (aa) substitutions ranged from 1 to 43 aa including insertions of 2 aa and deletions of 12 aa (Fischer et al., 2013a). In addition to these sequences generated directly from organ samples of malformed fetuses/newborns, similar mutation accumulation was observed within the hypervariable region after consecutive cell culture passages of viruses isolated from the central nervous system of stillborn lambs. The mutations obtained in two SBV isolates during in vitro passages were not randomly distributed, but accumulated in the glycoprotein Gc-coding region (Coupeau et al., 2013).

In contrast to the highly variable SBV genome in malformed newborns, the genome of viruses detected in the blood of acutely infected adult animals is much more stable (Wernike et al., 2015b) (Figs. 4 and 5). All M-segment sequences generated since 2011 which were deposited in the INSDC databases and two sequences from samples submitted to the German National Reference Laboratory for SBV in autumn 2016 cluster closely independent of the year or country of isolation.

Since the envelope glycoproteins Gn and Gc are major immunogens of orthobunyaviruses and the genomic region in which the mutation hot spot is located represents a key immunogenic domain of SBV (Roman-Sosa et al., 2016; Wernike et al., 2017), the mutation hot spot was supposed to be involved in immune evasion mechanisms. Another assumption could be

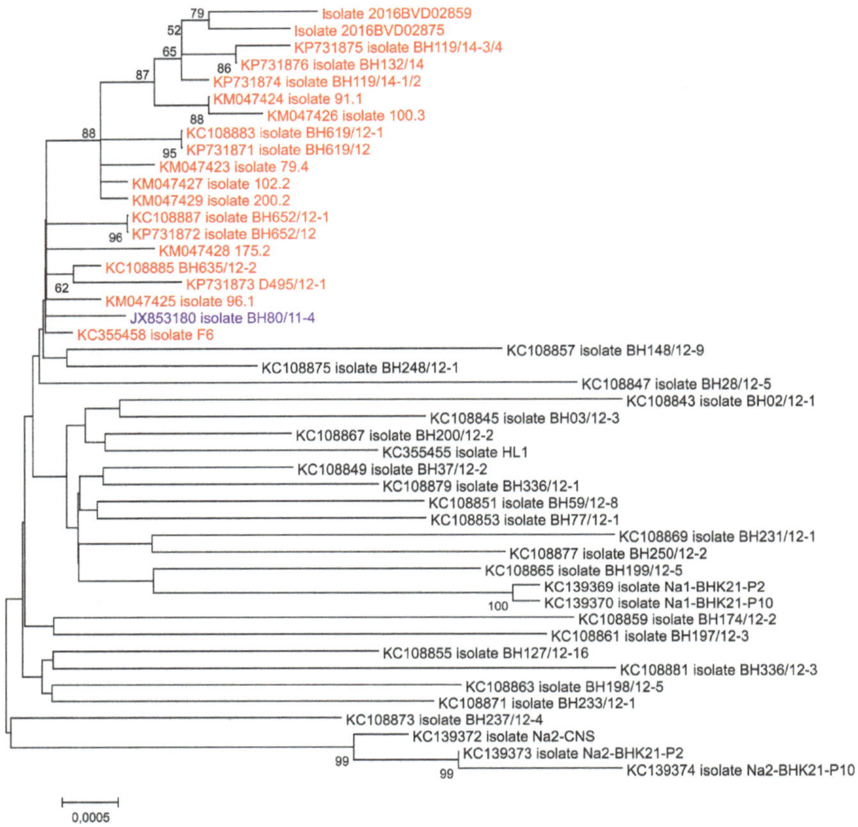

Fig. 4 Phylogenetic analysis based on the nucleotide sequences of the M-segment. The sequences were obtained from NCBI GenBank (labeled by accession number and isolate name) or are unpublished (isolates 2016BVD02859 and 2016BVD02875). Sequences labeled in *red* or *blue* (initial sample from a cattle farm located near the city of Schmallenberg, Germany) originate from the blood of acutely infected animals; sequences labeled in *black* were obtained from organ samples of malformed fetuses/newborns. Statistical support for nodes was obtained by bootstrapping (1000 replicates); only values ≥50% are shown. The *scale bar* indicates nucleotide substitutions per site.

the involvement in the adaption of the cell tropism within individual hosts (Coupeau et al., 2013; Fischer et al., 2013a). However, vector-transmitted viruses such as SBV have to adapt to two different hosts and undergo replication cycles in both hosts, i.e., the arthropod vector and the mammalian host. Thus, the high sequence stability of circulating SBV strains—detectable in the blood of acutely infected animals—might be necessary for the transmission to the vector, while the viruses present in malformed fetuses are not transmitted to the insect vectors.

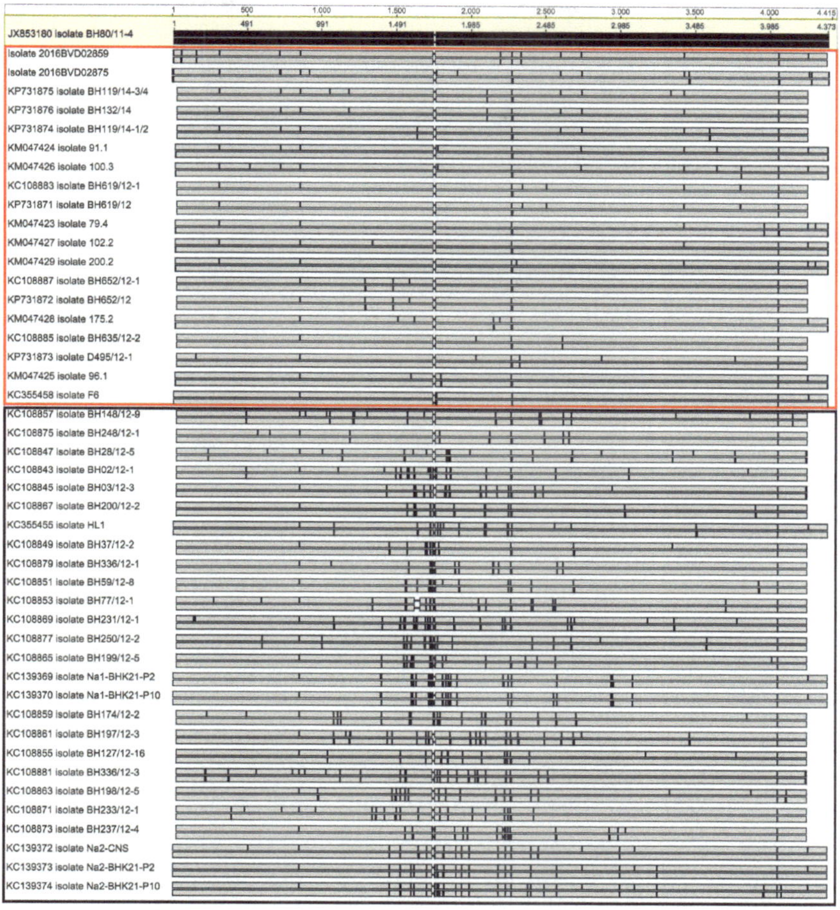

Fig. 5 Comparison of the M-segment sequences generated from acutely infected ruminants (framed in *red*) or from organ samples of malformed fetuses/newborns (framed in *black*). The sequences were obtained from NCBI GenBank (labeled by accession number and isolate name) or are unpublished (isolates 2016BVD02859 and 2016BVD02875). The first SBV isolate from 2011 is shaded in *yellow* and nucleotide (*upper gray bar* alongside the name of each isolate) and amino acid (*lower gray bar*) substitutions are highlighted as *vertical black lines*.

5. METHODS DEVELOPED FOR DIAGNOSIS AND TO PREVENT SBV INFECTION

The combination of a high sensitivity, the possibility of high-throughput screening, and the speed makes real-time RT-PCR to the most widely used method for the direct virus detection. Different PCR assays targeting either the L-, M-, or S-segment have been developed

(Bilk et al., 2012; Fischer et al., 2013b; Hoffmann et al., 2012) of which the S-segment-based system was shown to be most suitable for a specific and sensitive detection of SBV genome (Fischer et al., 2013b). Furthermore, a real-time RT-PCR assay for the detection of all members of the Simbu serogroup has been established which additionally enables a subsequent species classification via sequencing (Fischer et al., 2013b). However, the real-time RT-PCR as a tool to confirm an acute infection of adult animals is time restricted by the short viremia of a few days only; but viral genomes can be often detected for an extended period in the lymphoreticular system, particularly in the mesenteric lymph nodes (Wernike et al., 2012b, 2013a, b). In addition to the detection of viral RNA, the virus can be isolated on insect cell lines (e.g., *Culicoides variipennis* larvae cells), BHK, or African green monkey (Vero) cells (Coupeau et al., 2013; Hoffmann et al., 2012; Wernike et al., 2013b).

The different real-time RT-PCR assays can also be used to confirm fetal SBV infections, suitable sample materials include, e.g., brain samples, placenta, meconium, or hair swabs (Bilk et al., 2012; De Regge et al., 2012b, 2013). However, not all calves or lambs suspected of SBV infection test positive by PCR (De Regge et al., 2013; van Maanen et al., 2012), which has been also described for SBV-related viruses such as AKAV (Hartley et al., 1977). The most accepted explanation for this phenomenon is that the virus which has caused the initial lesions weeks or months before abortion, stillbirth, or birth is no longer present in the newborn.

Apart from samples from affected ruminants, real-time RT-PCR assays may also be used to screen insect vectors for the presence of viral RNA (Balenghien et al., 2014; Veronesi et al., 2013).

Owing to the short viremia of only a few days much more promising than real-time RT-PCR or virus isolation may be the detection of virus-specific antibodies, which is also a valuable tool for epidemiological investigations or to confirm a fetal infection, provided the sera were taken before the colostrum intake or fetal heart blood samples are tested.

Test systems for antibody detection include microneutralization, indirect immunofluorescence tests and various in-house or commercially available ELISAs (Bréard et al., 2013; Humphries and Burr, 2012; Lazutka et al., 2015; Loeffen et al., 2012; Wernike et al., 2013a). Although protocols are carried out with slight differences, the neutralization test was proven to be more sensitive than several ELISAs assessed in interlaboratory proficiency tests (van der Poel et al., 2014a; Wernike et al., 2016). In addition, the neutralization test is highly specific, while the S-segment-based commercial ELISA systems may led to crossreaction between SBV and other Simbu serogroup viruses (Wernike et al., 2016).

Owing to the impact of SBV infection on animal welfare, animal production, and the export of animals and their products (EFSA, 2012; Harris et al., 2014; Martinelle et al., 2014; Wüthrich et al., 2016), various preventive measures against SBV were discussed. Since it is an insect-transmitted virus, the application of repellents or insecticides could be considered to reduce the risk of exposure of susceptible animals to potentially infected vectors. However, a German case–control study conducted in 2012 provided no evidence for protective effects of such a treatment (Wernike et al., 2014a). Another possibility to prevent the transplacental transmission to the fetus may be an adjusted breeding system, i.e., the number of ewes and cows in the susceptible stage of gestation could be minimized during the season of the highest activity of the involved insect vectors. In addition, when grazing is practiced, the management systems could be adapted in ways that the young stock is kept outside which increases the likelihood of exposure to the insect vector. Immunity against SBV which is acquired before heifers conceive for the first time might prevent fetal infections in subsequent pregnancies. A much more reliable method to acquire immunity than an exposure to potentially infected insect vectors is admittedly vaccination against the virus. Hence, SBV-specific inactivated, live-attenuated, or subunit vaccines have been developed and successfully tested. For three inactivated whole virus vaccines marketing authorizations have been granted for the British, French, and European market, respectively (Anonymous, 2013b, 2015; Hechinger et al., 2014; Kraatz et al., 2015; Wernike et al., 2013c, 2017).

6. CURRENT SITUATION

Despite the very high SBV seroprevalences in Central European livestock in 2011, the virus still circulated in the 2012 and 2013 vector seasons, but at a low level (Elbers et al., 2015; Friedrich-Loeffler-Institut, 2015).

In summer and autumn 2014, SBV reappeared to a greater extent in continental Europe. The virus was repeatedly detected in blood samples of acutely infected animals (Wernike et al., 2015b) and, as a consequence, the frequency of offspring displaying SBV-induced malformation increased in the following winter in several Central European countries such as Belgium or Germany (Delooz et al., 2016; Wernike et al., 2015b). In contrast, the virus seemed not to be circulating in the south of England in that year as was deduced from serological analyses (Stokes et al., 2016).

In 2015, and the following winter, SBV cases were again reported only sporadically. However, in the 2016 vector season, the virus circulated once again on a larger scale in the European ruminant population and in the following winter an increasing number of malformed calves and lambs were born; cases have been reported from continental Europe to England, Wales, and Ireland (ProMED-mail, 2016a, b).

Germany, where SBV was detected for the first time and which represents besides the Netherlands and Belgium one of the initially most affected countries, was once more nationwide affected in the 2016 vector season as shown by the submission of PCR positive blood samples from various federal states to the German National Reference Laboratory for SBV infection (Table 1). With one exception, a blood sample from a mouflon, all samples originated from acutely infected cattle which were sampled in the context of preexport checks. Considering the very short viremia of only 3–6 days, a much larger number of infected adult animals had to be expected, which

Table 1 Origin of SBV-Positive Blood Samples Submitted to the German National Reference Laboratory for SBV in Summer and Autumn 2016

Submission Date	German Federal State	Number of Samples (Animal Species)	Quantification Cycle Value
11.08.2016	North Rhine-Westphalia	4 (cattle)	29.0–30.2
17.08.2016	Lower Saxony	1 (cattle)	28.3
18.08.2016	Rhineland-Palatinate	1 (cattle)	31.9
19.08.2016	North Rhine-Westphalia	1 (cattle)	39.5
23.08.2016	Rhineland-Palatinate	1 (cattle)	24.0
21.09.2016	Baden-Wuerttemberg	3 (cattle)	27.8–38.4
21.09.2016	Brandenburg	1 (mouflon), 4 (cattle)	24.7–28.4
27.09.2016	Baden-Wuerttemberg	5 (cattle)	31.4–37.2
04.10.2016	Mecklenburg-Hither Pomerania	1 (cattle)	32.8
13.10.2016	Baden-Wuerttemberg	2 (cattle)	22.7–27.3
20.10.2016	Saxony-Anhalt	3 (cattle)	22.5–26.3
23.11.2016	Lower Saxony	8 (cattle)	22.4–39.2

was once again confirmed by the increasing frequency of the birth of malformed offspring in Germany and further European countries in the following winter (Lewis, 2017; ProMED-mail, 2016b, 2017).

Such patterns of reemergence as seen for SBV during the last years are well known from other major members of the Simbu serogroup; there are, e.g., regular epidemics of AINOV in Japan or AKAV in Australia. Therefore, it has to be anticipated that SBV has established an endemic status in Europe and will reappear to a larger extent in regular intervals in the future.

7. SUMMARY

In late 2011, unspecific clinical symptoms such as fever, diarrhea, and decreased milk production were observed in dairy cattle in the Dutch/German border region. The causative agent was eventually detected by a metagenomic approach, isolated in cultured cells, and the etiological relationship between the virus and the observed disease has been proven in several experimental infection studies. Since the detection of SBV, substantial knowledge about the virus itself, but also about its transmission, the effects on the animal hosts and the epidemiology has been acquired. Furthermore, several molecular and serological test systems were developed for the diagnosis of SBV infections and multiple vaccine candidates have been tested successfully. The emergence and spread of SBV and the establishment of an endemic status in Europe is a constant reminder that new diseases can emerge at any time in our increasingly converging world, and that novel pathogens are able to spread very rapidly among naïve livestock.

ACKNOWLEDGMENT

Fig. 3 was kindly provided by M. Runge and S. Kleinschmidt (Hannover, Germany).

REFERENCES

Anonymous, 2012a. Findings in cases of SBV infection in sheep in the south-east. Vet. Rec. 170, 510.

Anonymous, 2012b. Geographical spread of UK cases of SBV 'within expectations'. Vet. Rec. 170, 217.

Anonymous, 2013a. First case of Schmallenberg disease confirmed in Scotland. Vet. Rec. 172, 435.

Anonymous, 2013b. VMD authorises SBV vaccine for use in the UK. Vet. Rec. 172, 543.

Anonymous, 2015. Schmallenberg virus vaccine. Vet. Rec. 177, 321.

Balenghien, T., Pages, N., Goffredo, M., Carpenter, S., Augot, D., Jacquier, E., Talavera, S., Monaco, F., Depaquit, J., Grillet, C., Pujols, J., Satta, G., Kasbari, M., Setier-Rio, M.L., Izzo, F., Alkan, C., Delecolle, J.C., Quaglia, M., Charrel, R., Polci, A., Breard, E., Federici, V., Cetre-Sossah, C., Garros, C., 2014. The emergence of Schmallenberg virus across Culicoides communities and ecosystems in Europe. Prev. Vet. Med. 116, 360.

Balmer, S., Vogtlin, A., Thur, B., Buchi, M., Abril, C., Houmard, M., Danuser, J., Schwermer, H., 2014. Serosurveillance of Schmallenberg virus in Switzerland using bulk tank milk samples. Prev. Vet. Med. 116, 370.

Barry, G., Varela, M., Ratinier, M., Blomstrom, A.L., Caporale, M., Seehusen, F., Hahn, K., Schnettler, E., Baumgärtner, W., Kohl, A., Palmarini, M., 2014. The NSs protein of Schmallenberg virus counteracts the antiviral response of the cell by inhibiting its transcriptional machinery. J. Gen. Virol. 95, 1640–1646.

Bilk, S., Schulze, C., Fischer, M., Beer, M., Hlinak, A., Hoffmann, B., 2012. Organ distribution of Schmallenberg virus RNA in malformed newborns. Vet. Microbiol. 159, 236.

Bradshaw, B., Mooney, J., Ross, P.J., Furphy, C., O'Donovan, J., Sanchez, C., Gomez-Parada, M., Toolan, D., 2012. Schmallenberg virus cases identified in Ireland. Vet. Rec. 171, 540.

Bréard, E., Lara, E., Comtet, L., Viarouge, C., Doceul, V., Desprat, A., Vitour, D., Pozzi, N., Cay, A.B., De Regge, N., Pourquier, P., Schirrmeier, H., Hoffmann, B., Beer, M., Sailleau, C., Zientara, S., 2013. Validation of a commercially available indirect ELISA using a nucleocapside recombinant protein for detection of Schmallenberg virus antibodies. PLoS One 8, e53446.

Chaintoutis, S.C., Kiossis, E., Giadinis, N.D., Brozos, C.N., Sailleau, C., Viarouge, C., Breard, E., Papanastassopoulou, M., Zientara, S., Papadopoulos, O., Dovas, C.I., 2014. Evidence of Schmallenberg virus circulation in ruminants in Greece. Tropl. Anim. Health Prod. 46, 251.

Chiari, M., Sozzi, E., Zanoni, M., Alborali, L.G., Lavazza, A., Cordioli, P., 2014. Serosurvey for Schmallenberg virus in alpine wild ungulates. Transbound. Emerg. Dis. 61, 1.

Chowdhary, R., Street, C., Travassos da Rosa, A., Nunes, M.R., Tee, K.K., Hutchison, S.K., Vasconcelos, P.F., Tesh, R.B., Lipkin, W.I., Briese, T., 2012. Genetic characterization of the Wyeomyia group of orthobunyaviruses and their phylogenetic relationships. J. Gen. Virol. 93, 1023.

Coupeau, D., Claine, F., Wiggers, L., Kirschvink, N., Muylkens, B., 2013. In vivo and in vitro identification of a hypervariable region in Schmallenberg virus. J. Gen. Virol. 94, 1168.

De Regge, N., Deblauwe, I., De Deken, R., Vantieghem, P., Madder, M., Geysen, D., Smeets, F., Losson, B., van den Berg, T., Cay, A.B., 2012a. Detection of Schmallenberg virus in different Culicoides spp. by real-time RT-PCR. Transbound. Emerg. Dis. 59, 471.

De Regge, N., Van Loo, H., Bertels, G., Saulmont, M., Vandenbussche, F., van den Berg, T., Cay, A.B., 2012b. Comparative analysis to determine the most appropriate tissue for Schmallenberg virus detection by RT-PCR in aborted lambs and calves. Proceedings of the EPIZONE 6th Annual Meeting. Brighton, June, 2012, p. 202.

De Regge, N., van den Berg, T., Georges, L., Cay, B., 2013. Diagnosis of Schmallenberg virus infection in malformed lambs and calves and first indications for virus clearance in the fetus. Vet. Microbiol. 162, 595.

Delooz, L., Saegerman, C., Quinet, C., Petitjean, T., De Regge, N., Cay, B., 2016. Resurgence of Schmallenberg virus in Belgium after 3 years of epidemiological silence. Transbound. Emerg. Dis.

Desmecht, D., Garigliany, M.M., Beer, M., Schirrmeier, H., Paternostre, J., Volpe, R., Linden, A., 2013. In: Detection of antibodies against Schmallenberg virus in wild boars, Belgium, 2010–2012. Proceedings of 31th Congress of the International Union of Game Biologists, Brussels, Belgium.

Dominguez, M., Gache, K., Touratier, A., Perrin, J.B., Fediaevsky, A., Collin, E., Breard, E., Sailleau, C., Viarouge, C., Zanella, G., Zientara, S., Hendrikx, P., Calavas, D., 2014. Spread and impact of the Schmallenberg virus epidemic in France in 2012–2013. BMC Vet. Res. 10, 248.
Dong, H., Li, P., Elliott, R.M., Dong, C., 2013. Structure of Schmallenberg orthobunyavirus nucleoprotein suggests a novel mechanism of genome encapsidation. J. Virol. 87, 5593.
EFSA, 2012. "Schmallenberg" Virus: Analysis of the Epidemiological Data (November 2012). EFSA Supporting Publications 2012. EN-360, Parma, Italy. http://www.efsa.europa.eu/en/supporting/doc/360e.pdf. accessed 15/07/2013.
Elbers, A.R., Loeffen, W.L., Quak, S., de Boer-Luijtze, E., van der Spek, A.N., Bouwstra, R., Maas, R., Spierenburg, M.A., de Kluijver, E.P., van Schaik, G., van der Poel, W.H., 2012. Seroprevalence of Schmallenberg virus antibodies among dairy cattle, the Netherlands, winter 2011-2012. Emerg. Infect. Dis. 18, 1065.
Elbers, A.R., Meiswinkel, R., van Weezep, E., Sloet van Oldruitenborgh-Oosterbaan, M.M., Kooi, E.A., 2013. Schmallenberg virus in Culicoides spp. biting midges, the Netherlands, 2011. Emerg. Infect. Dis. 19, 106.
Elbers, A.R., Meiswinkel, R., van Weezep, E., Kooi, E.A., van der Poel, W.H., 2015. Schmallenberg virus in Culicoides biting midges in the Netherlands in 2012. Transbound. Emerg. Dis. 62, 339.
Elliott, R.M., 2014. Orthobunyaviruses: recent genetic and structural insights. Nat. Rev. Microbiol. 12, 673–685.
Elliott, R.M., Blakqori, G., 2011. Molecular biology of orthobunyaviruses. In: Plyusnin, A., Elliott, R.M. (Eds.), Bunyaviridae. Molecular and Cellular Biology. Caister Academic Press, Norfolk, UK, p. 1.
Fernandez-Aguilar, X., Pujols, J., Velarde, R., Rosell, R., Lopez-Olvera, J.R., Marco, I., Pumarola, M., Segales, J., Lavin, S., Cabezon, O., 2014. Schmallenberg virus circulation in high mountain ecosystem, Spain. Emerg. Infect. Dis. 20, 1062.
Fischer, M., Hoffmann, B., Goller, K.V., Höper, D., Wernike, K., Beer, M., 2013a. A mutation 'hot spot' in the Schmallenberg virus M segment. J. Gen. Virol. 94, 1161.
Fischer, M., Schirrmeier, H., Wernike, K., Wegelt, A., Beer, M., Hoffmann, B., 2013b. Development of a pan-Simbu real-time reverse transcriptase PCR for the detection of Simbu serogroup viruses and comparison with SBV diagnostic PCR systems. Virol. J. 10, 327.
Friedrich-Loeffler-Institut, 2015. FLI: Schmallenberg virus. https://www.fli.de/en/news/animal-disease-situation/schmallenberg-virus/Friedrich-Loeffler-Institut, Greifswald-Insel Riems, Germany (last accessed: January 2017).
Garigliany, M.M., Bayrou, C., Kleijnen, D., Cassart, D., Desmecht, D., 2012a. Schmallenberg virus in domestic cattle, Belgium, 2012. Emerg. Infect. Dis. 18, 1512.
Garigliany, M.M., Hoffmann, B., Dive, M., Sartelet, A., Bayrou, C., Cassart, D., Beer, M., Desmecht, D., 2012b. Schmallenberg virus in calf born at term with porencephaly, Belgium. Emerg. Infect. Dis. 18, 1005.
Garigliany, M.M., Desmecht, D., Bayrou, C., Peeters, D., 2013. No serologic evidence for emerging Schmallenberg virus infection in dogs (Canis domesticus). Vector Borne Zoonotic Dis. 13, 830.
Goffredo, M., Monaco, F., Capelli, G., Quaglia, M., Federici, V., Catalani, M., Montarsi, F., Polci, A., Pinoni, C., Calistri, P., Savini, G., 2013. Schmallenberg virus in Italy: a retrospective survey in Culicoides stored during the bluetongue Italian surveillance program. Prev. Vet. Med. 111, 230.
Goller, K.V., Höper, D., Schirrmeier, H., Mettenleiter, T.C., Beer, M., 2012. Schmallenberg virus as possible ancestor of Shamonda virus. Emerg. Infect. Dis. 18, 1644.

Gouzil, J., Fablet, A., Lara, E., Caignard, G., Cochet, M., Kundlacz, C., Palmarini, M., Varela, M., Bréard, E., Sailleau, C., Viarouge, C., Coulpier, M., Zientara, S., Vitour, D., 2017. Nonstructural protein NSs of Schmallenberg virus is targeted to the nucleolus and induces nucleolar disorganization. J. Virol. 91, e01263-16.
Harris, K.A., Eglin, R.D., Hayward, S., Milnes, A., Davies, I., Cook, A.J., Downs, S.H., 2014. Impact of Schmallenberg virus on British sheep farms during the 2011/2012 lambing season. Vet. Rec. 175, 172.
Hartley, W.J., De Saram, W.G., Della-Porta, A.J., Snowdon, W.A., Shepherd, N.C., 1977. Pathology of congenital bovine epizootic arthrogryposis and hydranencephaly and its relationship to Akabane virus. Aust. Vet. J. 53, 319.
Hashiguchi, Y., Nanba, K., Kumagai, T., 1979. Congenital abnormalities in newborn lambs following Akabane virus infection in pregnant ewes. Natl. Inst. Anim. Health Q (Tokyo) 19, 1.
Hechinger, S., Wernike, K., Beer, M., 2014. Single immunization with an inactivated vaccine protects sheep from Schmallenberg virus infection. Vet. Res. 45, 79.
Helmer, C., Eibach, R., Tegtmeyer, P.C., Humann-Ziehank, E., Ganter, M., 2013. Survey of Schmallenberg virus (SBV) infection in German goat flocks. Epidemiol. Infect. 141, 2335.
Herder, V., Wohlsein, P., Peters, M., Hansmann, F., Baumgärtner, W., 2012. Salient lesions in domestic ruminants infected with the emerging so-called Schmallenberg virus in Germany. Vet. Pathol. 49, 588.
Hoffmann, B., Scheuch, M., Höper, D., Jungblut, R., Holsteg, M., Schirrmeier, H., Eschbaumer, M., Goller, K.V., Wernike, K., Fischer, M., Breithaupt, A., Mettenleiter, T.C., Beer, M., 2012. Novel orthobunyavirus in cattle, Europe, 2011. Emerg. Infect. Dis. 18, 469.
Hoffmann, B., Schulz, C., Beer, M., 2013. First detection of Schmallenberg virus RNA in bovine semen, Germany, 2012. Vet. Microbiol. 167, 289.
Humphries, D., Burr, P., 2012. Schmallenberg virus milk antibody ELISA. Vet. Rec. 171, 511.
Kirkland, P.D., Barry, R.D., Harper, P.A., Zelski, R.Z., 1988. The development of Akabane virus-induced congenital abnormalities in cattle. Vet. Rec. 122, 582.
Kobayashi, T., Yanase, T., Yamakawa, M., Kato, T., Yoshida, K., Tsuda, T., 2007. Genetic diversity and reassortments among Akabane virus field isolates. Virus Res. 130, 162.
Konno, S., Moriwaki, M., Nakagawa, M., 1982. Akabane disease in cattle: congenital abnormalities caused by viral infection. Spontaneous disease. Vet. Pathol. 19, 246.
Kraatz, F., Wernike, K., Hechinger, S., König, P., Granzow, H., Reimann, I., Beer, M., 2015. Deletion mutants of Schmallenberg virus are avirulent and protect from virus challenge. J. Virol. 89, 1825.
Kurogi, H., Inaba, Y., Takahashi, E., Sato, K., Satoda, K., 1977. Congenital abnormalities in newborn calves after inoculation of pregnant cows with Akabane virus. Infect. Immun. 17, 338.
Laloy, E., Riou, M., Barc, C., Belbis, G., Breard, E., Breton, S., Cordonnier, N., Crochet, D., Delaunay, R., Moreau, J., Pozzi, N., Raimbourg, M., Sarradin, P., Trapp, S., Viarouge, C., Zientara, S., Ponsart, C., 2015. Schmallenberg virus: experimental infection in goats and bucks. BMC Vet. Res. 11, 221.
Laloy, E., Braud, C., Breard, E., Kaandorp, J., Bourgeois, A., Kohl, M., Meyer, G., Sailleau, C., Viarouge, C., Zientara, S., Chai, N., 2016. Schmallenberg virus in zoo ruminants, France and the Netherlands. Emerg. Infect. Dis. 22, 2201.
Larska, M., Krzysiak, M., Smreczak, M., Polak, M.P., Zmudzinski, J.F., 2013a. First detection of Schmallenberg virus in elk (Alces alces) indicating infection of wildlife in Bialowieza National Park in Poland. Vet. J. 198, 279.
Larska, M., Polak, M.P., Grochowska, M., Lechowski, L., Zwiazek, J.S., Zmudzinski, J.F., 2013b. First report of Schmallenberg virus infection in cattle and midges in Poland. Transbound. Emerg. Dis. 60, 97.

Lazutka, J., Zvirbliene, A., Dalgediene, I., Petraityte-Burneikiene, R., Spakova, A., Sereika, V., Lelesius, R., Wernike, K., Beer, M., Sasnauskas, K., 2014. Generation of recombinant Schmallenberg virus nucleocapsid protein in yeast and development of virus-specific monoclonal antibodies. J. Immunol. Res. 2014, 160316.

Lazutka, J., Spakova, A., Sereika, V., Lelesius, R., Sasnauskas, K., Petraityte-Burneikiene, R., 2015. Saliva as an alternative specimen for detection of Schmallenberg virus-specific antibodies in bovines. BMC Vet. Res. 11, 237.

Lewis, C., 2017. Schmallenberg disease in UK flocks. Vet. Rec. 180, 100.

Lievaart-Peterson, K., Luttikholt, S.J.M., Van den Brom, R., Vellema, P., 2012. Schmallenberg virus infection in small ruminants—first review of the situation and prospects in Northern Europe. Small Rumin. Res. 106, 71.

Linden, A., Desmecht, D., Volpe, R., Wirtgen, M., Gregoire, F., Pirson, J., Paternostre, J., Kleijnen, D., Schirrmeier, H., Beer, M., Garigliany, M.M., 2012. Epizootic spread of Schmallenberg virus among wild Cervids, Belgium, Fall 2011. Emerg. Infect. Dis. 18, 2006.

Loeffen, W., Quak, S., De Boer-Luijtze, E., Hulst, M., van der Poel, W., Bouwstra, R., Maas, R., 2012. Development of a virus neutralisation test to detect antibodies against Schmallenberg virus and serological results in suspect and infected herds. Acta Vet. Scand. 54, 44.

Loeffler, F., 1884. Untersuchung über die Bedeutung der Mikroorganismen für die Entstehung der Diphtherie beim Menschen, bei der Taube und beim Kalbe. In: Mitteilungen aus dem kaiserlichen Gesundheitsamte, vol. 2, 421.

Luttikholt, S., Veldhuis, A., van den Brom, R., Moll, L., Lievaart-Peterson, K., Peperkamp, K., van Schaik, G., Vellema, P., 2014. Risk factors for malformations and impact on reproductive performance and mortality rates of Schmallenberg virus in sheep flocks in the Netherlands. PLoS One 9, e100135.

Martinelle, L., Dal Pozzo, F., Gauthier, B., Kirschvink, N., Saegerman, C., 2014. Field veterinary survey on clinical and economic impact of Schmallenberg virus in Belgium. Transbound. Emerg. Dis. 61, 285.

Martinelle, L., Poskin, A., Dal Pozzo, F., De Regge, N., Cay, B., Saegerman, C., 2015. Experimental infection of sheep at 45 and 60 days of gestation with Schmallenberg virus readily led to placental colonization without causing congenital malformations. PLoS One 10, e0139375.

Mason, C., Stevenson, H., Carty, H., Hosie, B., Caldow, G., Boyes, G., 2013. SBV in a dairy herd in Scotland. Vet. Rec. 172, 403.

Molenaar, F.M., La Rocca, S.A., Khatri, M., Lopez, J., Steinbach, F., Dastjerdi, A., 2015. Exposure of Asian elephants and other exotic ungulates to Schmallenberg virus. PLoS One 10, e0135532.

Mouchantat, S., Wernike, K., Lutz, W., Hoffmann, B., Ulrich, R.G., Borner, K., Wittstatt, U., Beer, M., 2015. A broad spectrum screening of Schmallenberg virus antibodies in wildlife animals in Germany. Vet. Res. 46, 99.

Parsonson, I.M., Della-Porta, A.B., Snowdon, W.A., Murray, M.D., 1975. Letter: congenital abnormalities in foetal lambs after inoculation of pregnant ewes with Akabane virus. Aust. Vet. J. 51, 585.

Parsonson, I.M., Della-Porta, A.J., Snowdon, W.A., 1977. Congenital abnormalities in newborn lambs after infection of pregnant sheep with Akabane virus. Infect. Immun. 15, 254.

Parsonson, I.M., McPhee, D.A., Della-Porta, A.J., McClure, S., McCullagh, P., 1988. Transmission of Akabane virus from the ewe to the early fetus (32 to 53 days). J. Comp. Pathol. 99, 215.

Peperkamp, K., Dijkman, R., van Maanen, C., Vos, J., Wouda, W., Holzhauer, M., van Wuijckhuise, L., Junker, K., Greijdanus, S., Roumen, M., 2012. Polioencephalomyelitis in a calf due to infection with Schmallenberg virus. Vet. Rec. 170, 570.

Plyusnin, A., Beaty, B.J., Elliott, R.M., Goldbach, R., Kormelink, R., Lundkvist, A., Schmaljohn, C.S., Tesh, R.B., 2012. Bunyaviridae. In: King, A.M.Q., Adams, M.J., Carstens, E.B., Lefkowits, E.J. (Eds.), Virustaxonomy: Ninth Report of the International Committee on Taxonomy of Viruses. Elsevier Academic Press, London, p. 725.

Ponsart, C., Pozzi, N., Breard, E., Catinot, V., Viard, G., Sailleau, C., Viarouge, C., Gouzil, J., Beer, M., Zientara, S., Vitour, D., 2014. Evidence of excretion of Schmallenberg virus in bull semen. Vet. Res. 45, 37.

Poskin, A., Van Campe, W., Mostin, L., Cay, B., De Regge, N., 2014. Experimental Schmallenberg virus infection of pigs. Vet. Microbiol. 170, 398.

ProMED-mail, 2012a. Schmallenberg Virus—Europe (62): Sweden (BL) pos. Serology. Archiv. 20121005.1326266, http://www.promedmail.org (posted 05.10.12).

ProMED-mail, 2012b. Schmallenberg Virus—Europe (73): Norway, Sweden, Update. Archiv. 20121128.1428668, http://www.promedmail.org (posted 28.11.12).

ProMED-mail, 2013. Schmallenberg Virus—Europe (04): Finland Confirmed, UK Update. Archiv. 20130115.1499846, http://www.promedmail.org (posted 15.01.13).

ProMED-mail, 2016a. Schmallenberg Virus—Europe (02): UK (England, Wales), Reemergence Susp. Archiv. 20161214.4698989, http://www.promedmail.org (posted 14.12.16).

ProMED-mail, 2016b. Schmallenberg Virus—Europe: Netherlands, Belgium, Reemergence. Archiv. 20161023.4578989, http://www.promedmail.org (posted 23.10.16).

ProMED-mail, 2017. Schmallenberg Virus—Europe: Ireland (CO, LS) Bovine, Reemergence. Archiv. 20170106.4747269, http://www.promedmail.org (posted 06.01.17).

Radostits, O.M., Gay, C.C., Hinchcliff, K.W., Constable, P.D., 2007. Congenital Arthrogryposis and Hydranencephaly; Akabane Disease; Cache Valley Virus Disease. Veterinary Medicine: A Textbook of the Diseases of Cattle, Horses, Sheep, Pigs and Goats, 10th ed. Saunders Ltd., Philadelphia, PA, p. 1207.

Rasmussen, L.D., Kristensen, B., Kirkeby, C., Rasmussen, T.B., Belsham, G.J., Bodker, R., Bøtner, A., 2012. Culicoids as vectors of Schmallenberg virus. Emerg. Infect. Dis. 18, 1204.

Rasmussen, L.D., Kirkeby, C., Bodker, R., Kristensen, B., Rasmussen, T.B., Belsham, G.J., Bøtner, A., 2014. Rapid spread of Schmallenberg virus-infected biting midges (Culicoides spp.) across Denmark in 2012. Transbound. Emerg. Dis. 61, 12.

Roman-Sosa, G., Brocchi, E., Schirrmeier, H., Wernike, K., Schelp, C., Beer, M., 2016. Analysis of the humoral immune response against the envelope glycoprotein Gc of Schmallenberg virus reveals a domain located at the amino terminus targeted by mAbs with neutralizing activity. J. Gen. Virol. 97, 571.

Saeed, M.F., Li, L., Wang, H., Weaver, S.C., Barrett, A.D., 2001. Phylogeny of the Simbu serogroup of the genus Bunyavirus. J. Gen. Virol. 82, 2173.

Saegerman, C., Martinelle, L., Dal Pozzo, F., Kirschvink, N., 2014. Preliminary survey on the impact of Schmallenberg virus on sheep flocks in south of Belgium. Transbound. Emerg. Dis. 61, 469.

Sailleau, C., Boogaerts, C., Meyrueix, A., Laloy, E., Breard, E., Viarouge, C., Desprat, A., Vitour, D., Doceul, V., Boucher, C., Zientara, S., Nicolier, A., Grandjean, D., 2013. Schmallenberg virus infection in dogs, France, 2012. Emerg. Infect. Dis. 19, 1896.

Scholte, E.J., Mars, M.H., Braks, M., Den Hartog, W., Ibanez-Justicia, A., Koopmans, M., Koenraadt, C.J., De Vries, A., Reusken, C., 2014. No evidence for the persistence of Schmallenberg virus in overwintering mosquitoes. Med. Vet. Entomol. 28, 110.

Schulz, C., Wernike, K., Beer, M., Hoffmann, B., 2014. Infectious Schmallenberg virus from bovine semen, Germany. Emerg. Infect. Dis. 20, 337.

Schulz, C., Beer, M., Hoffmann, B., 2015. Schmallenberg virus infection in South American camelids: field and experimental investigations. Vet. Microbiol. 180, 171.

St. George, T.D., Kirkland, P.D., 2004. Diseases caused by Akabane and related Simbu-group viruses. In: Coetzer, J.A.W., Tustin, R.C. (Eds.), Infectious Diseases of Livestock, second ed. Oxford University Press, Oxford, UK, p. 1029.

Steinrigl, A., Schiefer, P., Schleicher, C., Peinhopf, W., Wodak, E., Bago, Z., Schmoll, F., 2014. Rapid spread and association of Schmallenberg virus with ruminant abortions and foetal death in Austria in 2012/2013. Prev. Vet. Med. 116, 350.

Stokes, J.E., Baylis, M., Duncan, J.S., 2016. A freedom from disease study: Schmallenberg virus in the south of England in 2015. Vet. Rec. 179, 435.

van den Brom, R., Luttikholt, S.J., Lievaart-Peterson, K., Peperkamp, N.H., Mars, M.H., van der Poel, W.H., Vellema, P., 2012. Epizootic of ovine congenital malformations associated with Schmallenberg virus infection. Tijdschr. Diergeneeskd. 137, 106.

van der Poel, W.H., Cay, B., Zientara, S., Steinbach, F., Valarcher, J.F., Bøtner, A., Mars, M.H., Hakze-van der Honing, R., Schirrmeier, H., Beer, M., 2014a. Limited interlaboratory comparison of Schmallenberg virus antibody detection in serum samples. Vet. Rec. 174, 380.

van der Poel, W.H., Parlevliet, J.M., Verstraten, E.R., Kooi, E.A., Hakze-Van Der Honing, R., Stockhofe, N., 2014b. Schmallenberg virus detection in bovine semen after experimental infection of bulls. Epidemiol. Infect. 142, 1495.

van Maanen, C., van der Heijden, H., Wellenberg, G.J., Witteveen, G., Luttikholt, S., Bouwstra, R., Kooi, B., Vellema, P., Peperkamp, K., Mars, J., 2012. Schmallenberg virus antibodies in bovine and ovine fetuses. Vet. Rec. 171, 299.

Varela, M., Schnettler, E., Caporale, M., Murgia, C., Barry, G., McFarlane, M., McGregor, E., Piras, I.M., Shaw, A., Lamm, C., Janowicz, A., Beer, M., Glass, M., Herder, V., Hahn, K., Baumgartner, W., Kohl, A., Palmarini, M., 2013. Schmallenberg virus pathogenesis, tropism and interaction with the innate immune system of the host. PLoS Pathog. 9, e1003133.

Veldhuis, A.M., van Schaik, G., Vellema, P., Elbers, A.R., Bouwstra, R., van der Heijden, H.M., Mars, M.H., 2013. Schmallenberg virus epidemic in the Netherlands: spatiotemporal introduction in 2011 and seroprevalence in ruminants. Prev. Vet. Med. 112, 35.

Veldhuis, A.M., Carp-van Dijken, S., van Wuijckhuise, L., Witteveen, G., van Schaik, G., 2014. Schmallenberg virus in Dutch dairy herds: potential risk factors for high within-herd seroprevalence and malformations in calves, and its impact on productivity. Vet. Microbiol. 168, 281.

Veronesi, E., Henstock, M., Gubbins, S., Batten, C., Manley, R., Barber, J., Hoffmann, B., Beer, M., Attoui, H., Mertens, P.P., Carpenter, S., 2013. Implicating culicoides biting midges as vectors of Schmallenberg virus using semi-quantitative rt-PCR. PLoS One 8, e57747.

Walter, C.T., Barr, J.N., 2011. Recent advances in the molecular and cellular biology of bunyaviruses. J. Gen. Virol. 92, 2467.

Wensman, J.J., Blomqvist, G., Hjort, M., Holst, B.S., 2013. Presence of antibodies to Schmallenberg virus in a dog in Sweden. J. Clin. Microbiol. 51, 2802.

Wernike, K., Breithaupt, A., Keller, M., Hoffmann, B., Beer, M., Eschbaumer, M., 2012a. Schmallenberg virus infection of adult type I interferon receptor knock-out mice. PLoS One 7, e40380.

Wernike, K., Eschbaumer, M., Breithaupt, A., Hoffmann, B., Beer, M., 2012b. Schmallenberg virus challenge models in cattle: infectious serum or culture-grown virus? Vet. Res. 43, 84.

Wernike, K., Eschbaumer, M., Schirrmeier, H., Blohm, U., Breithaupt, A., Hoffmann, B., Beer, M., 2013a. Oral exposure, reinfection and cellular immunity to Schmallenberg virus in cattle. Vet. Microbiol. 165, 155.

Wernike, K., Hoffmann, B., Bréard, E., Bøtner, A., Ponsart, C., Zientara, S., Lohse, L., Pozzi, N., Viarouge, C., Sarradin, P., Leroux-Barc, C., Riou, M., Laloy, E., Breithaupt, A., Beer, M., 2013b. Schmallenberg virus experimental infection of sheep. Vet. Microbiol. 166, 461.

Wernike, K., Nikolin, V.M., Hechinger, S., Hoffmann, B., Beer, M., 2013c. Inactivated Schmallenberg virus prototype vaccines. Vaccine 31, 3558.

Wernike, K., Conraths, F., Zanella, G., Granzow, H., Gache, K., Schirrmeier, H., Valas, S., Staubach, C., Marianneau, P., Kraatz, F., Höreth-Böntgen, D., Reimann, I., Zientara, S., Beer, M., 2014a. Schmallenberg virus-two years of experiences. Prev. Vet. Med. 116, 423.

Wernike, K., Holsteg, M., Schirrmeier, H., Hoffmann, B., Beer, M., 2014b. Natural infection of pregnant cows with Schmallenberg virus—a follow-up study. PLoS One 9, e98223.

Wernike, K., Jöst, H., Becker, N., Schmidt-Chanasit, J., Beer, M., 2014c. Lack of evidence for the presence of Schmallenberg virus in mosquitoes in Germany, 2011. Parasit. Vectors 7, 402.

Wernike, K., Silaghi, C., Nieder, M., Pfeffer, M., Beer, M., 2014d. Dynamics of Schmallenberg virus infection within a cattle herd in Germany, 2011. Epidemiol. Infect. 142, 1501.

Wernike, K., Brocchi, E., Cordioli, P., Senechal, Y., Schelp, C., Wegelt, A., Aebischer, A., Roman-Sosa, G., Reimann, I., Beer, M., 2015a. A novel panel of monoclonal antibodies against Schmallenberg virus nucleoprotein and glycoprotein Gc allows specific orthobunyavirus detection and reveals antigenic differences. Vet. Res. 46, 27.

Wernike, K., Hoffmann, B., Conraths, F.J., Beer, M., 2015b. Schmallenberg virus recurrence, Germany, 2014. Emerg. Infect. Dis. 21, 1202.

Wernike, K., Beer, M., Hoffmann, B., 2016. Schmallenberg virus infection diagnosis: results of a German proficiency trial. Transbound. Emerg. Dis.

Wernike, K., Aebischer, A., Roman-Sosa, G., Beer, M., 2017. The N-terminal domain of Schmallenberg virus envelope protein Gc is highly immunogenic and can provide protection from infection. Sci. Rep. 7, 42500.

Wisloff, H., Nordvik, B.S., Sviland, S., Tonnessen, R., 2014. The first documented clinical case of Schmallenberg virus in Norway: fetal malformations in a calf. Vet. Rec. 174, 120.

Wüthrich, M., Lechner, I., Aebi, M., Vögtlin, A., Posthaus, H., Schüpbach-Regula, G., Meylan, M., 2016. A case–control study to estimate the effects of acute clinical infection with the Schmallenberg virus on milk yield, fertility and veterinary costs in Swiss dairy herds. Prev. Vet. Med. 126, 54.

Yanase, T., Kato, T., Aizawa, M., Shuto, Y., Shirafuji, H., Yamakawa, M., Tsuda, T., 2012. Genetic reassortment between Sathuperi and Shamonda viruses of the genus Orthobunyavirus in nature: implications for their genetic relationship to Schmallenberg virus. Arch. Virol. 157, 1611.

Zanella, G., Raballand, C., Durand, B., Sailleau, C., Pelzer, S., Benoit, F., Doceul, V., Zientara, S., Bréard, E., 2013. Likely introduction date of Schmallenberg virus into France according to monthly serological surveys in cattle. Transbound. Emerg. Dis. 62, e76–9.

CHAPTER FOUR

Assessing the Diversity of Rodent-Borne Viruses: Exploring of High-Throughput Sequencing and Classical Amplification/Sequencing Approaches

Stephan Drewes*, Petra Straková†, Jan F. Drexler‡,§, Jens Jacob¶, Rainer G. Ulrich*,#,1

*Institute of Novel and Emerging Infectious Diseases, Friedrich-Loeffler-Institut, Greifswald-Insel Riems, Germany

†Institute of Vertebrate Biology v.v.i., Academy of Sciences, Brno, Czech Republic

‡Charité—Universitätsmedizin Berlin, Corporate Member of Freie Universität Berlin, Humboldt-Universität zu Berlin, and Berlin Institute of Health, Institute of Virology, Berlin, Germany

§German Center for Infection Research (DZIF), Germany

¶Julius Kühn-Institute, Federal Research Centre for Cultivated Plants, Institute for Plant Protection in Horticulture and Forests, Vertebrate Research, Münster, Germany

#German Center for Infection Research (DZIF), Partner site Hamburg-Luebeck-Borstel-Insel Riems, Greifswald-Insel Riems, Germany

[1]Corresponding author: e-mail address: rainer.ulrich@fli.de

Contents

Abstract

Rodents are distributed throughout the world and interact with humans in many ways. They provide vital ecosystem services, some species are useful models in biomedical research and some are held as pet animals. However, many rodent species can have adverse effects such as damage to crops and stored produce, and they are of health concern because of the transmission of pathogens to humans and livestock.

The first rodent viruses were discovered by isolation approaches and resulted in break-through knowledge in immunology, molecular and cell biology, and cancer

Advances in Virus Research, Volume 99
ISSN 0065-3527
https://doi.org/10.1016/bs.aivir.2017.08.002

research. In addition to rodent-specific viruses, rodent-borne viruses are causing a large number of zoonotic diseases. Most prominent examples are reemerging outbreaks of human hemorrhagic fever disease cases caused by arena- and hantaviruses. In addition, rodents are reservoirs for vector-borne pathogens, such as tick-borne encephalitis virus and *Borrelia* spp., and may carry human pathogenic agents, but likely are not involved in their transmission to human.

In our days, next-generation sequencing or high-throughput sequencing (HTS) is revolutionizing the speed of the discovery of novel viruses, but other molecular approaches, such as generic RT-PCR/PCR and rolling circle amplification techniques, contribute significantly to the rapidly ongoing process. However, the current knowledge still represents only the tip of the iceberg, when comparing the known human viruses to those known for rodents, the mammalian taxon with the largest species number. The diagnostic potential of HTS-based metagenomic approaches is illustrated by their use in the discovery and complete genome determination of novel borna- and adenoviruses as causative disease agents in squirrels.

In conclusion, HTS, in combination with conventional RT-PCR/PCR-based approaches, resulted in a drastically increased knowledge of the diversity of rodent viruses. Future improvements of the used workflows, including bioinformatics analysis, will further enhance our knowledge and preparedness in case of the emergence of novel viruses. Classical virological and additional molecular approaches are needed for genome annotation and functional characterization of novel viruses, discovered by these technologies, and evaluation of their zoonotic potential.

ABBREVIATIONS

ANDV Andes virus
APHAEA Approaches in monitoring wildlife Population Health, and Ecology and Abundance
ASHV Arctic ground squirrel hepatitis B virus
cDNA complementary deoxyribonucleic acid
CPXV Cowpox virus
DOBV Dobrava-Belgrade virus
dsDNA double-stranded deoxyribonucleic acid
ECTV Ectromelia virus
GenBank genetic sequence database
GSHV Ground squirrel hepatitis virus
HAV Hepatitis A virus
HaPyV Hamster polyomavirus
HCV Hepatitis C virus
HEV Hepatitis E virus
HTNV Hantaan virus
HTS high-throughput sequencing
JTT Jones–Taylor–Thornton substitution model
JUNV Junin virus
LASV Lassa virus
LCMV Lymphocytic choriomeningitis virus
LNV Laguna Negra virus

MACV Machupo virus
MCMV Mouse cytomegalovirus
MEV Murine erythroblastosis virus
MHC major histocompatibility complex
MHV Murine hepatitis virus
MHV-68 Murine gammaherpesvirus 68
MLV Murine leukemia and lymphoma virus
MMTV Mouse mammary tumor virus
MOBV Mobala virus
MPtV Murine pneumotropic virus
MPV Murine parvovirus
MPyV Murine polyomavirus
NCBI National Center for Biotechnology Information
PARV Parana virus
PCR polymerase chain reaction
POC point-of-care
PUUV Puumala virus
RACE rapid amplification of cDNA ends
RCA rolling circle amplification
RT-PCR reverse transcription-polymerase chain reaction
RT-qPCR reverse transcription-quantitative polymerase chain reaction
SNV Sin Nombre virus
ssRNA positive (+) or negative (−) single-stranded ribonucleic acid
TBEV Tick-borne encephalitis virus
TULV Tula virus
UN United Nations
USA United States of America
VSBV-1 Variegated squirrel bornavirus 1
WHV Woodchuck hepatitis virus

1. RODENTS: IMPORTANT RESERVOIRS FOR ZOONOTIC PATHOGENS

Rodents are the most numerous mammalian taxon regarding the number of species and individuals. They are distributed throughout the world and populate diverse habitats including native forest, desert, agricultural land, and urban infrastructure. In Africa alone, the number of rodent species is close to 400 (Wilson and Reeder, 1993). As widespread as rodents occur as manifold are ecological features they have to cope with their environment morphologically (e.g., variation in adult body weight from about 4 g to 60 kg), physiologically (e.g., torpor and hibernation), biologically (e.g., reproduction along the gradient of r and K-strategies), and behaviorally (e.g., night vs day activity). Some rodent species are highly endangered and some are so abundant that they are significant agricultural pests (Singleton et al., 2007). Rodents are used for food and fur in several regions of the world, are highly relevant as experimental animals in biomedical research, and even held as pets.

Many parasites and bacterial and viral pathogens are present in rodents and some of them can be transmitted to humans and cause severe zoonotic disease (see later). Zoonotic diseases are defined as "infections that are naturally transmitted between vertebrate host and humans" (Hubálek and Rudolf, 2011). The epidemiology of zoonotic disease and human infection risk is strongly linked to the ecology of the associated wildlife pathogen host. This is particularly relevant for rodent-borne zoonotic disease because the abundance of many rodent host populations can fluctuate naturally at several orders of magnitude.

Fluctuation in rodent abundance is often driven seasonally by the vegetation period creating resources that allow rodents to reproduce successfully and hence populations to grow. Therefore, reproduction in small rodents is aligned to the availability of food resources (Eccard and Ylönen, 2001; Firquet et al., 1996; Phung et al., 2012). Additionally, and potentially much more relevant for human infection risk are multiannual fluctuations of rodent population dynamics. This phenomenon is certainly one of the most impressive ecological features of population dynamics and occurs throughout the world (Cornulier et al., 2013; Krebs and Myers, 1974; Leirs et al., 1997).

Some multiannual fluctuations of small mammal abundance are mainly driven bottom-up by occasional pulses of plentiful food provided by tree seed mast such as in the systems Asian rodent species (e.g., *Mus cervicolor, Rattus rattus, Berylmys berdmorei, Bandicota indica*)—bamboo (Douangboupha et al., 2003; Nag, 1999), bank vole (*Myodes glareolus*)—beech (*Fagus* spec.) (Linard et al., 2007), and house mouse (*Mus musculus*)—*Nothofagus* beech (Fitzgerald et al., 2004). In other species, or even the same species in other regions, similar multiannual fluctuations occur but mechanisms that drive them are different. For instance, multiannual fluctuation of bank vole population dynamics in Fennoscandia do not seem to be regulated by tree seed mast (Hörnfeldt, 1994). In these populations as well as in other small mammal species top-down effects may matter (Korpimäki et al., 2005) but other factors also contribute to variation (Korpimäki et al., 2004; Lambin et al., 2006). Such factors include dispersal abilities (Gliwicz, 1990), stress (Frank, 1957), genetic effects (Chitty, 1967), demographic features (Radchuk et al., 2016), as well as parasites and pathogens (Smith et al., 2008).

Human infection risk is often positively correlated to the abundance of the rodent host population—rodent population outbreaks are followed by a considerably increased number of human disease cases. Rodent density dependent (or delayed density dependent) transmission of zoonotic pathogens is mainly due to increased contact rates between infected rodents and humans (Thrall et al., 1993). This has been shown for a number of virus-rodent host-systems such as bank voles—Puumala virus (PUUV) (Reil et al., 2016; Tersago et al., 2009), *Oligoryzomys longicaudatus*—Andes virus (ANDV), and *Calomys laucha*—Laguna Negra virus (LNV) (review in Schlegel et al., 2014).

Rodents can carry a large variety of zoonotic pathogens (Meerburg et al., 2009), which can be classified in different ways, e.g., by taxonomy (see Table 1), transmission route, or disease caused. They can be transmitted to humans by different routes. On the one hand, pathogens can be shed by feces, urine, and saliva of persistently infected rodents and transmitted by inhalation of virus-contaminated dust, e.g., hantaviruses and leptospires. On the other hand, vectors, such as ticks, might be involved in the transmission with rodents being a reservoir, e.g., for tick-borne encephalitis virus, *Borrelia* spp. or *Rickettsia* spp. A direct transmission of rodent-borne zoonotic pathogens can be mediated by biting, e.g., for cowpox virus (CPXV), by alimentary route (rodents as "bush meat"), e.g., rats in low-income countries, e.g., monkeypox virus, or contact to the skin of infected rodent, e.g., lymphocytic choriomengitis virus (LCMV), or *Francisella tularensis*.

Zoonotic pathogens frequently cause diseases with unspecific flu-like symptoms and can therefore be misdiagnosed. Several hanta- and arenaviruses cause hemorrhagic fevers of different severity. Other zoonotic pathogens are causative agents of encephalitides or other neurological disorders or of skin rush symptoms (see Table 1).

In addition to zoonotic pathogens, rodents harbor-specific pathogens with no or still unknown zoonotic potential (see below: "3. PCR-mediated identification of novel viruses" and "4. Virus hunting by shot-gun high-throughput sequencing"). Furthermore, in rodents, e.g., urban rats, human pathogenic agents were detected, such as norovirus or extended spectrum beta-lactamase-producing enterobacteria (Guenther et al., 2012; Wolf et al., 2013).

In summary, rodents are a highly diverse group of mammals with reservoir function for zoonotic pathogens. In addition, rodents carry rodent-specific pathogens with most likely no zoonotic potential.

Table 1 Overview About Zoonotic Pathogens Associated With Rodents

Family	Genome	Species	Disease	Symptoms	References
Viruses					
Arenaviridae	−ssRNA, segmented	Junin virus (JUNV)	Argentine hemorrhagic fever (AHF)	Conjunctivitis, purpura, petechia, sepsis	Boxaca et al. (1965)
		Lassa virus (LASV)	Lassa fever	Fever, weakness, malaise, headache, throat ache	Fichet-Calvet et al. (2008) and Green et al. (1978)
		Lymphocytic Choriomeningitis virus (LCMV)	Lymphocytic choriomeningitis	Flu-like symptoms, severe headache, vomiting, delirium	Collis (1935)
		Machupo virus (MACV)	Bolivian hemorrhagic fever	Conjunctivitis, purpura, petechia, sepsis	Mackenzie (1965)
Flaviviridae	+ssRNA	Tick-borne encephalitis virus (TBEV)	Tick-borne encephalitis	Headache and severe pain in neck, fever, confusion, neurologic signs	Gritsun et al. (2003)
Hantaviridae	−ssRNA, segmented	Andes virus (ANDV)	Hantavirus cardiopulmonary syndrome (HCPS)	Flu-like symptoms, hemorrhages, cardiopulmonary, and/or renal impairment/failure	López et al. (1996)
		Sin Nombre virus (SNV)			Nichol et al. (1993)

		Dobrava-Belgrade virus (DOBV)	Hantavirus hemorrhagic fever with renal syndrome (HFRS)	Flu-like symptoms, hemorrhages, oliguria, proteinuria	Klempa et al. (2013)
		Hantaan virus (HTNV)			Lee et al. (1978)
		Puumala virus (PUUV)			Brummer-Korvenkontio et al. (1980)
		Seoul virus (SEOV)			Lee et al. (1982)
		Tula virus (TULV)			Plyusnin et al. (1994)
Poxviridae	dsDNA	Cowpox virus (CPXV)	Cowpox disease	Skin lesions	Moss and Flexner (1987)
Bacteria					
Coxiellaceae	dsDNA	*Coxiella burnetii*	Q fever	Flu-like symptoms, diarrhea, fatigue, headache, and myalgias	Fournier et al. (1998)
Enterobacteriaceae	dsDNA	*Yersinia pestis*	Plague	Septicemia, nausea, vomiting, diarrhea, abdominal pain, kidney failure	Perry and Fetherston (1997)
Francisellaceae	dsDNA	*Francisella tularensis*	Tularemia	Flu-like symptoms, lethargy, sepsis	Farlow et al. (2005)
Rickettsiaceae	dsDNA	*Rickettsia* spp.	Rickettsiosis	Fever, skin lesions, fatigue, exhaustion	Dobler and Wölfel (2009)

Continued

Table 1 Overview About Zoonotic Pathogens Associated With Rodents—cont'd

Family	Genome	Species	Disease	Symptoms	References
Spirochaetaceae	dsDNA	*Leptospira* spp.	Leptospirosis	Flu-like symptoms, conjunctival suffusion, skin rash, severe headache, renal failure	Bharti et al. (2003)
Parasites					
Babesiidae	dsDNA	*Babesia* spp.	Babesiosis	Asymptomatic or mild nonspecific illness	Homer et al. (2000)
Sarcocystidae	dsDNA	*Toxoplasma gondii*	Toxoplasmosis	Hydrocephalus, mental retardation, or spontaneous abortion	Van Knapen (1989)
Taeniidae	dsDNA	*Echinococcus multilocularis*	Alveolar echinococcosis	Headache, nausea, vomiting, abdominal pain	Eckert et al. (2000)

2. THE HISTORY OF RODENT-BORNE VIRUSES AND THEIR DISCOVERY

The change from a hunting to a domesticating and animal husbandries using civilization had profound consequences for mankind in the past 13,000 years. It was hypothesized that epidemic diseases arose at that time and zoonoses were established due to the permanent close contact between denser animal herds and in parallel a denser human population (Diamond, 2002; Fournié et al., 2017).

Some virus-caused rodent-borne zoonoses are known since long time, but the causative agents were discovered only in the 20th century. Due to the repeated appearance of a unique clinical picture in European and Asian inhabitants, some old records may be allocated to a certain rodent zoonosis. Ancient Chinese writers described a disease very similar to hemorrhagic fever with renal syndrome in 960 (Lee, 1982). In Russia, a hemorrhagic fever of men was apparently known since 1913 (Casals et al., 1970). Even during military conflicts of the 20th century, this rodent zoonosis had impact on opposing forces. During World War I (1914/15) overall 1455 cases of "Field nephritis" or "War nephritis" were recorded among British troops in the area of Flanders (Abercrombie, 1916; Bradford, 1916). Later on, in World War II (1942) Finnish and German soldiers suffered from the same syndrome in Scandinavia and Russia (Myhrman, 1951; Rüdisser, 1942; Stuhlfauth, 1943). All these cases were most likely attributed to a disease we know today as hantavirus disease (Johnson, 2001; Pilaski et al., 1986). The infectious agent was long time unknown, but the search was resumed during the Korean War (1950–1953), where more than 3000 UN troops showed a hemorrhagic fever in association with renal impairment (Johnson, 2001; Smadel, 1953). It lasted additional 25 years until the causative agent was successfully isolated, identified, and finally called hantavirus (Lee et al., 1978; see Table 1). However, there are indications that the hantavirus species PUUV was already present in Europe since the last deglaciation approx. 21,000 to 17,000 years before present (Drewes et al., 2017; Sironen et al., 2001). Nowadays, we know more than 50 hantaviruses detected not only in rodents but also in insectivores (shrews, moles) and bats (Witkowski et al., 2016; Yanagihara et al., 2015; see Fig. 1 and Table 1).

In addition to hemorrhagic fevers, rodents can harbor also viruses that cause other zoonotic diseases including encephalitis (see Table 1). In 1933, such an epidemic occurred in the region of St. Louis County, USA. Based on samples from this epidemic, in the course of infection experiments a new virus was

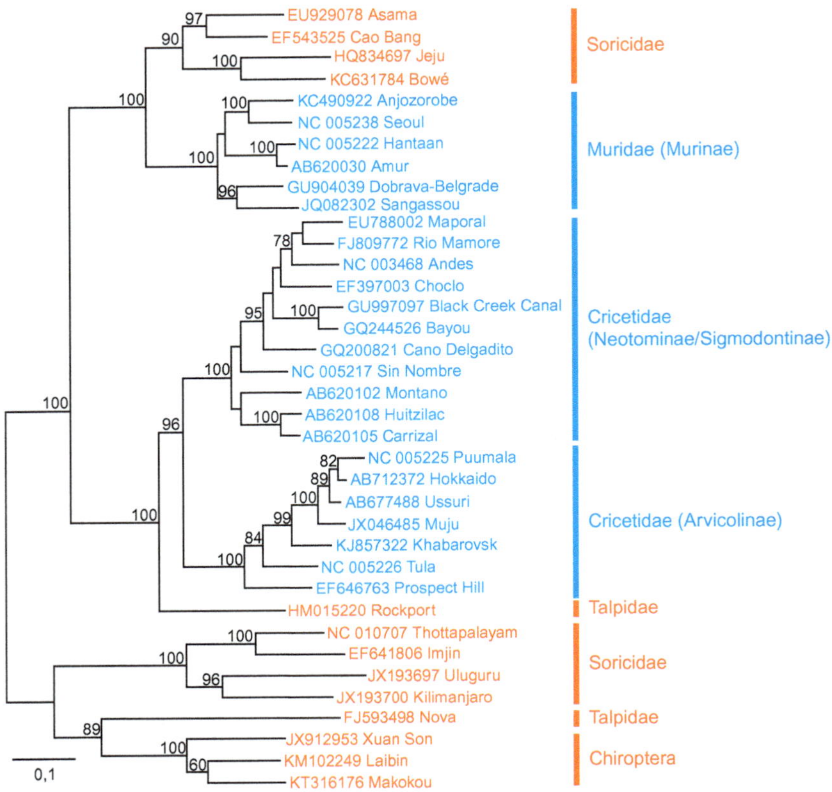

Fig. 1 Phylogenetic relationship of rodent-borne and other (shrew-, mole-, and bat-borne) hantaviruses. The shown tree is based on available complete ORF amino acid sequences of hantaviral L segment (MEGA v6) (Tamura et al., 2013) Maximum Likelihood method based on the Whelan and Goldman models with six independent Gamma rate categories. The tree with the highest log likelihood (−51,228.9665) is shown. The percentage of trees from 1000 repetitions in which the associated taxa clustered together is shown next to the branches for values ≥60%. Initial tree(s) for the heuristic search were obtained by applying the Neighbor-Joining method to a matrix of pairwise distances estimated using a JTT model. The scale bar indicates an evolutionary distance of given substitutions per position in the sequence. Host animal families are given on the right side in *blue* for Euarchontoglires (rodent families Muridae and Cricetidae with corresponding subfamilies) and *orange* for Laurasiatheria (shrew, mole, and bat taxa). *Taken from Witkowski, P.T., Drexler, J.F., Kallies, R., Ličková, M., Bokorová, S., Mananga, G.D., Szemes, T., Leroy, E.M., Krüger, D.H., Drosten, C., Klempa, B., 2016. Phylogenetic analysis of a newfound bat-borne hantavirus supports a laurasiatherian host association for ancestral mammalian hantaviruses. Infect. Genet. Evol. 41, 113–119, with permission of the corresponding publishers.*

isolated from spinal fluid and brain tissue of monkeys that differed from the etiologic agent of St. Louis encephalitis in microscopic pathologic anatomy (Armstrong and Lillie, 1934; Beeman, 2007). A bit later, this unknown virus could be attributed to lymphocytic choriomeningitis in humans (Rivers and McNair Scott, 1935). Moreover, the finding of chronically infected "white mice" in a breeding colony, often without symptoms, confirmed the house mouse (*M. musculus*) as natural reservoir of LCMV (Traub, 1936; see Fig. 2). Although the house mouse is believed to be the primary host, this virus has been found in other rodent species as well (Blasdell et al., 2008; Kallio-Kokko et al., 2006). In 1970, the virus family *Arenaviridae* was established with LCMV as prototype species (Rowe et al., 1970). Each member of this family is strongly associated with a certain rodent species. The Old World lineage comprises viruses such as LCMV, Lassa virus (LASV), Mobala virus (MOBV) and their hosts are rodents from the family Muridae, subfamily Murinae, while the New World lineage includes viruses such as Junin virus (JUNV), Machupo virus (MACV), and Parana virus (PARV) and their hosts are rodents from the family Cricetidae, subfamily Sigmodontinae (Gonzalez et al., 2007; see Table 1). Some of these viruses were already discovered in the 1950s to 1970s (see Fig. 2) and are known to cause severe hemorrhagic fevers in Western Africa (LASV), or South America, namely Argentina and Bolivia (JUNV, MACV) (Hubálek and Rudolf, 2011).

The early history of virus discovery was based on isolation of the viruses in cell culture or by animal passages. Some rodent-borne viruses are among viruses that were used in the pioneering period of molecular and cell

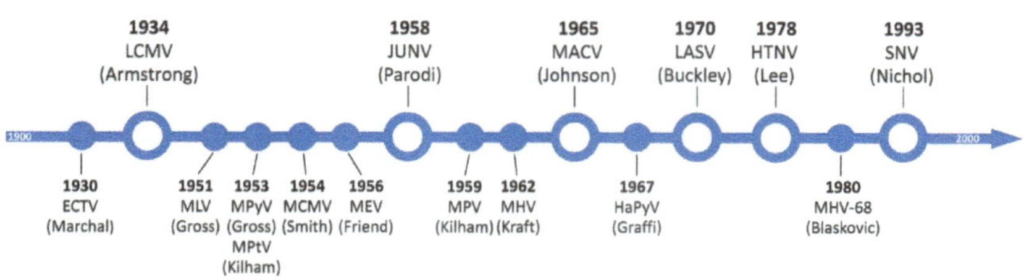

Fig. 2 History of the discoveries of rodent-associated viruses (information for this scheme was taken from the powerpoint compilation "Foundations of Virology" (http://www.utmb.edu/virusimages/) kindly provided by Frederick A. Murphy). ECTV, Ectromelia virus; HaPyV, *Mesocricetus auratus* polyomavirus (Hamster polyomavirus, previously designated Hamster papovavirus); HTNV, Hantaan virus; LASV, Lassa virus; LCMV, Lymphocytic choriomeningitis virus; MACV, Machupo virus; MCMV, Mouse cytomegalovirus; MEV, Murine erythroblastosis virus; MHV, Murine hepatitis virus; MHV-68, Murine gammaherpesvirus 68; MLV, Murine leukemia and lymphoma virus; MPtV, Murine pneumotropic virus; MPyV, Murine polyomavirus; MPV, Murine parvovirus; SNV, Sin Nombre virus. Zoonotic viruses are given above the timeline and are labeled by a *hollow circle*.

biology, cancer research, and immunology. The timechart of the discovery of zoonotic and nonzoonotic viruses is given in Fig. 2. As mentioned earlier, LCMV was discovered in the 1930s and became a model virus for basic studies in virology and particularly immunology, finally resulting in the discovery of major histocompatibility complex (MHC) restriction of CD8-positive T lymphocytes (Zhou et al., 2012).

The tumor virology has played a central role in cancer research since the middle of the 20th century (Javier and Butel, 2008). The Bittner virus, also known as mouse mammary tumor virus (MMTV), was found to cause spontaneous mammary tumors in mice (Bittner, 1942). It could be demonstrated that this virus (at that time designated as an "infectious filterable agent") was present in the milk of cancer-prone mice. By this circumstance, the route of transmission to newborn mice was revealed (Javier and Butel, 2008). Gross (1951) identified the first mouse leukemia virus (murine leukemia virus, MLV, the first known oncogenic mammalian retrovirus) and a mouse polyomavirus (MPyV) that induced multiple solid tumors in mice (Gross, 1953; Stewart and Eddy, 1959). The MPyV, a DNA tumor virus, was discovered as the first member of the previous family Papovaviridae, later separated into the recent family *Polyomaviridae*. In addition, it was one of the first viruses where the complete genome was determined and its organization characterized (Tooze, 1981). After these discoveries, more mouse leukemia viruses were found (Friend, 1957; Graffi, 1957; Moloney, 1960). These investigations suggested a role of viruses in cancer in animals and supported the idea that human tumors might be caused by viruses, and thereby encouraged scientists in the whole world to search for new human cancer viruses. Many years later, these findings also led to the discovery and description of the reverse transcriptase enzyme, viral-encoded integrase enzyme, and the cellular origin of viral oncogenes (Baltimore, 1970; Butel, 2000; Javier and Butel, 2008; Temin, 1964a,b, Temin and Mizutani, 1970). Because of the small genome sizes of RNA and DNA tumor viruses, they were used as model systems in cancer research, where they proved powerful tools revealing fundamental molecular events that trigger cancer induction and progression (Butel, 2000; Javier and Butel, 2008). Most known cellular oncogenes were revealed through studies of RNA tumor viruses, whereas DNA tumor viruses were involved in the discovery of tumor suppressor proteins, such as p53 (Braithwaite and Prives, 2006; Tegtmeyer, 1975).

In summary, several rodent-associated viruses were discovered in the past by classical isolation approaches and either detected as causative agents of zoonotic diseases and/or used in break-through basic research investigations in biomedicine.

3. PCR-MEDIATED IDENTIFICATION OF NOVEL VIRUSES

The development of the RT-PCR and PCR technology enhances the speed of the discovery of novel viruses. For virus discovery, generic (nested) RT-PCR/PCR assays were developed and successfully applied. Based on a generic PCR targeting the DNA polymerase gene of herpesviruses a large number of novel beta- and gammaherpesviruses were discovered in rodents (Ehlers et al., 2007; see Table 2). The discovery of novel double-stranded DNA viruses with a circular genome was based on the development of the rolling circle amplification (RCA) technique. This technique is based on a random hexamer priming and strand displacement reaction mediated by the phage phi29 DNA polymerase. Application of this methodology resulted in the identification of novel polyoma- and papilloma viruses in different rodent species (Fig. 3 and Table 2).

The discovery of novel RNA viruses was also based on generic RT-PCR approaches. The development of a L gene-based pan-hantavirus RT-PCR (Klempa et al., 2006) resulted not only in the identification of novel hantaviruses in rodents, but also in shrews, moles, and even bats (see, e.g., Klempa et al., 2007; Pounder et al., 2013; Straková et al., 2017; see Table 1 and Fig. 1).

Furthermore, the discovery of a novel hepatitis E virus, member of the family *Hepeviridae*, in Norway rats was grounded on the development of a broad-spectrum nested RT-PCR targeting all at that time known hepeviruses (Johne et al., 2010a). Subsequent investigations resulted in the primer walking-based determination of the complete genome of rat hepatitis E virus (ratHEV; Johne et al., 2010b) and the determination of its broad, most likely worldwide geographical distribution in different rat species (see Ryll et al., 2017; Table 2). Currently, ratHEV is still the only rodent-associated hepevirus known (see Fig. 4A). The search for additional human hepatitis virus-related agents in a large number of globally sampled small mammals resulted in the identification of novel hepaci- and hepatoviruses, relatives of hepatitis C virus (HCV) and hepatitis A virus (HAV), respectively (Drexler et al., 2013, 2015). The studies detected these novel viruses only in a subset of rodent species, such as bank vole hepacivirus and South African four-striped mouse hepacivirus, and hepatoviruses in 13 rodent species (see Table 2 and Fig. 4B and D). Similarly, a screening of a similar sample of 9000 small mammals including about 4300 rodent individuals of 33 species detected morbillivirus-related agents as almost the only group of paramyxoviruses in rodents (Drexler et al., 2012). As illustrated in

Table 2 Overview of Nonzoonotic Rodent-Associated Viruses

Family	Genome	Species	Host (Common Name)	Host (Scientific Name)	References
Adenoviridae	dsDNA	Murine adenovirus 1	House mouse	*Mus musculus*	Hashimoto et al. (1966)
		Murine adenovirus 3	Striped field mouse	*Apodemus agrarius*	Klempa et al. (2009)
		Squirrel adenovirus 1	Eurasian red squirrel	*Sciurus vulgaris*	Abendroth et al. (2017) and Everest et al. (2010)
Astroviridae	+ssRNA	Rodent astrovirus cluster A	Chestnut white-bellied rat	*Niviventer fulvescens*	To et al. (2017)
			Sikkim rat	*Rattus andamanensis*	
		Rodent astrovirus cluster B	Norway rat	*Rattus norvegicus*	
		Rodent astrovirus cluster C	Coxing's white-bellied rat	*Niviventer coninga*	
		Rodent astrovirus cluster D	Greater bandicoot rat	*Bandicota indica*	
			Sikkim rat	*Rattus andamanensis*	
Flaviviridae	+ssRNA	Bank vole hepacivirus	Bank vole	*Myodes glareolus*	Drexler et al. (2013)
		South African four-striped mouse hepacivirus	Four-striped mouse	*Rhabdomys pumilio*	
Hepeviridae	+ssRNA	Rat hepatitis E virus	Norway rat	*Rattus norvegicus*	Johne et al. (2010a,b) and Ryll et al. (2017)
			Black rat	*Rattus rattus*	

Herpesviridae	dsDNA	*Apodemus flavicollis* cytomegalovirus 1, 2, 3	Yellow-necked mouse	*Apodemus flavicollis*	Ehlers et al. (2007)
		Apodemus sylvaticus cytomegalovirus 1	Wood mouse	*Apodemus sylvaticus*	
		Arvicola terrestris cytomegalovirus 1	European water vole	*Arvicola terrestris*	
		Bandicota indica cytomegalovirus 1, 2, 3, 4	Greater bandicoot rat	*Bandicota indica*	
		Microtus agrestis cytomegalovirus 1	Field vole	*Microtus agrestis*	
		Microtus arvalis cytomegalovirus 1	Common vole	*Microtus arvalis*	
		Murine cytomegalovirus	House mouse	*Mus musculus*	
		Mus cervicolor cytomegalovirus 1	Fawn-colored mouse	*Mus cervicolor*	
		Myodes glareolus cytomegalovirus 1	Bank vole	*Myodes glareolus*	

Continued

Table 2 Overview of Nonzoonotic Rodent-Associated Viruses—cont'd

Family	Genome	Species	Host (Common Name)	Host (Scientific Name)	References
		Ondatra zibethicus cytomegalovirus 1	Muskrat	*Ondatra zibethicus*	
		Rat cytomegalovirus	Norway rat	*Rattus norvegicus*	
		Rattus exulans cytomegalovirus 1	Polynesian rat	*Rattus exulans*	
		Rattus rattus cytomegalovirus 1	Black rat	*Rattus rattus*	
		Rattus tiomanicus cytomegalovirus 1	Malaysian field rat	*Rattus tiomanicus*	
		Mouse cytomegalovirus	House mouse	*Mus musculus*	Smith (1954)
		Murine gammaherpesvirus 68	Striped field mouse, Yellow-necked mouse, Wood mouse	*Apodemus agrarius, Apodemus flavicollis, Apodemus sylvaticus*	Ehlers et al. (2007)
Herpesviridae	dsDNA	*Apodemus agrarius* rhadinovirus 1	Striped field mouse	*Apodemus agrarius*	Ehlers et al. (2007)
		Apodemus flavicollis rhadinovirus 1	Yellow-necked mouse	*Apodemus flavicollis*	
		Apodemus sylvaticus rhadinovirus 1	Wood mouse	*Apodemus sylvaticus*	

	Bandicota indica rhadinovirus 1, 2, 3, 4	Greater bandicoot rat	*Bandicota indica*
	Bandicota savilei rhadinovirus 1	Lesser bandicoot rat	*Bandicota savilei*
	Microtus agrestis rhadinovirus 1	Field vole	*Microtus agrestis*
	Mus cervicolor rhadinovirus 1	Fawn-colored mouse	*Mus cervicolor*
	Mus musculus rhadinovirus 1	House mouse	*Mus musculus*
	Myodes glareolus rhadinovirus 1	Bank vole	*Myodes glareolus*
	Rattus exulans rhadinovirus 1, 2	Polynesian rat	*Rattus exulans*
	Rattus norvegicus rhadinovirus 1, 2	Norway rat	*Rattus norvegicus*
	Rattus rattus rhadinovirus 1, 2, 3	Black rat	*Rattus rattus*
	Rattus tiomanicus rhadinovirus 1, 2	Malaysian field rat	*Rattus tiomanicus*

Continued

Table 2 Overview of Nonzoonotic Rodent-Associated Viruses—cont'd

Family	Genome	Species	Host (Common Name)	Host (Scientific Name)	References
Papillomaviridae	dsDNA	*Apodemus sylvaticus* papilloma virus 1	Wood mouse	*Apodemus sylvaticus*	Schulz et al. (2009, 2012)
		Mus musculus papilloma virus 1	House mouse	*Mus musculus*	
		Rattus norvegicus papillomavirus 1, 2	Norway rat	*Rattus norvegicus*	
		Erethizon dorsatum papilloma virus Type 1	North American porcupine	*Erethizon dorsatum*	Rector et al. (2005)
Paramyxoviridae	−ssRNA	*Apodemus flavicollis* paramyxovirus	Yellow-necked mouse	*Apodemus flavicollis*	Drexler et al. (2012)
		Microtus agrestis paramyxovirus	Field vole	*Microtus agrestis*	
		Microtus arvalis paramyxovirus	Common vole	*Microtus arvalis*	
		Myodes glareolus paramyxovirus	Bank vole	*Myodes glareolus*	
		Rattus rattus paramyxovirus	Black rat	*Rattus rattus*	
		Rhabdomys pumilio paramyxovirus	Four-striped mouse	*Rhabdomys pumilio*	

Picornaviridae	+ssRNA	Aichi virus 1	Mice	*Apodemus* spp.	Farkas et al. (2012)
		Apodemus agrarius hepatovirus	Striped field mouse	*Apodemus agrarius*	Drexler et al. (2015)
		Baiomys musculus hepatovirus	Southern pygmy mouse	*Baiomys musculus*	
		Cricetulus migratorius hepatovirus	Gray dwarf hamster	*Cricetulus migratorius*	
		Liomys pictus hepatovirus	Painted spiny pocket mouse	*Liomys pictus*	
		Microtus agrestis hepatovirus	Field vole	*Microtus agrestis*	
Picornaviridae	+ssRNA	*Microtus arvalis* hepatovirus	Common vole	*Microtus arvalis*	Drexler et al. (2015)
		Myodes glareolus hepatovirus	Bank vole	*Myodes glareolus*	
		Oryzomys couesi hepatovirus	Coues' rice rat	*Oryzomys couesi*	
		Peromyscus hepatovirus	Deer mouse	*Peromyscus* sp.	
		Reithrodontomys fulvescens hepatovirus	Fulvous harvest mouse	*Reithrodontomys fulvescens*	
		Sigmodon mascotensis hepatovirus	Jaliscan cotton rat	*Sigmodon mascotensis*	

Continued

Table 2 Overview of Nonzoonotic Rodent-Associated Viruses—cont'd

Family	Genome	Species	Host (Common Name)	Host (Scientific Name)	References
		Rosavirus B	Norway rat	*Rattus norvegicus*	Lau et al. (2016)
		Rosavirus C	Black rat	*Rattus rattus*	
			Chestnut spiny rat	*Maxomys ochraceiventer*	
			Coxing's white-bellied rat	*Niviventer coninga*	
			Greater bandicoot rat	*Bandicota indica*	
			Sikkim rat	*Rattus andamanensis*	
Polyomaviridae	dsDNA	Bank vole polyoma virus	Bank vole	*Myodes glareolus*	Nainys et al. (2015)
		Common vole polyoma virus	Common vole	*Microtus arvalis*	
		Mesocricetus auratus polyoma virus	Golden hamster	*Mesocricetus auratus*	Graffi et al. (1967)
		Mouse polyoma virus	House mouse	*Mus musculus*	Gross (1953)
		Murine pneumotropic virus	House mouse	*Mus musculus*	Kilham and Murphy (1953)
		Rattus norvegicus polyomavirus 1	Norway rat	*Rattus norvegicus*	Ehlers et al. (2015)
		Rattus norvegicus polyomavirus 2	Norway rat	*Rattus norvegicus*	Rigatti et al. (2016)
Retroviridae	+ssRNA reverse transcriptase	Murine erythroblastosis virus	House mouse	*Mus musculus*	Friend (1956)
		Mouse mammary tumor virus	House mouse	*Mus musculus*	Bittner (1942)

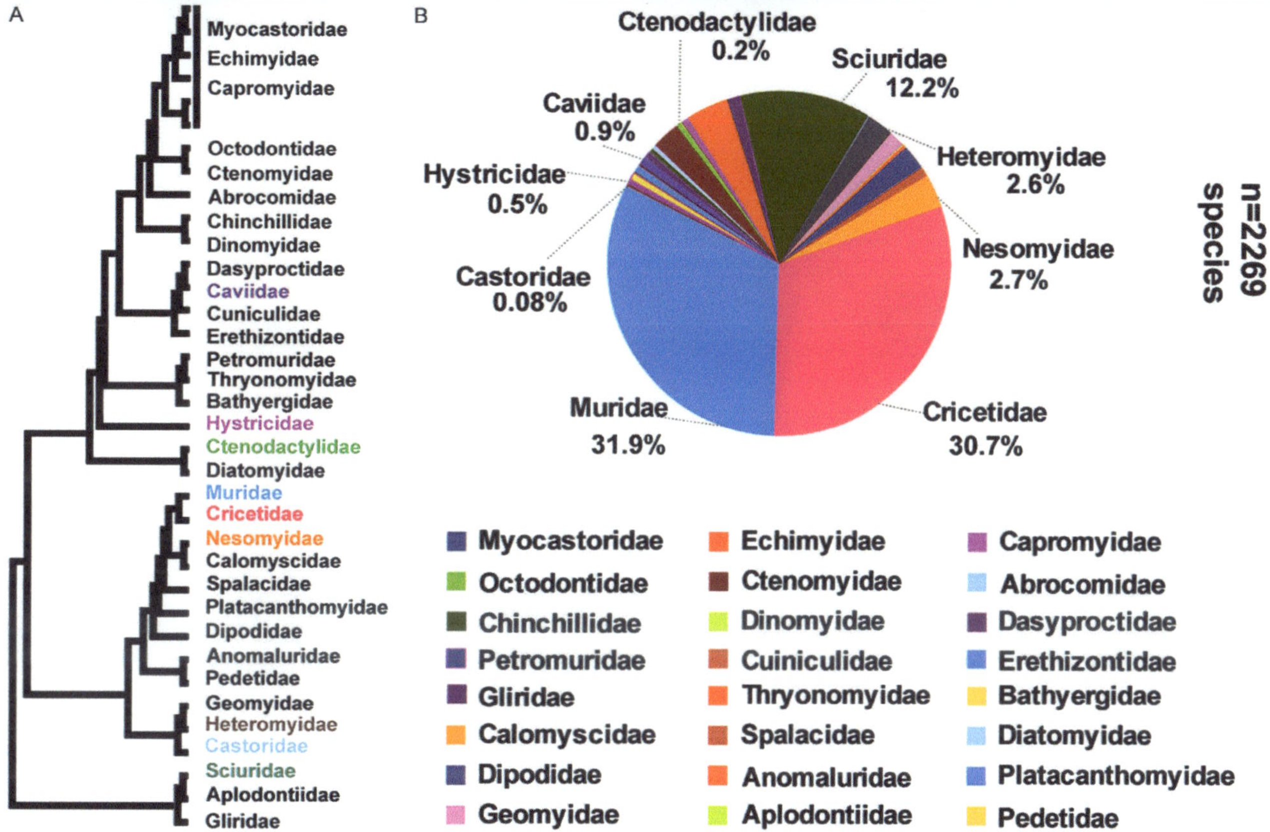

Fig. 3 Potential bias in systematic studies on virus diversity in rodents. (A) Phylogeny of rodent families according to Fabre et al. 2012. Unresolved families are summarized by *vertical line*. Families showing virus detections in systematic PCR/RT-PCR-based studies in the last 10 years are shown in *color*. (B) Species distribution in rodent families according to the red list from the International Union for Conservation of Nature—IUCN, available at: https://www.iucn.org/, accessed 2nd July 2017. (C) Studies providing recent virus detections in systematic PCR/RT-PCR-based studies by rodent family given to the right and viral family given on *top*, identified by first author, family name and publication year. dsDNA, double-stranded DNA viruses; ssRNA, single-stranded RNA viruses.

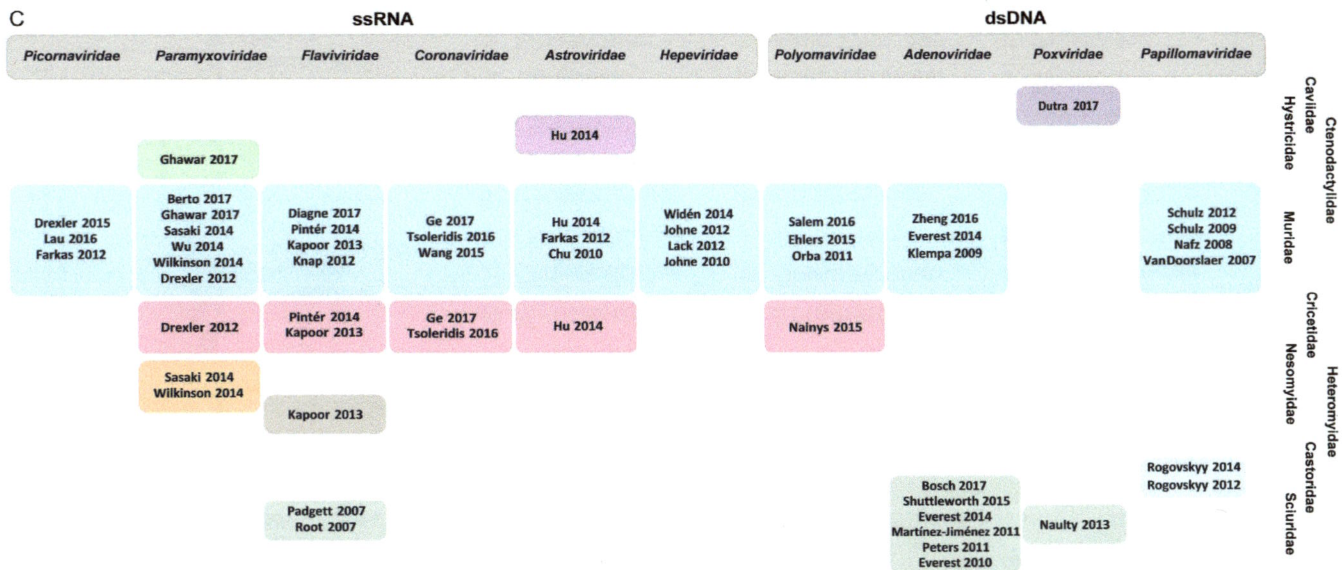
C
ssRNA
dsDNA
Picornaviridae
Paramyxoviridae
Flaviviridae
Coronaviridae
Astroviridae
Hepeviridae
Polyomaviridae
Adenoviridae
Poxviridae
Papillomaviridae
Dutra 2017
Hu 2014
Ghawar 2017
Drexler 2015
Lau 2016
Farkas 2012
Berto 2017
Ghawar 2017
Sasaki 2014
Wu 2014
Wilkinson 2014
Drexler 2012
Diagne 2017
Pintér 2014
Kapoor 2013
Knap 2012
Ge 2017
Tsoleridis 2016
Wang 2015
Hu 2014
Farkas 2012
Chu 2010
Widén 2014
Johne 2012
Lack 2012
Johne 2010
Salem 2016
Ehlers 2015
Orba 2011
Zheng 2016
Everest 2014
Klempa 2009
Schulz 2012
Schulz 2009
Nafz 2008
VanDoorslaer 2007
Drexler 2012
Pintér 2014
Kapoor 2013
Ge 2017
Tsoleridis 2016
Hu 2014
Nalnys 2015
Sasaki 2014
Wilkinson 2014
Kapoor 2013
Rogovskyy 2014
Rogovskyy 2012
Bosch 2017
Shuttleworth 2015
Everest 2014
Martínez-Jiménez 2011
Peters 2011
Everest 2010
Naulty 2013
Padgett 2007
Root 2007
Ctenodactylidae
Caviidae
Hystricidae
Muridae
Cricetidae
Nesomyidae
Heteromyidae
Castoridae
Sciuridae

Fig. 3—Cont'd

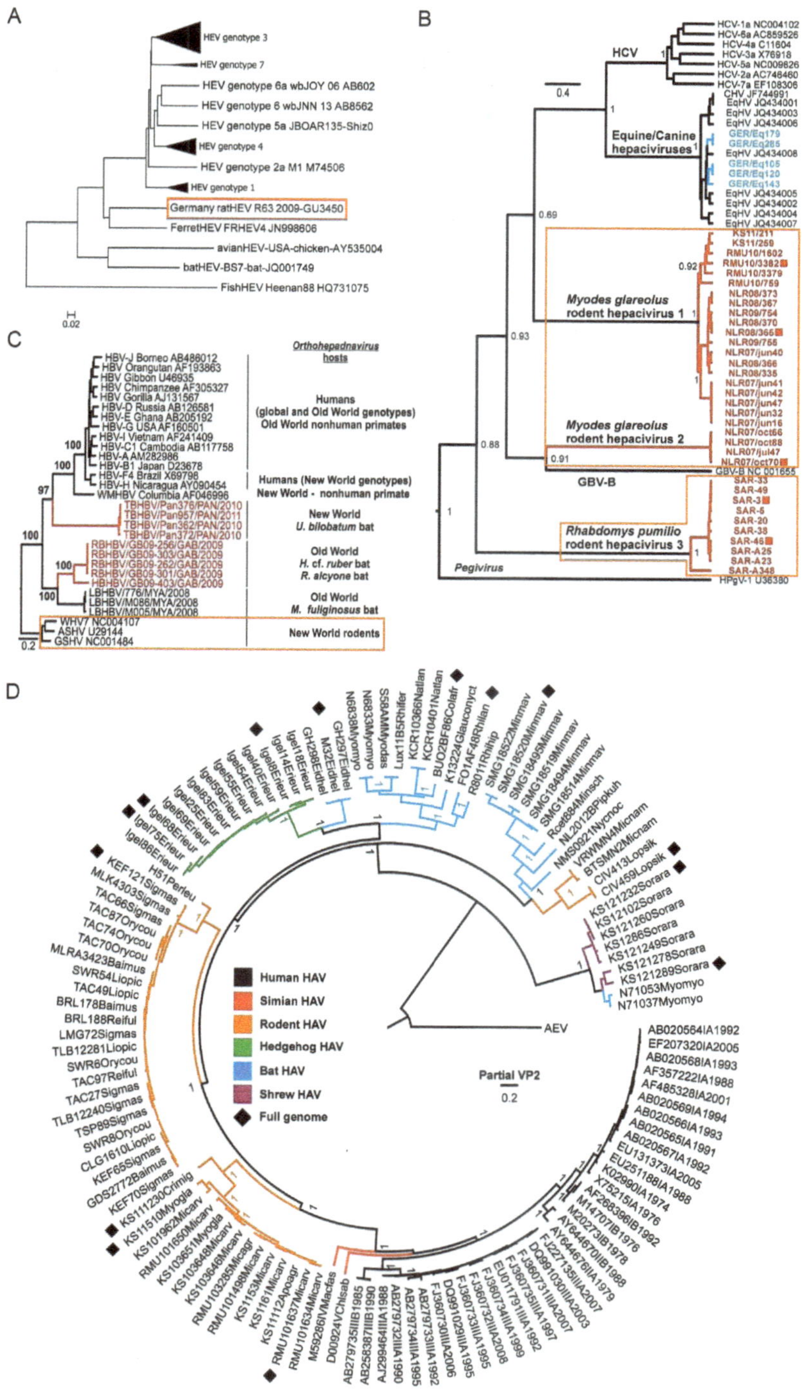

Fig. 4 See legend on next page.

Fig. 4 Nucleotide sequence-based phylogenetic trees of hepeviruses (A), hepaciviruses (B), hepadnaviruses (C), and hepatoviruses (D), illustrating the currently known rodent-associated viruses labeled by orange frames. (A) The multiple sequence alignment of complete hepevirus-reference sequences according to Smith et al., 2014, 2016 was created with ClustalW and the phylogenetic reconstruction was performed with the Neighbor-Joining-Method, including 1000 bootstraps. (B) Partial NS3 gene phylogeny. The analysis comprised a 978 nucleotide fragment of the hepatitis C virus (HCV) NS3 gene corresponding to positions 3912–4889 in HCV 1a H77 (GenBank, NC_004102). GenBank accession numbers of reference hepaciviruses are indicated to the *right* of taxon names. Tree topology was inferred using BEAST with a GTR nucleotide substitution model as described in the methods section. Rodent hepaciviruses from this study are shown in *red and boldface*, equine hepaciviruses from this study are shown in *blue and boldface*. *Red squares* indicate those viruses whose near full-length genomes were generated. Statistical support of grouping is shown as posterior probabilities at deep nodes. Scale bar corresponds to genetic distance. To the right, 5000 tree replicates of the same analysis are rendered using Densitree (initial 5000 trees discarded as burn-in). (C) Phylogenetic analysis of hepadnaviruses including the New World rodent viruses. Maximum-likelihood (ML) full genome phylogeny. Unique bat viruses from this study are in *red*. Values at deep nodes represent Bayesian posterior probabilities or percentage ML bootstrap replicates; scale bars represent genetic distance. *ASHV*, Arctic ground squirrel hepatitis B virus; *GSHV*, Ground squirrel hepatitis virus; *WHV*, Woodchuck hepatitis virus. (D) Host relationships, hepatovirus relationships, viral species prediction, and genomic properties. Phylogenetic relationships of hepatitis A virus (HAV)-related hepatoviruses in the partial VP2 encoding domain. Values at nodes show support of grouping from Bayesian posterior probabilities. *Trees in panels B, C, and D were taken from Drexler, J.F., Corman, V.M., Müller, M.A., Lukashev, A.N., Gmyl, A., Coutard, B., Adam, A., Ritz, D., Leijten, L.M., van Riel, D., Kallies, R., Klose, S.M., Gloza-Rausch, F., Binger, T., Annan, A., Adu-Sarkodie, Y., Oppong, S., Bourgarel, M., Rupp, D., Hoffmann, B., Schlegel, M., Kümmerer, B.M., Krüger, D.H., Schmidt-Chanasit, J., Setién, A.A., Cottontail, V.M., Hemachudha, T., Wacharapluesadee, S., Osterrieder, K., Bartenschlager, R., Matthee, S., Beer, M., Kuiken, T., Reusken, C., Leroy, E.M., Ulrich, R.G., Drosten, C., 2013a. Evidence for novel hepaciviruses in rodents. PLoS Pathog. 9, e1003438; Drexler, J.F., Geipel, A., König, A., Corman, V.M., van Riel, D., Leijten, L.M., Bremer, C.M., Rasche, A., Cottontail, V.M., Maganga, G.D., Schlegel, M., Müller, M.A., Adam, A., Klose, S.M., Carneiro, A.J., Stöcker, A., Franke, C.R., Gloza-Rausch, F., Geyer, J., Annan, A., Adu-Sarkodie, Y., Oppong, S., Binger, T., Vallo, P., Tschapka, M., Ulrich, R.G., Gerlich, W.H., Leroy, E., Kuiken, T., Glebe, D., Drosten, C., 2013b. Bats carry pathogenic hepadnaviruses antigenically related to hepatitis B virus and capable of infecting human hepatocytes. Proc. Natl. Acad. Sci. U. S. A. 110, 16151–16156; Drexler, J.F., Corman, V.M., Lukashev, A.N., van den Brand, J.M., Gmyl, A.P., Brunink, S., Rasche, A., Seggewibeta, N., Feng, H., Leijten, L.M., Vallo, P., Kuiken, T., Dotzauer, A., Ulrich, R.G., Lemon, S.M., Drosten, C., Hepatovirus Ecology Consortium, 2015. Evolutionary origins of hepatitis A virus in small mammals. Proc. Natl. Acad. Sci. U. S. A. 112, 15190–15195, with permission of the corresponding publishers.*

Fig. 3, currently only for a minority of rodent families (9 of 24) viruses were discovered. The best knowledge of viruses exists for rodents of the family Muridae (Fig. 3C). This is most likely due to detection of viruses in highly abundant and/or commensal murine rodent species, such as Norway rat, house mouse, or yellow-necked mouse. Similarly, certain vole species (common vole, bank vole; both family Cricetidae) are also highly abundant and were therefore evaluated for several virus taxa. In addition, the currently known virus taxa were only detected in a few of the rodent species, even of these two aforementioned families (see Fig. 4). Therefore, our current knowledge of the viral diversity represents only the tip of the iceberg. In addition, the large-scale screenings of rodents for selected viruses suggest also lineage extinction events during the evolutionary history of these virus groups (see e.g., Drexler et al., 2013, 2015 and Fig. 4B and D).

Another source of further knowledge of the origin of viruses is based on the increasing number of rodent genomes determined. Genome-wide screens of these genomes resulted in the identification of endogeneous virus-derived sequences, mainly of retroviral but also nonretroviral origin (Belyi et al., 2010; Bock and Stoye, 2000; Horie et al., 2010). In a recent study, an endogeneous bornavirus element was experimentally shown to inhibit the replication of an exogeneous bornavirus in a squirrel (Fujino et al., 2014).

In summary, application of conventional RT-PCR and PCR approaches resulted in a drastic increase of our knowledge on the diversity of RNA and DNA viruses. The large number of rodent families or species without detection of any virus and a comparison of the number of human and rodent viruses highlight large gaps in our knowledge of rodent viruses (see Figs. 3 and 4).

4. VIRUS HUNTING BY SHOT-GUN HIGH-THROUGHPUT SEQUENCING

Conventional and real-time RT-PCR/PCR methods operate on the basis of specific primer pairs and require therefore sequence information of the virus to be detected (or related ones). The above-described generic or pan-RT-PCR/PCR approaches allow a broader detection of viruses, including novel within or in a higher taxonomic unit (Anthony et al., 2013). In contrast, shot-gun high-throughput sequencing (HTS) methods are "open view" methods that allow the detection of completely novel virus species, genera and families without knowing target sequences. In case of the Illumina and IonTorrent technology, this is possible due to a primer ligation to all termini of DNA fragments and a subsequent clonal amplification and sequencing of millions of individual clones in parallel (Garrido-Cardenas et al., 2017). The final and limiting step comprises the analysis of huge metagenomic sequence datasets by software algorithms to filter the relevant information (see, e.g., Scheuch et al., 2015). So far, there are only few studies available dealing with HTS-based approaches in finding novel rodent viruses.

In an initial survey, feces samples of 105 feral rodents of eight species from California and Virginia, USA, were collected and analyzed by 454 pyrosequencing (Phan et al., 2011). The most abundant sequences belonged to the *Circoviridae*, *Picobirnaviridae*, *Picornaviridae*, *Astroviridae*, *Parvoviridae*, *Papillomaviridae*, *Adenoviridae*, and *Coronaviridae* (Table 3). Among them were a novel complete mouse kobuvirus genome (M-5) that is in close relationship to human gastroenteritis causing aichivirus, and members of two new genera of the family *Picornaviridae*. Various complete circovirus-like sequences were found suggesting a membership to new virus families. The first mouse sapelovirus and picobirnavirus-related sequences were detected, whereas the first complete murine astrovirus genome was described. Further, a complete novel mouse papillomavirus genome and a novel mouse adenovirus genome were sequenced. In feces, also several sequences of plant and insect viruses occurred demonstrating the omnivorous diet of rodents.

Three HTS studies investigated feces samples from wild urban rats collected in Berlin, Copenhagen, Hong Kong, Kuala Langat, Kuala Lumpur, and New York and showed similar findings in regard of detected virus families (Table 3). In comparison to Phan et al. (2011), who focused on rural wild rodents, the urban rats carried already known and until now unknown viruses with zoonotic potential, but mostly so far unknown zoonotic potential, including bocavirus, boone cardiovirus, rat rotavirus A and B, rat hepacivirus, rat hepatitis E virus, rat sapovirus, rat norovirus, Seoul virus, Theiler's encephalomyelitis virus, and Theiler's virus of rats (Firth et al., 2014; Hansen et al., 2016; Sachsenröder et al., 2014).

In conclusion, shot-gun HTS was up to now used only in a few examples on wild rodents, but impressively demonstrates the potential of these technologies to uncover the diversity of viruses harbored by these small mammals. The multiple detection of novel viruses in rodents raises questions on their zoonotic potential.

Table 3 Viruses Detected in Rodent Specimens by High-Throughput Sequencing Approaches

Family	Category	Virus	Rodent Species	References
Adenoviridae	Mammalian virus	Adenovirus	*Sciurus vulgaris, Rattus norvegicus, Peromyscus maniculatus*	Abendroth et al. (2017), Firth et al. (2014), and Phan et al. (2011)
	Mammalian virus	n.g.	n.g.	Hansen et al. (2016) and Sachsenröder et al. (2014)
Alphaflexiviridae	Plant virus	n.g.	n.g.	Hansen et al. (2016) and Phan et al. (2011)
Arteriviridae	Mammalian virus	Arterivirus	*Rattus norvegicus*	Firth et al. (2014)
Astroviridae	Mammalian virus	Murine astrovirus	*Rattus norvegicus, Mus musculus*	Firth et al. (2014) and Phan et al. (2011)
	Mammalian virus	Rat astrovirus[a]	*Rattus norvegicus*	Firth et al. (2014)
Baculoviridae	Insect virus	n.g.	n.g.	Hansen et al. (2016)
Bornaviridae	Mammalian virus	Variegated squirrel bornavirus 1	*Sciurus variegatoides*	Hoffmann et al. (2015a)
Bromoviridae	Insect virus	n.g.	n.g.	Hansen et al. (2016) and Phan et al. (2011)
Caliciviridae	Mammalian virus	Rat norovirus like	*Rattus norvegicus*	Sachsenröder et al. (2014)
	Mammalian virus	Sapovirus	*Rattus norvegicus*	Firth et al. (2014) and Sachsenröder et al. (2014)
	Mammalian virus	n.g.	n.g.	Hansen et al. (2016)
Carmotetraviridae	Insect virus	n.g.	n.g.	Hansen et al. (2016)

Continued

Table 3 Viruses Detected in Rodent Specimens by High-Throughput Sequencing Approaches—cont'd

Family	Category	Virus	Rodent Species	References
Circoviridae	Mammalian virus	Rodent circo-like viruses	*Neotoma cinerea*, *Peromyscus maniculatus*, *Mus musculus*, *Microtus pennsylvanicus*	Phan et al. (2011)
	Mammalian virus	Stool-associated circular ssDNA virus	*Rattus norvegicus*	Firth et al. (2014) and Sachsenröder et al. (2014)
	Mammalian virus	n.g.	n.g.	Hansen et al. (2016)
Coronaviridae	Mammalian virus	n.g.	n.g.	Phan et al. (2011)
Densoviridae	Insect virus	n.g.	n.g.	Phan et al. (2011)
Dicistroviriade	Insect virus	Chronic bee paralysis virus	n.g.	Phan et al. (2011)
	Insect virus	Kelp fly virus	n.g.	Phan et al. (2011)
	Insect virus	Nora virus	n.g.	Phan et al. (2011)
	Insect virus	n.g.	n.g.	Hansen et al. (2016)
Flaviviridae	Mammalian virus	Norway rat hepacivirus	*Rattus norvegicus*	Firth et al. (2014)
	Mammalian virus	Norway rat pegivirus	*Rattus norvegicus*	Firth et al. (2014)
	Mammalian virus	Norway rat pestivirus	*Rattus norvegicus*	Firth et al. (2014)
Geminiviridae	Plant virus	n.g.	n.g.	Phan et al. (2011)
Hantaviridae	Mammalian virus	Seoul virus	*Rattus norvegicus*	Firth et al. (2014)

Hepeviridae	Mammalian virus	Hepatitis E virus	*Rattus norvegicus*	Firth et al. (2014)
Iridoviridae	Insect virus	n.g.	n.g.	Hansen et al. (2016) and Phan et al. (2011)
Mimiviridae	Protist virus	n.g.	n.g.	Hansen et al. (2016)
Nanoviridae	Plant virus	n.g.	n.g.	Phan et al. (2011)
Nodaviridae	Mammalian virus, insect virus	n.g.	n.g.	Hansen et al. (2016)
Papillomaviridae	Mammalian virus	Mouse papillomavirus	*Peromyscus maniculatus*	Phan et al. (2011)
	Mammalian virus	n.g.	n.g.	Hansen et al. (2016)
Partitiviridae	Plant virus	n.g.	n.g.	Hansen et al. (2016) and Phan et al. (2011)
Parvoviridae	Mammalian virus	Bocavirus	*Rattus norvegicus*	Firth et al. (2014) and Sachsenröder et al. (2014)
	Mammalian virus	Kilham rat virus[a]	*Rattus norvegicus*	Firth et al. (2014)
	Mammalian virus	Rodent adeno-associated virus	*Peromyscus maniculatus*	Phan et al. (2011)
	Mammalian virus	n.g.	n.g.	Hansen et al. (2016) and Sachsenröder et al. (2014)
Phycodnaviridae	Algal virus	n.g.	n.g.	Hansen et al. (2016) and Phan et al. (2011)

Continued

Table 3 Viruses Detected in Rodent Specimens by High-Throughput Sequencing Approaches—cont'd

Family	Category	Virus	Rodent Species	References
Picobirnaviridae	Mammalian virus	Rodent picobirnavirus	*Mus musculus, Microtus pennsylvanicus*	Phan et al. (2011)
	Mammalian virus	n.g.	n.g.	Sachsenröder et al. (2014)
Picornaviridae	Mammalian virus	Boone cardiovirus	*Rattus norvegicus, Rattus* spp.	Firth et al. (2014) and Hansen et al. (2016)
	Mammalian virus	Kobuvirus	*Rattus norvegicus, Peromyscus crinitus, Peromyscus maniculatus*	Firth et al. (2014) and Phan et al. (2011)
	Mammalian virus	Manhattan parechovirus	*Rattus norvegicus*	Firth et al. (2014)
	Mammalian virus	Mosavirus	*Peromyscus crinitus*	Phan et al. (2011)
	Mammalian virus	Mouse sapelovirus	*Mus musculus*	Phan et al. (2011)
	Mammalian virus	Norway rat hunnivirus	*Rattus norvegicus*	Firth et al. (2014)
	Mammalian virus	Rat picornavirus	*Rattus norvegicus*	Firth et al. (2014) and Sachsenröder et al. (2014)
	Mammalian virus	Rat theilovirus	*Rattus norvegicus, Rattus* spp.	Firth et al. (2014) and Hansen et al. (2016)
	Mammalian virus	Rosavirus	*Rattus norvegicus, Peromyscus crinitus*	Firth et al. (2014) and Phan et al. (2011)

	Mammalian virus	Saffold virus	*Rattus* spp.	Hansen et al. (2016)
	Mammalian virus	Theiler's encephalomyelitis virus	*Rattus* spp.	Hansen et al. (2016)
	Mammalian virus	Theiler's-like virus	*Rattus* spp.	Hansen et al. (2016)
	Mammalian virus	Thera virus	*Rattus norvegicus*	Firth et al. (2014)
Polydnaviridae	Insect virus	n.g.	n.g.	Phan et al. (2011)
Polyomaviridae	Mammalian virus	n.g.	n.g.	Hansen et al. (2016)
Poxviridae	Mammalian virus, insect virus	n.g.	n.g.	Hansen et al. (2016)
Reoviridae	Mammalian virus	Rat Rotavirus A	*Rattus norvegicus*	Sachsenröder et al. (2014)
	Mammalian virus	Rat Rotavirus B	*Rattus norvegicus*	Sachsenröder et al. (2014)
Retroviridae	Mammalian virus	n.g.	n.g.	Hansen et al. (2016)

Continued

Table 3 Viruses Detected in Rodent Specimens by High-Throughput Sequencing Approaches—cont'd

Family	Category	Virus	Rodent Species	References
Secoviridae	Plant virus	n.g.	n.g.	Hansen et al. (2016) and Phan et al. (2011)
Tombusviridae	Plant virus	n.g.	n.g.	Hansen et al. (2016) and Phan et al. (2011)
Totiviridae	Fungal virus, protist virus	n.g.	n.g.	Hansen et al. (2016)
Tymoviridae	Plant virus	n.g.	n.g.	Phan et al. (2011)
Virgaviridae	Insect virus	n.g.	n.g.	Hansen et al. (2016) and Phan et al. (2011)

[a]Only virus sequences assembled into contigs whose sequence identity exceeded 90% were included. Single virus reads were excluded.
In this table given viruses were characterized/confirmed by full or nearly full genome assembly, phylogenetic analysis, or other pathogen-specific assays.
n.g. not given in detail by cited publication.

5. HTS APPLICATION FOR DIAGNOSTICS OF RODENT-BORNE VIRUSES

Recently, shot-gun HTS-based metagenomic analyses were also used for identification of a novel zoonotic bornavirus, its complete genome determination and molecular epidemiologal evaluation (Hoffmann et al., 2015a). Starting in the end of 2011, three fatal human encephalitis cases among squirrel breeders occurred in Germany over a duration of 2 years. Classical serological and molecular detection methods all failed in finding the causative agent. The following HTS-based approach revealed a novel bornavirus, tentatively designated Variegated squirrel bornavirus 1 (VSBV-1). Moreover, the use of the HTS technology enabled a rapid generation of two complete coding sequences, one from a patient and one from an in-contact variegated squirrel (*Sciurus variegatoides*). The very high identity of the VSBV-1 sequences from the squirrel and the patient confirmed the etiological role and zoonotic potential of the novel bornavirus. Based on the almost complete genome sequences, novel VSBV-1-specific real-time RT-PCR (RT-qPCR) was established and used for the detection of VSBV-1 in the other two patients and for determination of the tissue distribution in the affected squirrel. A subsequent VSBV-1-specific RT-qPCR screening of additional 796 squirrels of different species from Germany, the Netherlands, United Kingdom, Croatia, Belgium, and Thailand resulted in the detection of VSBV-1 RNA in six additional Variegated squirrels, one Red-tailed squirrel (*Sciurus granatensis*) (both subfamily Sciurinae), in 18 Prevost's squirrels (*Callosciurus prevostii*), one Finlayson's squirrel (*Callosciurus finlaysonii*), and one Swinhoe's striped squirrel (*Tamiops swinhoei*) (all subfamily Callosciurinae), originating from Germany, the Netherlands, and Croatia (Schlottau et al., 2017a,b). In contrast, in indigenous Eurasian red squirrels (*Sciurus vulgaris*) no indication for VSBV-1 RNA was found. Complete genome determination for 19 VSBV-1 strains from seven holdings indicated a highly conserved genome and the absence of obvious indications for host adaptation (Schlottau et al., 2017a,b).

Further, in a Eurasian red squirrel, found dead in an aviary in Germany, the complete genome of a novel adenovirus was determined by combination of a microarray and HTS-based approach (Abendroth et al., 2017). The initial microarray analysis already suggested the presence of mastadenovirus sequences, which was confirmed by HTS-based complete genome determination. Interestingly, the squirrel adenovirus genome was highly divergent from the genomes of adenoviruses associated with house mouse

(*M. musculus*) and striped field mouse (*Apodemus agrarius*), respectively. The high sequence identity to sequences determined earlier in squirrels in United Kingdom and Germany indicated that the novel virus represents all these previously identified strains, therefore, tentatively designated Squirrel adenovirus 1. Based on the novel hexon gene sequence determined by the HTS approach, a RT-qPCR assay was developed, validated, and used for investigations during a disease outbreak in Eurasian red squirrels in Germany (Wernike, Beer, Ulrich et al., in preparation).

Additionally, the complete sequence of a first common vole (*Microtus arvalis*) derived CPXV isolate was determined by HTS and the open reading frames (ORFs) were annotated (Hoffmann et al., 2015b). The full-length DNA sequence of this novel CPXV strain showed a high sequence identity to the sequence of a highly virulent rat CPXV strain. In addition, the genome of the novel strain contains a large additional ORF of about 6000 nucleotides, absent in the reference CPXV strain Brighton Red, which might contain virulence markers to be characterized.

In summary, HTS-based approaches can allow a rapid identification of novel viruses in clinical specimens and their complete genome determination. However, a general application of the HTS technology might be hampered by a lower sensitivity compared to convential or real-time RT-PCR/PCR assays. The combination of HTS with microarray screening (or other methods, such as hybrid-capture) may enhance the sensitivity of HTS.

6. WORKFLOWS FOR VIRUS DISCOVERY IN RODENT SPECIMENS

The selection of a workflow for a virus discovery study mainly depends on its objective: (i) diagnostics and complete genome determination on a clinical specimen; (ii) identification and genome characterization of a cell culture virus isolate; (iii) metagenome analysis for virus hunting. For the letter, the available starting material is also crucial for the selection of the workflow. The sample collection on the one hand depends on the tropism of the virus to be discovered, but if not known, frequently blood or feces samples are used, in order to reduce the host cell background.

In case of a sample with a pathological finding or a cell culture isolate of a cytopathogenic virus a shot-gun HTS approach might be appropriate. The cell culture-based open view-approach might be also accompanied by electron microscopy, another open view-approach. If gaps in the contigs arise after assembly, the HTS approach might be accompanied by a conventional, perhaps primer-walking-based, sequencing approach.

In case of a more complex sample, an enrichment step might improve the results of the subsequent HTS analysis. When a novel virus sequence was discovered by conventional generic RT-PCR/PCR or by HTS approach, a pathogen-specific real-time RT-PCR/PCR could be established identifying the target tissue with the highest viral load and thereby allowing the selection of a tissue sample for further genome sequencing via HTS.

One of the key factors for the discovery of a new virus is proper collection of specimen and transportation to the laboratory. For the discovery of new viruses in rodents, different samples can be used (organs, tissues, blood, swabs). A noninvasive sampling, e.g., collection of feces samples or collection of oral swabs, may allow a surveillance study in rodent populations, but may work only during virus shedding and might therefore depend on the timepoint of sample collection. Similarly, the collection of blood samples will yield only virus sequences in case of a viremia, causing potential problems in latency or oscillating patterns of virus replication during persistent infections. Alternatively, a snap trapping of rodents might be more safe for the personnel. For this purpose, harmonized methods for abundance estimation of rodents and for diagnosis of rodent-borne pathogens were defined within the project "harmonized Approaches in monitoring wildlife Population Health, and Ecology and Abundance" (APHAEA) (Drewes et al., 2016; Sonnenburg et al., 2017).

All samples can be temporarily stored at 4°C in the fridge, for longer at −80°C in the freezer (typically for RNA viruses). For virus isolation, samples should be placed on cell cultures immediately after they were taken or should be deeply frozen at −80°C until used (Forman and Valsamakis, 2011).

One of the final bottlenecks in HTS analysis is the assembly of the generated sequence reads. For this procedure, different pipelines have been established (see, e.g., Del Chierico et al., 2015; Scheuch et al. 2015; Torri et al., 2012). First, reads must be extracted from the whole dataset, joined together into contigs and contigs into scaffolds. If possible, a template for genome assembly already described related virus from GenBank, NCBI may help to enhance the speed of assembly. When the number of the reads and their length are not sufficient, gaps between contigs can appear. To fill these gaps, the primer-walking method with designed primers from already assembled contigs can be applied (Volpicella et al. 2012). For obtaining 3′ and 5′ends of viral genomes, RACE (rapid amplification of cDNA ends) method is usually used. Sometimes, just complete coding regions of newly described viruses were used for an initial phylogenetic analysis.

In summary, the discovery and characterization of novel viruses need case-dependent workflows that may combine HTS approaches with conventional RT-PCR/PCR use.

7. CONCLUSIONS

Recently, the first HTS-driven metagenomics analyses as well as the conventional RT-PCR/PCR-mediated approaches drastically increased our knowledge of the viral diversity existing in rodents. Nevertheless, this currently known virus diversity still indicates large gaps in knowledge of rodent viruses compared to those discovered in humans and the knowledge of the rodent virome is still only fragmentary. A combination of the aforementioned approaches, incorporating also hybrid-capture methods, as well as the further determination of rodent genomes and potential endogenous viral elements will enable to fill-in the gaps of our current knowledge within the next years. A major driving force in this respect is the development of novel or improvement of existing sequencing and analysis workflows, including rapid and reliable bioinformatics analysis, for the discovery of novel viruses. The information on the newly described viruses and virus sequences can be used for the development of a point-of-care (POC)-applicable technology to enable the rapid identification of novel viruses during new outbreak scenarios. Nevertheless, the molecular methods must be accompanied by classical virological techniques allowing, e.g., also the fulfillment of the Henle–Loeffler–Koch postulates, where appropriate, and the evaluation of the zoonotic potential of newly discovered viruses. To fulfill these postulates in the era of metagenomics, the comparison in metagenomics libraries must show a significant difference between diseased and healthy control animal (Mokili et al., 2012): (1) The metagenomics traits in animal with disease must be significantly different from healthy animal (i.e., some specific traits are found in sick animal but not in the healthy control. (2) Inoculation of samples from the sick animal into the healthy animal must lead to the induction of the disease state. Comparison of metagenomes before and after inoculation should show the acquisition or increase of new metagenomics traits. (3) New traits can be purified (serial dilution or time-point sampling of specimens from a sick animal). (4) Inoculation of the suspected purified traits into a healthy animal should induce disease if the specific traits form the etiology of disease (taken and modified from Mokili et al. 2012).

ACKNOWLEDGMENTS

HTS investigations in the laboratory of R.G.U. were supported by contract-research-project for the Bundeswehr Medical Service E/U2AD/CF512/DF557 (MetalnfRisk) and by the German Center of Infection Research (DZIF). In addition, the investigations of the Network "Rodent-borne pathogens" were funded in part by the German Federal Ministry of Education and Research (BMBF) through the German Research Platform for Zoonoses (FKZ 01KI1018 and 01KI1303) to R.G.U. The support of Stefan Fischer and Anja Glatter in preparing the tables, of René Ryll for provision of a hepevirus phylogenetic tree, and of Martin Beer and Dirk Höper for critical reading and helpful comments is kindly acknowledged. I am very grateful to Prof. Frederick A. Murphy for creating and providing the powerpoint compilation "Foundations of Virology" (http://www.utmb.edu/virusimages/) which was used to generate Fig. 2 of this manuscript.

REFERENCES

Abendroth, B., Höper, D., Ulrich, R.G., Larres, G., Beer, M., 2017. A red squirrel associated adenovirus identified by a combined microarray and deep sequencing approach. Arch. Virol. https://doi.org/10.1007/s00705-017-3463-5 (Epub ahead of print).

Abercrombie, R.G., 1916. Observations on the acute phase of five hundred cases of war nephritis. J. R. Army Med. Corps 27, 131–157.

Anthony, S.J., Epstein, J.H., Murray, K.A., Navarrete-Macias, I., Zambrana-Torrelio, C.M., Solovyov, A., Ojeda-Flores, R., Arrigo, N.C., Islam, A., Ali Khan, S., Hosseini, P., Bogich, T.L., Olival, K.J., Sanchez-Leon, M.D., Karesh, W.B., Goldstein, T., Luby, S.P., Morse, S.S., Mazet, J.A., Daszak, P., Lipkin, W.I., 2013. A strategy to estimate unknown viral diversity in mammals. MBio 4, e00598–13.

Armstrong, C., Lillie, R.D., 1934. Experimental lymphocytic choriomeningitis of monkeys and mice produced by a virus encountered in studies of the 1933 St. Louis encephalitis epidemic. Public Health Rep. 49, 1019–1027.

Baltimore, D., 1970. RNA-dependent DNA polymerase in virions of RNA tumour viruses. Nature 226, 1209–1211.

Beeman, E.A., 2007. "Green thumb virologist": Saint Louis encephalitis; lymphocytic choriomeningitis. In: Charles Armstrong, M.D. (Ed.), A Biography. National Institutes of Health, Bethesda, MD, USA, pp. 167–207.

Belyi, V.A., Levine, A.J., Skalka, A.M., 2010. Unexpected inheritance: multiple integrations of ancient bornavirus and ebolavirus/marburgvirus sequences in vertebrate genomes. PLoS Pathog. 6, e1001030.

Bharti, A.R., Nally, J.E., Ricaldi, J.N., Matthias, M.A., Diaz, M.M., Lovett, M.A., Levett, P.N., Gilman, R.H., Willig, M.R., Gotuzzo, E., Vinetz, J.M., Peru-United States Leptospirosis Consortium, 2003. Leptospirosis: a zoonotic disease of global importance. Lancet Infect. Dis. 3, 757–771.

Bittner, J.J., 1942. Possible relationship of the estrogenic hormones genetic susceptibility and milk influence in the production of mammary cancer in mice. Cancer Res. 2, 710–721.

Blasdell, K.R., Becker, S.D., Hurst, J., Begon, M., Bennett, M., 2008. Host range and genetic diversity of arenaviruses in rodents, United Kingdom. Emerg. Infect. Dis. 14, 1455–1458.

Bock, M., Stoye, J.P., 2000. Endogenous retroviruses and the human germline. Curr. Opin. Genet. Dev. 10, 651–655.
Boxaca, M.C., de Guerrero, L.B., Parodi, A.S., Rugiero, H.R., Gonzalez Cappa, S., 1965. Viremia in patients with Argentinean hemorrhagic fever. Rev. Asoc. Med. Argent. 79, 230–238.
Bradford, J.R., 1916. Nephritis in the British troops in Flanders. A preliminary note. J. R. Army Med. Corps. 27, 445–459.
Braithwaite, A.W., Prives, C.L., 2006. p53: more research and more questions. Cell Death Differ. 13, 877–880.
Brummer-Korvenkontio, M., Vaheri, A., Hovi, T., von Bonsdorff, C.H., Vuorimies, J., Manni, T., Penttinen, K., Oker-Blom, N., Lähdevirta, J., 1980. Nephropathia epidemica: detection of antigen in bank voles and serologic diagnosis of human infection. J. Infect. Dis. 141, 131–134.
Butel, J.S., 2000. Viral carcinogenesis: revelation of molecular mechanisms and etiology of human disease. Carcinogenesis 21, 405–426.
Casals, J., Henderson, B.E., Hoogstraal, H., Johnson, K.M., Shelokov, A., 1970. A review of Soviet viral hemorrhagic fevers, 1969. J. Infect. Dis. 122, 437–453.
Chitty, D., 1967. The natural selection of self-regulatory behaviour in animal populations. Proc. Ecol. Soc. Aust. 2, 78.
Collis, W.R., 1935. Acute benign lymphocytic meningitis (acute aseptic meningitis). Br. Med. J. 2, 1148–1150.
Cornulier, T., Yoccoz, N.G., Bretagnolle, V., Brommer, J.E., Butet, A., Ecke, F., Elston, D.A., Framstad, E., Henttonen, H., Hörnfeldt, B., Huitu, O., Imholt, C., Ims, R.A., Jacob, J., Jędrzejewska, B., Millon, A., Petty, S.J., Pietiäinen, H., Tkadlec, E., Zub, K., Lambin, X., 2013. Europe-wide dampening of population cycles in keystone herbivores. Science 340, 63–66.
Del Chierico, F., Ancora, M., Marcacci, M., Camma, C., Putignani, L., Conti, S., 2015. Choice of next-generation sequencing pipelines. Methods Mol. Biol. 1231, 31–47.
Diamond, J., 2002. Evolution, consequences and future of plant and animal domestication. Nature 418, 700–707.
Dobler, G., Wölfel, R., 2009. Typhus and other rickettsioses: emerging infections in Germany. Dtsch. Arztebl. Int. 106, 348–354.
Douangboupha, B., Aplin, K.P., Singleton, G.R., 2003. Rodent outbreaks in the uplands of Laos: analysis of historical patterns and the identity of nuu khii. ACIAR Monogr. Ser. 96, 103–111.
Drewes, S., Schmidt, S., Jacob, J., Imholt, C., Ulrich, R.G., 2016. APHAEA/EWDA Species Card: Voles and Mouses. http://www.aphaea.org/cards/species/voles.
Drewes, S., Ali, H.S., Saxenhofer, M., Rosenfeld, U.M., Binder, F., Cuypers, F., Schlegel, M., Röhrs, S., Heckel, G., Ulrich, R.G., 2017. Host-associated absence of human Puumala virus infections in northern and eastern Germany. Emerg. Infect. Dis. 23, 83–86.
Drexler, J.F., Corman, V.M., Müller, M.A., Maganga, G.D., Vallo, P., Binger, T., Gloza-Rausch, F., Cottontail, V.M., Rasche, A., Yordanov, S., Seebens, A., Knörnschild, M., Oppong, S., Adu Sarkodie, Y., Pongombo, C., Lukashev, A.N., Schmidt-Chanasit, J., Stöcker, A., Carneiro, A.J., Erbar, S., Maisner, A., Fronhoffs, F., Buettner, R., Kalko, E.K., Kruppa, T., Franke, C.R., Kallies, R., Yandoko, E.R., Herrler, G., Reusken, C., Hassanin, A., Krüger, D.H., Matthee, S., Ulrich, R.G., Leroy, E.M., Drosten, C., 2012. Bats host major mammalian paramyxoviruses. Nat. Commun. 3. https://doi.org/10.1038/ncomms1796.
Drexler, J.F., Corman, V.M., Müller, M.A., Lukashev, A.N., Gmyl, A., Coutard, B., Adam, A., Ritz, D., Leijten, L.M., van Riel, D., Kallies, R., Klose, S.M., Gloza-Rausch, F., Binger, T., Annan, A., Adu-Sarkodie, Y., Oppong, S., Bourgarel, M., Rupp, D.,

Hoffmann, B., Schlegel, M., Kümmerer, B.M., Krüger, D.H., Schmidt-Chanasit, J., Setién, A.A., Cottontail, V.M., Hemachudha, T., Wacharapluesadee, S., Osterrieder, K., Bartenschlager, R., Matthee, S., Beer, M., Kuiken, T., Reusken, C., Leroy, E.M., Ulrich, R.G., Drosten, C., 2013. Evidence for novel hepaciviruses in rodents. PLoS Pathog. 9, e1003438.

Drexler, J.F., Corman, V.M., Lukashev, A.N., van den Brand, J.M., Gmyl, A.P., Brunink, S., Rasche, A., Seggewibeta, N., Feng, H., Leijten, L.M., Vallo, P., Kuiken, T., Dotzauer, A., Ulrich, R.G., Lemon, S.M., Drosten, C., Hepatovirus Ecology Consortium, 2015. Evolutionary origins of hepatitis A virus in small mammals. Proc. Natl. Acad. Sci. U. S. A. 112, 15190–15195.

Eccard, J.A., Ylönen, H., 2001. Initiation of breeding after winter in bank voles: effects of food and population density. Can. J. Zool. 79, 1743–1753.

Eckert, J., Conraths, F.J., Tackmann, K., 2000. Echinococcosis: an emerging or re-emerging zoonosis? Int. J. Parasitol. 30, 1283–1294.

Ehlers, B., Küchler, J., Yasmum, N., Dural, G., Voigt, S., Schmidt-Chanasit, J., Jäkel, T., Matuschka, F.R., Richter, D., Essbauer, S., Hughes, D.J., Summers, C., Bennett, M., Stewart, J.P., Ulrich, R.G., 2007. Identification of novel rodent herpesviruses, including the first gammaherpesvirus of *Mus musculus*. J. Virol. 81, 8091–8100.

Ehlers, B., Richter, D., Matuschka, F.R., Ulrich, R.G., 2015. Genome sequences of a rat polyomavirus related to murine polyomavirus, Rattus norvegicus polyomavirus 1. Genome Announc. 3, e00997–15.

Everest, D.J., Grierson, S.S., Meredith, A.L., Milne, E.M., 2010. Adenovirus in a red squirrel (*Sciurus vulgaris*) from Scotland. Vet. Rec. 167, 184. https://doi.org/10.1136/vr.c3928.

Fabre, P.H., Hautier, L., Dimitrov, D., Douzery, E.J., 2012. A glimpse on the pattern of rodent diversification: a phylogenetic approach. BMC Evol. Biol. 12. https://doi.org/10.1186/1471-2148-12-88.

Farkas, T., Fey, B., Keller, G., Martella, V., Egyed, L., 2012. Molecular detection of novel astroviruses in wild and laboratory mice. Virus Genes 45, 518–525.

Farlow, J., Wagner, D.M., Dukerich, M., Stanley, M., Chu, M., Kubota, K., Petersen, J., Keim, P., 2005. *Francisella tularensis* in the United States. Emerg. Infect. Dis. 11, 1835–1841.

Fichet-Calvet, E., Lecompte, E., Koivogui, L., Daffis, S., ter Meulen, J., 2008. Reproductive characteristics of *Mastomys natalensis* and Lassa virus prevalence in Guinea, West Africa. Vector Borne Zoonotic Dis. 8, 41–48.

Firquet, E., Leirs, H., Bronner, G., 1996. Germinating grasses and reproductive seasonality of *Mastomys*-species (Rodentia, Muridae). Mammalia 60, 775–779.

Firth, C., Bhat, M., Firth, M.A., Williams, S.H., Frye, M.J., Simmonds, P., Conte, J.M., Ng, J., Garcia, J., Bhuva, N.P., Lee, B., Che, X., Quan, P.L., Lipkin, W.I., 2014. Detection of zoonotic pathogens and characterization of novel viruses carried by commensal *Rattus norvegicus* in New York City. MBio 5, e01933–01914.

Fitzgerald, B., Efford, M., Karl, B., 2004. Breeding of house mice and the mast seeding of southern beeches in the Orongorongo Valley, New Zealand. New Zeal. J. Zool. 31, 167–184.

Forman, M.S., Valsamakis, A., 2011. Specimen collection, transport, and processing: virology. In: Versalovic, J., Carroll, K., Funke, G., Jorgensen, J., Landry, M., Warnock, D. (Eds.), Manual of Clinical Microbiology, 10th ed. American Society of Microbiology, Washington, DC, pp. 1276–1288.

Fournié, G., Pfeiffer, D.U., Bendrey, R., 2017. Early animal farming and zoonotic disease dynamics: modelling brucellosis transmission in Neolithic goat populations. R. Soc. Open Sci. 4. https://doi.org/10.1098/rsos.160943.

Fournier, P.E., Marrie, T.J., Raoult, D., 1998. Diagnosis of Q fever. J. Clin. Microbiol. 36, 1823–1834.
Frank, F., 1957. The causality of microtine cycles in Germany (second preliminary research report). J. Wildl. Manag. 21, 113–121.
Friend, C., 1956. The isolation of a virus causing a malignant disease of the hematopoietic system in adult Swiss mice. Proc. Am. Assoc. Cancer Res. 2, 106.
Friend, C., 1957. Cell-free transmission in adult Swiss mice of a disease having the character of a leukemia. J. Exp. Med. 105, 307–318.
Fujino, K., Horie, M., Honda, T., Merriman, D.K., Tomonaga, K., 2014. Inhibition of Borna disease virus replication by an endogenous bornavirus-like element in the ground squirrel genome. Proc. Natl. Acad. Sci. U. S. A. 111, 13175–13180.
Garrido-Cardenas, J., Garcia-Maroto, F., Alvarez-Bermejo, J., Manzano-Agugliaro, F., 2017. DNA sequencing sensors: an overview. Sensors 17. https://doi.org/10.3390/s17030588.
Gliwicz, J., 1990. The first born, their dispersal, and vole cycles. Oecologia 83, 519–522.
Gonzalez, J.P., Emonet, S., de Lamballerie, X., Charrel, R., 2007. Arenaviruses. Curr. Top. Microbiol. Immunol. 315, 253–288.
Graffi, A., 1957. Chloroleukemia of mice. Ann. N. Y. Acad. Sci. 68, 540–558.
Graffi, A., Schramm, T., Bender, E., Bierwolf, D., Graffi, I., 1967. On a new virus containing skin tumor in golden hamster. Arch. Geschwulstforsch. 30, 277–283.
Green, C.A., Gordon, D.H., Lyons, N.F., 1978. Biological species in *Praomys* (*Mastomys*) *natalensis* (Smith), a rodent carrier of Lassa virus and bubonic plague in Africa. Am. J. Trop. Med. Hyg. 27, 627–629.
Gritsun, T.S., Lashkevich, V.A., Gould, E.A., 2003. Tick-borne encephalitis. Antiviral Res. 57, 129–146.
Gross, L., 1951. "Spontaneous" leukemia developing in C3H mice following inoculation in infancy, with AK-leukemic extracts, or AK-embrvos. Proc. Soc. Exp. Biol. Med. 76, 27–32.
Gross, L., 1953. A filterable agent, recovered from Ak leukemic extracts, causing salivary gland carcinomas in C3H mice. Proc. Soc. Exp. Biol. Med. 83, 414–421.
Guenther, S., Bethe, A., Fruth, A., Semmler, T., Ulrich, R.G., Wieler, L.H., Ewers, C., 2012. Frequent combination of antimicrobial multiresistance and extraintestinal pathogenicity in Escherichia coli isolates from urban rats (*Rattus norvegicus*) in Berlin, Germany. PLoS One 7, e50331.
Hansen, T.A., Mollerup, S., Nguyen, N.P., White, N.E., Coghlan, M., Alquezar-Planas, D.E., Joshi, T., Jensen, R.H., Fridholm, H., Kjartansdóttir, K.R., Mourier, T., Warnow, T., Belsham, G.J., Bunce, M., Willerslev, E., Nielsen, L.P., Vinner, L., Hansen, A.J., 2016. High diversity of picornaviruses in rats from different continents revealed by deep sequencing. Emerg. Microbes Infect. 5. https://doi.org/10.1038/emi.2016.90.
Hashimoto, K., Sugiyama, T., Sasaki, S., 1966. An adenovirus isolated from the feces of mice I. Isolation and identification. Jpn. J. Microbiol. 10, 115–125.
Hoffmann, D., Franke, A., Jeckel, M., Tamosiunaite, A., Schluckebier, J., Granzow, H., Hoffmann, B., Fischer, S., Ulrich, R.G., Höper, D., Goller, K., Osterrieder, N., Beer, M., 2015a. Out of the reservoir: phenotypic and genotypic characterization of a novel cowpox virus isolated from a common vole. J. Virol. 89, 10959–10969.
Hoffmann, B., Tappe, D., Höper, D., Herden, C., Boldt, A., Mawrin, C., Niederstrasser, O., Müller, T., Jenckel, M., van der Grinten, E., Lutter, C., Abendroth, B., Teifke, J.P., Cadar, D., Schmidt-Chanasit, J., Ulrich, R.G., Beer, M., 2015b. A variegated squirrel bornavirus associated with fatal human encephalitis. N. Engl. J. Med. 373, 154–162.
Homer, M.J., Bruinsma, E.S., Lodes, M.J., Moro, M.H., Telford 3rd, S., Krause, P.J., Reynolds, L.D., Mohamath, R., Benson, D.R., Houghton, R.L., Reed, S.G.,

Persing, D.H., 2000. A polymorphic multigene family encoding an immunodominant protein from *Babesia microti*. J. Clin. Microbiol. 38, 362–368.
Horie, M., Honda, T., Suzuki, Y., Kobayashi, Y., Daito, T., Oshida, T., Ikuta, K., Jern, P., Gojobori, T., Coffin, J.M., Tomonaga, K., 2010. Endogenous non-retroviral RNA virus elements in mammalian genomes. Nature 463, 84–87.
Hörnfeldt, B., 1994. Delayed density dependence as a determinant of vole cycles. Ecology 75, 791–806.
Hubálek, Z., Rudolf, I., 2011. Microbial Zoonoses and Sapronoses. Springer, Dordrecht-Heidelberg-London-New York, pp. 1–457.
Javier, R.T., Butel, J.S., 2008. The history of tumor virology. Cancer Res. 68, 7693–7706.
Johne, R., Plenge-Bönig, A., Hess, M., Ulrich, R.G., Reetz, J., Schielke, A., 2010a. Detection of a novel hepatitis E-like virus in faeces of wild rats using a nested broad-spectrum RT-PCR. J. Gen. Virol. 91, 750–758.
Johne, R., Heckel, G., Plenge-Bönig, A., Kindler, E., Maresch, C., Reetz, J., Schielke, A., Ulrich, R.G., 2010b. Novel hepatitis E virus genotype in Norway rats, Germany. Emerg. Infect. Dis. 16, 1452–1455.
Johnson, K.M., 2001. Hantaviruses: history and overview. Curr. Top. Microbiol. Immunol. 256, 1–14.
Kallio-Kokko, H., Laakkonen, J., Rizzoli, A., Tagliapietra, V., Cattadori, I., Perkins, S.E., Hudson, P.J., Cristofolini, A., Versini, W., Vapalahti, O., Vaheri, A., Henttonen, H., 2006. Hantavirus and arenavirus antibody prevalence in rodents and humans in Trentino, Northern Italy. Epidemiol. Infect. 134, 830–836.
Kilham, L., Murphy, H.W., 1953. A pneumotropic virus isolated from C3H mice carrying the Bittner Milk Agent. Proc. Soc. Exp. Biol. Med. 82, 133–137.
Klempa, B., Fichet-Calvet, E., Lecompte, E., Auste, B., Aniskin, V., Meisel, H., Denys, C., Koivogui, L., ter Meulen, J., Krüger, D.H., 2006. Hantavirus in African wood mouse, Guinea. Emerg. Infect. Dis. 12, 838–840.
Klempa, B., Fichet-Calvet, E., Lecompte, E., Auste, B., Aniskin, V., Meisel, H., Barriere, P., Koivogui, L., ter Meulen, J., Krüger, D.H., 2007. Novel hantavirus sequences in Shrew, Guinea. Emerg. Infect. Dis. 13, 520–522.
Klempa, B., Krüger, D.H., Auste, B., Stanko, M., Krawczyk, A., Nickel, K.F., Überla, K., Stang, A., 2009. A novel cardiotropic murine adenovirus representing a distinct species of mastadenoviruses. J. Virol. 83, 5749–5759.
Klempa, B., Avšič-Županc, T., Clement, J., Dzagurova, T.K., Henttonen, H., Heyman, P., Jakab, F., Krüger, D.H., Maes, P., Papa, A., Tkachenko, E.A., Ulrich, R.G., Vapalahti, O., Vaheri, A., 2013. Complex evolution and epidemiology of Dobrava-Belgrade hantavirus: definition of genotypes and their characteristics. Arch. Virol. 158, 521–529.
Korpimäki, E., Brown, P.R., Jacob, J., Pech, R.P., 2004. The puzzles of population cycles and outbreaks of small mammals solved? Bioscience 54, 1071–1079.
Korpimäki, E., Norrdahl, K., Huitu, O., Klemola, T., 2005. Predator-induced synchrony in population oscillations of coexisting small mammal species. Proc. R. Soc. B 272, 193–202.
Krebs, C.J., Myers, J.H., 1974. Population cycles in small mammals. Adv. Ecol. Res. 8, 267–399.
Lambin, X., Bretagnolle, V., Yoccoz, N.G., 2006. Vole population cycles in northern and southern Europe: is there a need for different explanations for single pattern? J. Anim. Ecol. 75, 340–349.
Lau, S.K., Woo, P.C., Li, K.S., Zhang, H.J., Fan, R.Y., Zhang, A.J., Chan, B.C., Lam, C.S., Yip, C.C., Yuen, M.C., Chan, K.H., Chen, Z.W., Yuen, K.Y., 2016. Identification of novel Rosavirus species that infects diverse rodent species and causes multisystemic dissemination in mouse model. PLoS Pathog. 12, e1005911.

Lee, H.W., 1982. Hemorrhagic fever with renal syndrome. History of Hantaan virus and epidemiological features. Scand. J. Infect. Dis. 36, 82–85.
Lee, H.W., Lee, P.W., Johnson, K.M., 1978. Isolation of the etiologic agent of Korean Hemorrhagic fever. J. Infect. Dis. 137, 298–308.
Lee, H.W., Baek, L.J., Johnson, K.M., 1982. Isolation of Hantaan virus, the etiologic agent of Korean hemorrhagic fever, from wild urban rats. J. Infect. Dis. 146, 638–644.
Leirs, H., Stenseth, N.C., Nichols, J.D., Hines, J.E., Verhagen, R., Verheyen, W., 1997. Stochastic seasonality and nonlinear density-dependent factors regulate population size in an African rodent. Nature 389, 176–180.
Linard, C., Lamarque, P., Heyman, P., Ducoffre, G., Luyasu, V., Tersago, K., Vanwambeke, S.O., Lambin, E.F., 2007. Determinants of the geographic distribution of Puumala virus and Lyme borreliosis infections in Belgium. Int. J. Health Geogr. 6. https://doi.org/10.1186/1476-072X-6-15.
López, N., Padula, P., Rossi, C., Lázaro, M.E., Franze-Fernández, M.T., 1996. Genetic identification of a new hantavirus causing severe pulmonary syndrome in Argentina. Virology 220, 223–226.
Mackenzie, R.B., 1965. Epidemiology of Machupo virus infection: I. Pattern of human infection, San Joaquín, Bolivia, 1962-1964. Am. J. Trop. Med. Hyg. 14, 808–813.
Meerburg, B.G., Singleton, G.R., Kijlstra, A., 2009. Rodent-borne diseases and their risks for public health. Crit. Rev. Microbiol. 35, 221–270.
Mokili, J.L., Rohwer, F., Dutilh, B.E., 2012. Metagenomics and future perspectives in virus discovery. Curr. Opin. Virol. 2, 63–77.
Moloney, J.B., 1960. Biological studies on a lymphoid-leukemia virus extracted from sarcoma 37. I. Origin and introductory investigations. J. Natl. Cancer Inst. 24, 933–951.
Moss, B., Flexner, C., 1987. Vaccinia virus expression vectors. Annu. Rev. Immunol. 5, 305–324.
Myhrman, G., 1951. Nephropathia epidemica a new infectious disease in northern Scandinavia. Acta Med. Scand. 140, 52–56.
Nag, S., 1999. Bamboo, rats and famines: famine relief and perceptions of British paternalism in the Mizo Hills (India). Environ. Hist. 5, 245–252.
Nainys, J., Timinskas, A., Schneider, J., Ulrich, R.G., Gedvilaite, A., 2015. Identification of two novel members of the tentative genus Wukipolyomavirus in wild rodents. PLoS One 10, e0140916.
Nichol, S.T., Spiropoulou, C.F., Morzunov, S., Rollin, P.E., Ksiazek, T.G., Feldmann, H., Sanchez, A., Childs, J., Zaki, S., Peters, C.J., 1993. Genetic identification of a hantavirus associated with an outbreak of acute respiratory illness. Science 262, 914–917.
Perry, R.D., Fetherston, J.D., 1997. *Yersinia pestis*—etiologic agent of plague. Clin. Microbiol. Rev. 10, 35–66.
Phan, T.G., Kapusinszky, B., Wang, C., Rose, R.K., Lipton, H.L., Delwart, E.L., 2011. The fecal viral flora of wild rodents. PLoS Pathog. 7, e1002218.
Phung, N.T.M., Brown, P.R., Leung, L.K.P., 2012. Changes in population abundance, reproduction and habitat use of the rice-field rat, *Rattus argentiventer*, in relation to rice-crop growth stage in a lowland rice agroecosystem in Vietnam. Wildl. Res. 39, 250–257.
Pilaski, J., Zöller, L., Blenk, H., 1986. Hämorrhagisches Fieber mit renalem Syndrom (HFRS): eine durch Nagetiere übertragene Nephropathie des Menschen. Wehrmed. Mschr. 30, 435–444.
Plyusnin, A., Vapalahti, O., Lankinen, H., Lehväslaiho, H., Apekina, N., Myasnikov, Y., Kallio-Kokko, H., Henttonen, H., Lundkvist, Å., Brummer-Korvenkontio, M., Gavrilovskaya, I., Vaheri, A., 1994. Tula virus: a newly detected hantavirus carried by European common voles. J. Virol. 68, 7833–7839.

Pounder, K.C., Begon, M., Sironen, T., Henttonen, H., Watts, P.C., Voutilainen, L., Vapalahti, O., Klempa, B., Fooks, A.R., McElhinney, L.M., 2013. Novel hantavirus in field vole, United Kingdom. Emerg. Infect. Dis. 19, 673–675.

Radchuk, V., Ims, R.A., Andreassen, H.P., 2016. From individuals to population cycles: the role of extrinsic and intrinsic factors in rodent populations. Ecology 97, 720–732.

Rector, A., Tachezy, R., Van Doorslaer, K., MacNamara, T., Burk, R.D., Sundberg, J.P., Van Ranst, M., 2005. Isolation and cloning of a papillomavirus from a North American porcupine by using multiply primed rolling-circle amplification: the Erethizon dorsatum papillomavirus type 1. Virology 331, 449–456.

Reil, D., Imholt, C., Drewes, S., Ulrich, R.G., Eccard, J.A., Jacob, J., 2016. Environmental conditions in favour of a hantavirus outbreak in 2015 in Germany? Zoonoses Public Health 63, 83–88.

Rigatti, L.H., Toptan, T., Newsome, J.T., Moore, P.S., Chang, Y., 2016. Identification and characterization of novel rat polyomavirus 2 in a colony of X-SCID rats by P-PIT assay. mSphere 1, e00334-16.

Rivers, T.M., McNair Scott, T.F., 1935. Meningitis in man caused by a filterable virus. Science 81, 439–440.

Rowe, W.P., Murphy, F.A., Bergold, G.H., Casals, J., Hotchin, J., Johnson, K.M., Lehmann-Grube, F., Mims, C.A., Traub, E., Webb, P.A., 1970. Arenoviruses: proposed name for a newly defined virus group. J. Virol. 5, 651–652.

Rüdisser, E., 1942. Beobachtungen zur Pathogenese und Aetiologie von Nierenentzündungen im Feldzug im Osten Winter 1941/42. Münch. Med. Wochenschr. 41, 863–866.

Ryll, R., Bernstein, S., Heuser, E., Schlegel, M., Dremsek, P., Zumpe, M., Wolf, S., Pépin, M., Bajomi, D., Müller, G., Heiberg, A.-N., Spahr, C., Lang, J., Groschup, M.H., Ansorge, H., Freise, J., Guenther, S., Baert, C., Ruiz-Fons, F., Pikula, J., Knap, N., Tsakmakidis, I., Dovas, C., Zanet, S., Imholt, C., Heckel, G., Johne, R., Ulrich, R.G., 2017. Detection of rat hepatitis E virus in wild Norway rats (*Rattus norvegicus*) and Black rats (*Rattus rattus*) from 11 European countries. Vet. Microbiol 208, 58–68.

Sachsenröder, J., Braun, A., Machnowska, P., Ng, T.F., Deng, X., Guenther, S., Bernstein, S., Ulrich, R.G., Delwart, E., Johne, R., 2014. Metagenomic identification of novel enteric viruses in urban wild rats and genome characterization of a group A rotavirus. J. Gen. Virol. 95, 2734–2747.

Scheuch, M., Höper, D., Beer, M., 2015. RIEMS: a software pipeline for sensitive and comprehensive taxonomic classification of reads from metagenomics datasets. BMC Bioinformatics 16. https://doi.org/10.1186/s12859-015-0503-6.

Schlegel, M., Jacob, J., Krüger, D.H., Rang, A., Ulrich, R.G., 2014. Hantavirus emergence in rodents, insectivores and bats. What comes next? In: Johnson, N. (Ed.), Role of Animals in Emerging Viral Diseases. Academic Press, Waltham, MA, pp. 235–291.

Schlottau, K., Jenckel, M., van den Brand, J., Fast, C., Herden, C., Höper, D., Homeier-Bachmann, T., Thielebein, J., Mensing, N., Diender, B., Hoffmann, D., Ulrich, R.G., Mettenleiter, T.C., Koopmans, M., Tappe, D., Schmidt-Chanasit, J., Reusken, C.B., Beer, M., Hoffmann, B., 2017a. Variegated Squirrel Bornavirus 1 in Squirrels, Germany and the Netherlands. Emerg. Infect. Dis. 23, 477–481.

Schlottau, K., Hoffmann, B., Homeier-Bachmann, T., Fast, C., Ulrich, R.G., Beer, M., Hoffmann, D., 2017b. Multiple detection of zoonotic variegated squirrel bornavirus 1 RNA in different squirrel species suggests a possible unknown origin for the virus. Arch. Virol. https://doi.org/10.1007/s00705-017-3432-z. (Epub ahead of print).

Schulz, E., Gottschling, M., Wibbelt, G., Stockfleth, E., Nindl, I., 2009. Isolation and genomic characterization of the first Norway rat (*Rattus norvegicus*) papillomavirus and its phylogenetic position within Pipapillomavirus, primarily infecting rodents. J. Gen. Virol. 90, 2609–2614.

Schulz, E., Gottschling, M., Ulrich, R.G., Richter, D., Stockfleth, E., Nindl, I., 2012. Isolation of three novel rat and mouse papillomaviruses and their genomic characterization. PLoS One 7, e47164.

Singleton, G.R., Brown, P.R., Jacob, J., Aplin, K.P., 2007. Unwanted and unintended effects of culling: a case for ecologically-based rodent management. Integr. Zool. 2, 247–259.

Sironen, T., Vaheri, A., Plyusnin, A., 2001. Molecular evolution of Puumala hantavirus. J. Virol. 75, 11803–11810.

Smadel, J.E., 1953. Epidemic hemorrhagic fever. Am. J. Public Health Nations Health 43, 1327–1330.

Smith, M.G., 1954. Propagation of salivary gland virus of the mouse in tissue cultures. Proc. Soc. Exp. Biol. Med. 86, 435–440.

Smith, M.J., White, A., Sherratt, J.A., Telfer, S., Begon, M., Lambin, X., 2008. Disease effects on reproduction can cause population cycles in seasonal environments. J. Anim. Ecol. 77, 378–389.

Smith, D.B., Simmonds, P., Members of the International Committee on the Taxonomy of Viruses Hepeviridae Study Group, Jameel, S., Emerson, S.U., Harrison, T.J., Meng, X.J., Okamoto, H., Van der Poel, W.H., Purdy, M.A., 2014. Consensus proposals for classification of the family Hepeviridae. J. Gen. Virol. 95, 2223–2232.

Smith, D.B., Simmonds, P., Izopet, J., Oliveira-Filho, E.F., Ulrich, R.G., Johne, R., Koenig, M., Jameel, S., Harrison, T.J., Meng, X.J., Okamoto, H., Van der Poel, W.H., Purdy, M.A., 2016. Proposed reference sequences for hepatitis E virus subtypes. J. Gen. Virol. 97, 537–542.

Sonnenburg, J., Ryser-Degiorgis, M.P., Kuiken, T., Ferroglio, E., Ulrich, R.G., Conraths, F.J., Gortázar, C., Staubach, C., APHAEA project partners, 2017. Harmonizing methods for wildlife abundance estimation and pathogen detection in Europe-a questionnaire survey on three selected host-pathogen combinations. BMC Vet. Res. 13. https://doi.org/10.1186/s12917-016-0935-x.

Stewart, S.E., Eddy, B.E., 1959. Tumor induction by SE polyoma virus and the inhibition of tumors by specific neutralizing antibodies. Am. J. Public Health Nations Health 49, 1493–1496.

Straková, P., Dufkova, L., Širmarová, J., Salát, J., Bartonička, T., Klempa, B., Pfaff, F., Höper, D., Hoffmann, B., Ulrich, R.G., Růžek, D., 2017. Novel hantavirus identified in European bat species *Nyctalus noctula*. Infect. Genet. Evol. 48, 127–130.

Stuhlfauth, K., 1943. Bericht über ein neues schlammfieberähnliches Krankheitsbild bei deutschen Truppen in Lappland. Dtsch. Med. Wochenschr. 69, 439–443.

Tamura, K., Stecher, G., Peterson, D., Filipski, A., Kumar, S., 2013. MEGA6: molecular evolutionary genetics analysis version 6.0. Mol. Biol. Evol. 30, 2725–2729.

Tegtmeyer, P., 1975. Function of simian virus 40 gene A in transforming infection. J. Virol. 15, 613–618.

Temin, H.M., 1964a. The participation of DNA in Rous sarcoma virus production. Virology 23, 486–494.

Temin, H.M., 1964b. Nature of the provirus of Rous sarcoma. Natl. Cancer Inst. Monogr. 17, 557–570.

Temin, H.M., Mizutani, S., 1970. RNA-dependent DNA polymerase in virions of Rous sarcoma virus. Nature 226, 1211–1213.

Tersago, K., Verhagen, R., Servais, A., Heyman, P., Ducoffre, G., Leirs, H., 2009. Hantavirus disease (nephropathia epidemica) in Belgium: effects of tree seed production and climate. Epidemiol. Infect. 137, 250–256.

Thrall, P.H., Antonovics, J., Hall, D.W., 1993. Host and pathogen coexistence in sexually transmitted and vector-borne diseases characterized by frequency-dependent disease transmission. Am. Nat. 142, 543–552.

To, K.K.W., Chan, W.M., Li, K.S.M., Lam, C.S.F., Chen, Z., Tse, H., Lau, S.K.P., Woo, P.C.Y., Yuen, K.Y., 2017. High prevalence of four novel astrovirus genotype species identified from rodents in China. J. Gen. Virol. 98, 1004–1015.

Tooze, J. (Ed.), 1981. DNA Tumor Viruses. Molecular biology of tumor viruses, second ed. Cold Spring Harbor Laboratory Press, New York, pp. 1–1073.

Torri, F., Dinov, I.D., Zamanyan, A., Hobel, S., Genco, A., Petrosyan, P., Clark, A.P., Liu, Z., Eggert, P., Pierce, J., Knowles, J.A., Ames, J., Kesselman, C., Toga, A.W., Potkin, S.G., Vawter, M.P., Macciardi, F., 2012. Next generation sequence analysis and computational genomics using graphical pipeline workflows. Genes 3, 545–575.

Traub, E., 1936. The epidemiology of lymphocytic choriomeningitis in white mice. J. Exp. Med. 64, 183–200.

Van Knapen, F., 1989. Toxoplasmosis, old stories and new facts. Int. Ophthalmol. 13, 371–375.

Volpicella, M., Leoni, C., Costanza, A., Fanizza, I., Placido, A., Ceci, L.R., 2012. Genome walking by next generation sequencing approaches. Biology 1, 495–507.

Wilson, D.E., Reeder, D.M., 1993. Mammal Species of the World: A Taxonomic and Geographic Reference. Smithsonian Institution Press, Washington and London, pp. 1–1001.

Witkowski, P.T., Drexler, J.F., Kallies, R., Ličková, M., Bokorová, S., Mananga, G.D., Szemes, T., Leroy, E.M., Krüger, D.H., Drosten, C., Klempa, B., 2016. Phylogenetic analysis of a newfound bat-borne hantavirus supports a laurasiatherian host association for ancestral mammalian hantaviruses. Infect. Genet. Evol. 41, 113–119.

Wolf, S., Reetz, J., Johne, R., Heiberg, A.C., Petri, S., Kanig, H., Ulrich, R.G., 2013. The simultaneous occurrence of human norovirus and hepatitis E virus in a Norway rat (*Rattus norvegicus*). Arch. Virol. 158, 1575–1578.

Yanagihara, R., Gu, S.H., Song, J.W., 2015. Expanded host diversity and global distribution of hantaviruses: implications for identifying and investigating previously unrecognized hantaviral diseases. In: Global Virology I–Identifying and Investigating Viral Diseases. Springer, New York, pp. 161–198.

Zhou, X., Ramachandran, S., Mann, M., Popkin, D.L., 2012. Role of lymphocytic choriomeningitis virus (LCMV) in understanding viral immunology: past, present and future. Viruses 4, 2650–2669.

FURTHER READING

Berto, A., Anh, P.H., Carrique-Mas, J.J., Simmonds, P., Van Cuong, N., Tue, N.T., Van Dung, N., Woolhouse, M.E., Smith, I., Marsh, G.A., Bryant, J.E., Thwaites, G.E., Baker, S., Rabaa, M.A., VIZIONS Consortium, 2017. Detection of potentially novel paramyxovirus and coronavirus viral RNA in bats and rats in the Mekong Delta region of southern Viet Nam. Zoonoses Public Health. https://doi.org/10.1111/zph.12362 (Epub ahead of print).

Blaskovic, D., Stancekova, M., Svobodova, J., Mistrikova, J., 1980. Isolation of five strains of herpesviruses from two species of free living small rodents. Acta Virol. 24, 468.

Bosch, S., Lurz, P.W.W., Seybold, K., Abendroth, B., Larres, G., Schlottau, K., Wylezich, C., Schröder, C., Wernike, K., Beer, M., Ulrich, R.G., 2017. Nachweis eines neuen Squirrel-Adenovirus bei einem Krankheitsausbruch bei Eichhörnchen (*Sciurus vulgaris*) in Deutschland 2013–2016. Mitteilungen aus unserer Säugetierwelt 20, 27–35.

Buckley, S.M., Casals, J., 1970. Lassa fever, a new virus disease of man from West Africa. 3. Isolation and characterization of the virus. Am. J. Trop. Med. Hyg. 19, 680–691.
Chu, D.K.W., Chin, A.W.H., Smith, G.J., Chan, K.H., Guan, Y., Malik Peiris, J.S., Poon, L.L.M., 2010. Detection of novel astroviruses in urban brown rats and previously known astroviruses in humans. J. Gen. Virol. 91, 2457–2462.
Diagne, M.M., Faye, M., Faye, O., Sow, A., Balique, F., Sembène, M., Granjon, L., Handschumacher, P., Faye, O., Diallo, M., Sall, A.A., 2017. Emergence of Wesselsbron virus among black rat and humans in Eastern Senegal in 2013. One Health 3, 23–28.
Dutra, L.A., de Freitas Almeida, G.M., Oliveira, G.P., Abrahao, J.S., Kroon, E.G., Trindade, G.S., 2017. Molecular evidence of orthopoxvirus DNA in capybara (*Hydrochoerus hydrochaeris*) stool samples. Arch. Virol. 162, 439–448.
Everest, D.J., Shuttleworth, C.M., Stidworthy, M.F., Grierson, S.S., Duff, J.P., Kenward, R.E., 2014. Adenovirus: an emerging factor in red squirrel *Sciurus vulgaris* conservation. Mammal. Rev. 44, 225–233.
Ge, X.Y., Yang, W.H., Zhou, J.H., Li, B., Zhang, W., Shi, Z.L., Zhang, Y.Z., 2017. Detection of alpha- and betacoronaviruses in rodents from Yunnan, China. Virol. J. 14. https://doi.org/10.1186/s12985-017-0766-9.
Ghawar, W., Pascalis, H., Bettaieb, J., Mélade, J., Gharbi, A., Snoussi, M.A., Laouini, D., Goodman, S.M., Ben Salah, A., Dellagi, K., 2017. Insight into the global evolution of Rodentia associated Morbilli-related paramyxoviruses. Sci. Rep. 7. https://doi.org/10.1038/s41598-017-02206-0.
Hu, B., Chmura, A.A., Li, J., Zhu, G., Desmond, J.S., Zhang, Y., Zhang, W., Epstein, J.H., Daszak, P., Shi, Z., 2014. Detection of diverse novel astroviruses from small mammals in China. J. Gen. Virol. 95, 2442–2449.
Johne, R., Dremsek, P., Kindler, E., Schielke, A., Plenge-Bönig, A., Gregersen, H., Wessels, U., Schmidt, K., Rietschel, W., Groschup, M.H., Guenther, S., Heckel, G., Ulrich, R.G., 2012. Rat hepatitis E virus: geographical clustering within Germany and serological detection in wild Norway rats (*Rattus norvegicus*). Infect. Genet. Evol. 12, 947–956.
Johnson, K.M., 1965. Epidemiology of Machupo virus infection. 3. Significance of virological observations in man and animals. Am. J. Trop. Med. Hyg. 14, 816–818.
Kapoor, A., Simmonds, P., Scheel, T.K., Hjelle, B., Cullen, J.M., Burbelo, P.D., Chauhan, L.V., Duraisamy, R., Sanchez Leon, M., Jain, K., Vandegrift, K.J., Calisher, C.H., Rice, C.M., Lipkin, W.I., 2013. Identification of rodent homologs of hepatitis C virus and pegiviruses. MBio 4, e00216–13.
Kilham, L., Olivier, L.J., 1959. A latent virus of rats isolated in tissue culture. Virology 7, 428–437.
Knap, N., Korva, M., Dolinšek, V., Sekirnik, M., Trilar, T., Avšič-Županc, T., 2012. Patterns of tick-borne encephalitis virus infection in rodents in Slovenia. Vector Borne Zoonotic Dis. 12, 236–242.
Kraft, L.M., 1962. An apparently new lethal virus disease of infant mice. Science 137, 282–283.
Lack, J.B., Volk, K., Van Den Bussche, R.A., 2012. Hepatitis E virus genotype 3 in wild rats, United States. Emerg. Infect. Dis. 18, 1268–1273.
Marchal, J., 1930. Infectious ectromelia. A hitherto undescribed virus disease of mice. J. Pathol. 33, 713–728.
Martínez-Jiménez, D., Graham, D., Couper, D., Benko, M., Schoniger, S., Gurnell, J., Sainsbury, A.W., 2011. Epizootiology and pathologic findings associated with a newly described adenovirus in the red squirrel, *Sciurus vulgaris*. J. Wildl. Dis. 47, 442–454.
Müller, H., Gissmann, L., 1978. Mastomys natalensis papilloma virus (MnPV), the causative agent of epithelial proliferations: characterization of the virus particle. J. Gen. Virol. 41, 315–323.

Murphy, F.A., Foundations of Virology. Available at http://www.utmb.edu/virusimages/. Accessed 3 April 2017.
Nafz, J., Schäfer, K., Chen, S.F., Bravo, I.G., Ibberson, M., Nindl, I., Stockfleth, E., Rösl, F., 2008. A novel rodent papillomavirus isolated from anogenital lesions in its natural host. Virology 374, 186–197.
Naulty, F., Everest, D., Warnock, N.D., Phelan, K., Callanan, J.J., 2013. Squirrelpox virus in red squirrels (*Sciurus vulgaris*) in the Republic of Ireland. J. Wildl. Dis. 49, 1070–1073.
O'Banion, M.K., Reichmann, M.E., Sundberg, J.P., 1988. Cloning and characterization of a papillomavirus associated with papillomas and carcinomas in the European harvest mouse (*Micromys minutus*). J. Virol. 62, 226–233.
Orba, Y., Kobayashi, S., Nakamura, I., Ishii, A., Hang'ombe, B.M., Mweene, A.S., Thomas, Y., Kimura, T., Sawa, H., 2011. Detection and characterization of a novel polyomavirus in wild rodents. J. Gen. Virol. 92, 789–795.
Padgett, K.A., Reisen, W.K., Kahl-Purcell, N., Fang, Y., Cahoon-Young, B., Carney, R., Anderson, N., Zucca, L., Woods, L., Husted, S., Kramer, V.L., 2007. West Nile virus infection in tree squirrels (Rodentia: Sciuridae) in California, 2004-2005. Am. J. Trop. Med. Hyg. 76, 810–813.
Parodi, A.S., Greenway, D.J., Rugiero, H.R., Frigerio, M., De La Barrera, J.M., Mettler, N., Garzon, F., Boxaca, M., Guerrero, L., Nota, N., 1958. Concerning the epidemic outbreak in Junin. Dia Med. 30, 2300–2301.
Peters, M., Vidovszky, M.Z., Harrach, B., Fischer, S., Wohlsein, P., Kilwinski, J., 2011. Squirrel adenovirus type 1 in red squirrels (*Sciurus vulgaris*) in Germany. Vet. Rec. 169. https://doi.org/10.1136/vr.d2610.
Pintér, R., Madai, M., Horváth, G., Németh, V., Oldal, M., Kemenesi, G., Dallos, B., Bányai, K., Jakab, F., 2014. Molecular detection and phylogenetic analysis of tick-borne encephalitis virus in rodents captured in the transdanubian region of Hungary. Vector Borne Zoonotic Dis. 14, 621–624.
Rogovskyy, A.S., Baszler, T.V., Bradway, D.S., Bruning, D.L., Davitt, C.M., Evermann, J.F., Burk, R.D., Chen, Z., Mansfield, K.G., Haldorson, G.J., 2012. A novel papillomavirus isolated from proliferative skin lesions of a wild American beaver (*Castor canadensis*). J. Vet. Diagn. Invest. 24, 750–754.
Rogovskyy, A.S., Chen, Z., Burk, R.D., Bankhead, T., 2014. Characterization of the North American beaver (*Castor canadensis*) papillomavirus genome. Vet. Microbiol. 168, 214–220.
Root, J.J., Oesterle, P.T., Sullivan, H.J., Hall, J.S., Marlenee, N.L., McLean, R.G., Montenieri, J.A., Clark, L., 2007. Fox squirrel (*Sciurus niger*) associations with West Nile virus. Am. J. Trop. Med. Hyg. 76, 782–784.
Salem, N.B., Moens, U., Ehlers, B., 2016. Genome sequences of murine pneumotropic virus (Polyomaviridae) detected in wild house mice (*Mus musculus*). Genome Announc. 4, e01545–15.
Sasaki, M., Muleya, W., Ishii, A., Orba, Y., Hang'ombe, B.M., Mweene, A.S., Moonga, L., Thomas, Y., Kimura, T., Sawa, H., 2014. Molecular epidemiology of paramyxoviruses in Zambian wild rodents and shrews. J. Gen. Virol. 95, 325–330.
Shuttleworth, C.M., Signorile, A.L., Everest, D.J., Lurz, P.W.W., 2015. Assessing causes and significance of red squirrel (*Sciurus vulgaris*) mortality during regional population restoration: an applied conservation perspective. HYSTRIX J. 26, 69–75.
Tsoleridis, T., Onianwa, O., Horncastle, E., Dayman, E., Zhu, M., Danjittrong, T., Wachtl, M., Behnke, J.M., Chapman, S., Strong, V., Dobbs, P., Ball, J.K., Tarlinton, R.E., McClure, C.P., 2016. Discovery of novel alphacoronaviruses in European rodents and shrews. Viruses 8. https://doi.org/10.3390/v8030084.
Van Doorslaer, K., Rector, A., Jenson, A.B., Sundberg, J.P., Van Ranst, M., Ghim, S.J., 2007. Complete genomic characterization of a murine papillomavirus isolated from

papillomatous lesions of a European harvest mouse (*Micromys minutus*). J. Gen. Virol. 88, 1484–1488.

Wang, W., Lin, X.D., Guo, W.P., Zhou, R.H., Wang, M.R., Wang, C.Q., Ge, S., Mei, S.H., Li, M.H., Shi, M., Holmes, E.C., Zhang, Y.Z., 2015. Discovery, diversity and evolution of novel coronaviruses sampled from rodents in China. Virology 474, 19–27.

Widén, F., Ayral, F., Artois, M., Olofson, A.S., Lin, J., 2014. PCR detection and analysis of potentially zoonotic hepatitis E virus in French rats. Virol. J. 11. https://doi.org/10.1186/1743-422X-11-90.

Wilkinson, D.A., Mélade, J., Dietrich, M., Ramasindrazana, B., Soarimalala, V., Lagadec, E., le Minter, G., Tortosa, P., Heraud, J.M., de Lamballerie, X., Goodman, S.M., Dellagi, K., Pascalis, H., 2014. Highly diverse morbillivirus-related paramyxoviruses in wild fauna of the southwestern Indian Ocean Islands: evidence of exchange between introduced and endemic small mammals. J. Virol. 88, 8268–8277.

Wu, Z., Yang, L., Yang, F., Ren, X., Jiang, J., Dong, J., Sun, L., Zhu, Y., Zhou, H., Jin, Q., 2014. Novel henipa-like virus, Mojiang paramyxovirus, in rats, China, 2014. Emerg. Infect. Dis. 20, 1064–1066.

Zheng, X.Y., Qiu, M., Ke, X.M., Guan, W.J., Li, J.M., Huo, S.T., Chen, S.W., Zhong, X.S., Zhou, W., Xiong, Y.Q., Ge, J., Chen, Q., 2016. Detection of novel adenoviruses in fecal specimens from rodents and shrews in southern China. Virus Genes 52, 417–421.

CHAPTER FIVE

The Expanding Field of Mammalian Astroviruses: Opportunities and Challenges in Clinical Virology

Céline L. Boujon*, Michel C. Koch*, Torsten Seuberlich*,[1]
*University of Bern, Bern, Switzerland
[1]Corresponding author: e-mail address: torsten.seuberlich@vetsuisse.unibe.ch

Contents

Abstract

Astroviruses are best known as being one of the leading causes of diarrhea in infants and were first described in this context in 1975. In its first years, astrovirus research was mainly restricted to electron microscopy and serology studies. The ability to culture

Advances in Virus Research, Volume 99
ISSN 0065-3527
https://doi.org/10.1016/bs.aivir.2017.07.002

some of these viruses in vitro allowed a first consequent step forward, especially at the molecular level. Since the emergence of more powerful genetic methods, though, the face of this research field has dramatically changed and evolved. From the exponential number of discoveries of new astrovirus strains in the most varied of animal species to their association with atypical diseases, these viruses revealed a lot of surprises, and many more are probably still waiting to be uncovered. This chapter summarizes the most important knowledge about astroviruses and discusses the implication of the latest findings in this area of research.

1. INTRODUCTION

Astroviruses appeared in literature for the first time in 1975 (Appleton and Higgins, 1975; Madeley and Cosgrove, 1975a,b). At that time, the study of astroviruses was mostly restricted to electron microscopy investigations. In the following years through to the beginning of this millennium, research gradually focused on molecular biology characterizations. Since the emergence of next-generation sequencing (NGS) methods, however, our knowledge about astroviruses has dramatically increased and changed. In the last two decades, initial presumptions about the *Astroviridae* family have been continuously challenged due to technological progresses.

The family *Astroviridae* belongs to group IV of the Baltimore classification system, which comprises viruses with a genome consisting of single-stranded RNA of positive polarity, typically between 6.2 and 7.7 kb in length. Virions are nonenveloped and have an icosahedral structure, and common to the genome of all astroviruses is the presence of three open reading frames (ORFs). ORF1a and 1b encode nonstructural proteins, whereas capsid proteins are translated from ORF2.

Astroviruses are classified into two genera: Mamastrovirus (MAstV) and Avastrovirus (AAstV), affecting mammalian and avian species, respectively. AAstV are associated with a panel of diseases and syndromes in poultry livestock, and cause enteric or systemic infections and pathologies. By contrast, disease association in mammalian species is less evident, with enteric virus infections being mainly asymptomatic, i.e., subclinical. In humans, however, a certain genotype species (*MAstV 1*) is one of the main causes of gastrointestinal disease in children; nevertheless, in such cases, clinical symptoms are often mild, and the infection is self-limiting.

With new virus discoveries in animals and humans, the spectrum of astrovirus genotype species has increased significantly, and there is a growing body of evidence that MAstV are involved in the pathogenesis of nongastrointestinal diseases. This chapter intends to provide a general overview of advances in the expanding field of MAstV research and discusses their clinical significance for human and veterinary virology.

2. TAXONOMY OF *ASTROVIRIDAE*

Originally, the taxonomy of astroviruses depended on the species in which a specific strain had been isolated for the first time. Nowadays, it relies on the complete capsid precursor protein sequence. Within the two genera MAstV and AAstV, individual genotype species are thus recognized by the International Committee on Taxonomy of Viruses (ICTV) and present a genetic amino acid p-distance of the capsid precursor protein of at least 0.338. Not all astroviruses that have been reported have been officially classified yet, but to date there are 19 genotype species of MAstV (*MAstV 1–19*), and another 14 species being proposed (*MAstV 20–33*). For AAstV, three genotype species (*AAstV 1–3*) are being recognized and approval is pending for four (*AAstV 4–7*). As in the majority of published work the old terminology has been (and still continues to be) used, this text will refer to both old and new nomenclatures.

3. MOLECULAR BIOLOGY OF ASTROVIRUSES

Molecular biology studies have been performed mostly with classic human astroviruses. Basic principles of viral replication have also been extrapolated from those of related positive-strand RNA viruses, such as *Picornaviridae* and *Caliciviridae*, and have not been fully reinvestigated for astroviruses. The three astroviral ORFs encode as much polyproteins, which are then proteolytically cleaved by cellular and viral proteases (Fig. 1). In some astrovirus strains, an additional ORF, termed ORFX, has been recognized. Its function and significance, however, is not known. Cell entry is assumed to involve a clathrin-mediated endocytosis pathway, but virus-specific receptors are still unknown. Following uncoating, viral nonstructural proteins are transcribed from the positive-sense single-stranded RNA genome. The ORF1b-encoded RNA-dependent RNA polymerase (RdRp) replicates the genomic RNA via an intermediate negative-sense RNA template, producing full-length RNA copies and a subgenomic RNA that serves for the translation of the structural capsid precursor protein VP90 from ORF2 (for review, see Bosch et al., 2014).

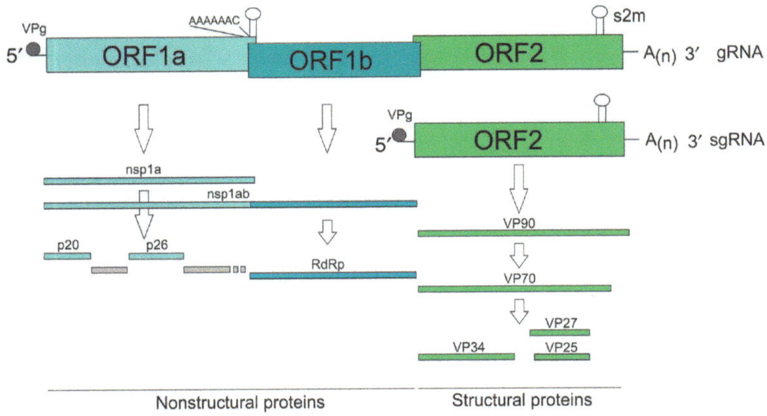

Fig. 1 Molecular biology of *Astroviridae*. The viral genome consists of a single strand of positive-sense RNA (gRNA) between 6.2 and 7.7 kb in size. It has short 5′ and 3′ untranslated regions, is polyadenylated, and encodes viral proteins in three open reading frames (ORF1a, ORF1b, and ORF2). A prominent stem–loop structure (s2m) near the 3′ end is conserved among most members of the *Astroviridae* family and is likely involved in RNA replication. The 5′ end supposedly carries a virus-encoded viral genome-linked protein (VPg). During RNA replication, a subgenomic RNA (sgRNA) is generated and is used for the translation of the structural capsid precursor protein (VP90) encoded by ORF2. VP90 is processed by cellular proteases to VP70, which is then incorporated into viral particles. Extracellular cleavage of VP70 results in the mature structural proteins VP34, VP27, and VP25. Nonstructural proteins (nsp) are translated from ORF1a and ORF1b, resulting in nsp1a and after a ribosomal frameshift at the 3′ end of ORF1a in nsp1ab, respectively. The ribosomal frameshift occurs at a ribosome slippery sequence (AAAAAAC) and a conserved stem–loop structure. These nonstructural precursor proteins are further processed to the RNA-dependent RNA polymerase (RdRp), as well as diverse intermediate and final protein products, among others p20 and p26 (of which the exact function is still unknown). For further information, we refer to the review of Bosch et al. (2014).

In ORF1a, different motives have been recognized: a protease, transmembrane helices and, in some cases, a viral genome-linked protein (VPg) (Méndez et al., 2013). After its translation, the resulting polyprotein, nsp1a, is gradually cleaved into several proteins. The precursor protein from ORF1b, which is translated together with nsp1a through a ribosomal frameshift mechanism and termed nsp1ab, is cleaved from it in a second phase, finally generating the RdRp. The capsid precursor protein VP90 displays certain features: the N-terminal part is rather conserved among astroviruses and thought to form the virion core, whereas the C-terminal domain is more variable and obviously constitutes its spikes (Méndez et al., 2013). After translation of VP90, it is first processed to VP70, which is then assembled to immature virions. The mechanism of virion release from infected cells is still not characterized in detail, but cellular caspases seem to hold an important role in that matter (Méndez et al., 2004). Finally, extracellular trypsin cleavage, which gives rise to three distinct capsid proteins, is required so that particles become infectious (Bass and Qiu, 2000). Even if many aspects of the molecular biology of astroviruses have been elucidated since they were discovered, a detailed understanding of the RNA replication complex and of molecular determinants for host and tissue specificity is still missing. This type of research is largely hampered by the lack of cell culture systems and detection tools for most astrovirus genotype species.

4. TOOLS FOR ASTROVIRUS DETECTION AND RESEARCH

One of the main obstacles in astrovirus research is the limited availability of diagnostic reagents, in vitro tools as well as in vivo models. Many efforts have been put into that direction for classic human astroviruses (*MAstV 1*); conversely, investigations about novel human and animal astroviruses are often hindered by the lack of appropriate antibodies and cell culture systems. In the early years of astrovirus research, these viruses were studied through the use of electron microscopy (EM) and recognized thanks to the star-like appearance that was thought to be typical for them; however, it was soon discovered that in some cases and under certain sample processing conditions, only a small proportion of virions display this characteristic feature (Madeley, 1979). Yet, also in recent years, EM proved to be a valuable component in virus discovery (Rivera et al., 2010) and in the morphological confirmation of evidence for astrovirus by NGS (Li et al., 2013).

Various polyclonal and monoclonal antibodies have been developed and applied, mainly in the field of classic human astroviruses (*MAstV 1*) research, in order to identify viral antigens in cell culture, and in clinical specimens by ELISA techniques. Recombinant antigens have been used in serological studies, but all these tools are still missing for other virus genotype species.

The first cells in which astroviruses could be successfully propagated in vitro were primary human embryo kidney cells (Lee and Kurtz, 1981). The addition of trypsin to the culture medium turned out to be essential for virion maturation and infectivity. Since then, bovine and feline embryonic cell culture types were also found to support replication of bovine and feline astroviruses, respectively (Aroonprasert et al., 1989; Harbour et al., 1987; Woode and Bridger, 1978), and a porcine astrovirus strain replicated in an established embryonic swine kidney cell line (Shimizu et al., 1990).

Moreover, a canine astrovirus from Italy was successfully isolated in a canine kidney cell line (MDCK) (Martella et al., 2011a). The most common cell line used in human astrovirus (*MAstV 1*) research and diagnostics is human colon carcinoma type 2 (CaCo-2) (Perot et al., 2017). However, the vast majority of animal astroviruses and the new human astrovirus genotype species have not yet been successfully propagated in vitro.

Reverse genetics systems have been described only for classic human astroviruses (*MAstV 1*), based either on the transfection of susceptible cells with in vitro transcribed RNA (Geigenmüller et al., 1997) or with plasmids containing the virus genome as complementary DNA (Chapellier et al., 2015). The advantage of such strategies is the possibility of genetic manipulations through mutagenesis, an approach that can therefore be useful in decoding the functions of specific genome parts. However, these possibilities have not yet been extensively exploited in the past years, mainly due to still relatively inefficient production of infectious viral particles.

A serious issue in astrovirus research is the lack of a representative small animal model. Currently, turkeys are often used as such, but the fact that turkey astroviruses are genetically distant from their human counterparts weakens the translational strength of such studies. In laboratory mice some astroviruses have been identified (Ng et al., 2013; Yokoyama et al., 2012), but again they do not seem to be associated with disease, making them poorly suited to the study of astroviral pathogenesis in vivo.

From 2008 on, past assumptions about astrovirus diversity have been shaken up, when NGS technologies in combination with bioinformatics have begun to be applied increasingly in pathogen discovery. This so-called viral metagenomics consists of massive parallel and unbiased sequencing of the entire genetic information in a given clinical specimen. Sequence reads are then compared to virus genome databases and/or assembled to contiguous sequences (contigs), which are then screened for sequence similarities to known sequences of viral genomes and to viral proteins after in silico translation. New virus candidates can be identified as long as they reveal a certain level of sequence similarity to virus database entries (Höper et al., 2016). While this approach has been shown to be highly effective, it requires specific equipment and bioinformatics expertise. Besides NGS, the use of conserved sequence primers in RT-PCR assays has significantly contributed to the discovery of new astroviruses, because of their suitability for more economic mass screening of samples compared to viral metagenomics and the relative ease in data generation and interpretation.

Most of these primers are aimed at ORF1b, which is the best conserved ORF among members of the *Astroviridae* family (Chu et al., 2008; Finkbeiner et al., 2009b; Mittelholzer et al., 2003a; Noel et al., 1995). However, while some of these conserved primer pairs revealed a broad reactivity for viruses of different genotype species, none of them is fully universal (Perot et al., 2017). Nowadays, discovery of new astroviruses mostly relies either on the use of these pan-astroviral primers, on viral metagenomics or on strategies combining both approaches.

5. ASTROVIRUS DISCOVERIES IN THE PREGENOMICS ERA

The story of astroviruses as we know them started in 1975, when small round particles with a diameter of 29–30 nm were described in stool samples of children suffering from diarrhea by EM (Appleton and Higgins, 1975). Shortly after, it was found that these viruses displayed a distinct star-shaped surface structure (Madeley and Cosgrove, 1975b), a feature that gave this virus family its name (*ástron* means star in Ancient Greek) (Madeley and Cosgrove, 1975a). In the following years, similar viruses were also reported from the feces of lambs (Snodgrass and Gray, 1977), calves (Woode and Bridger, 1978), dogs (Williams, 1980), cats (Hoshino et al., 1981), pigs (Bridger, 1980), red deer (Tzipori et al., 1981), and mice (Kjeldsberg and Hem, 1985), mostly based on virus particle detection by EM and propagation in primary cell cultures.

Originally based on serological assays, eight serotypes of human astroviruses were described (HuAstV-1 to -8) (Belliot et al., 1997; Lee and Kurtz, 1994). Genetic studies then showed that these serotypes correlated well with the molecular sequence of the capsid proteins of the viruses (Noel et al., 1995), which is now used to determine astrovirus genotype species (Bosch et al., 2014; De Benedictis et al., 2011). The first whole-genome sequence of an astrovirus to be determined was that of HuAstV-2 in 1993 (Jiang et al., 1993), shortly followed by HuAstV-1 (Lewis et al., 1994; Willcocks et al., 1994). These so-called Oxford strains are still considered to be the reference sequences for those serotypes (Geigenmüller et al., 1997).

The first AAstV to be described was duck astrovirus 1 (DAstV-1, formerly called duck hepatitis virus 2; Smyth, 2017), which was found to cause fatal hepatitis in these birds (Gough et al., 1984). In turkeys, two genetically different astroviruses (TAstV-1 and -2) have been associated with enteritis and the so-called poult enteritis mortality syndrome. Avian nephritis virus (ANV) was first observed by EM in fecal samples of apparently healthy chicken broilers, and infections usually result in subclinical disease, with histological changes in the kidneys (for review see, Koci and Schultz-Cherry, 2002).

Overall, until the year 2005 astroviruses had been described in only 12 species, with a mere 18 complete ORF2 sequences being available on Genbank (Fig. 2).

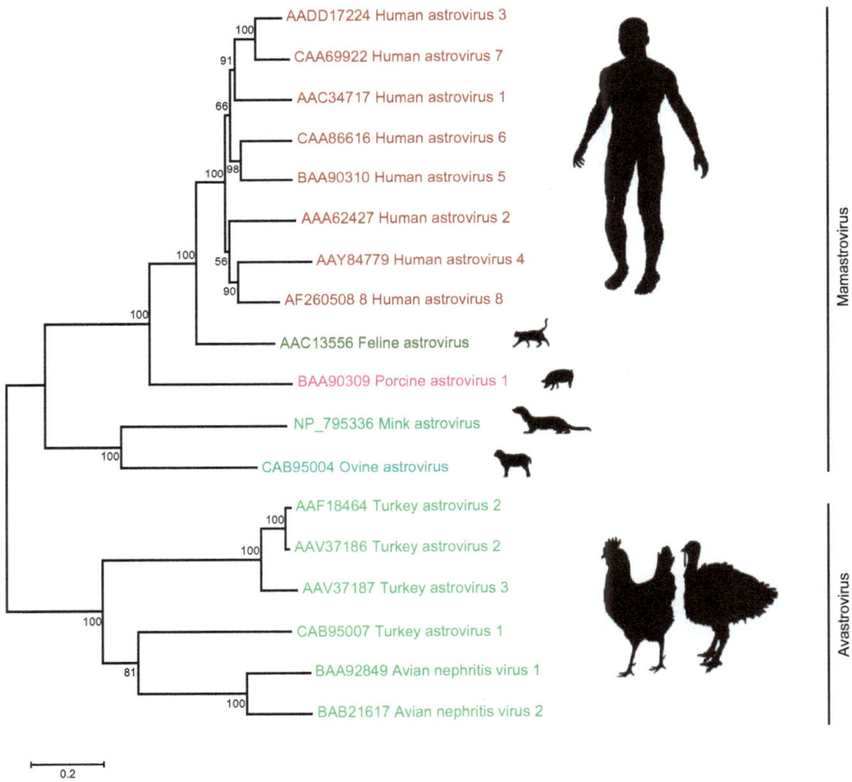

Fig. 2 Phylogenetic relationships of astrovirus strains as known in the pregenomics era, in the year 2005. The analysis is based on the entire amino acid sequences of the capsid protein. Sequences were obtained from Genbank, filtered by submission date, aligned with MUSCLE (Edgar, 2004), and the tree was constructed with MEGA 6 (Tamura et al., 2013) using the neighbor-joining method with 1000 bootstrap replications and the pairwise deletion option. The bar indicates amino acid sequence p-distances.

6. NOVEL ASTROVIRUSES IN HUMANS

The first novel human astrovirus to be detected by viral metagenomics was HAstV-MLB1, which was genetically highly divergent from the eight serotypes of classic human astroviruses (*MAstV 1*) described earlier. It was identified for the first time in a stool sample from a boy suffering from diarrhea (Finkbeiner et al., 2008a,b). The sample was collected in Melbourne, Australia, which explains the virus name. Shortly afterward, the same authors reported HuAstV-VA1, which was discovered in an outbreak of gastroenteritis in a child care center in Virginia, USA (Finkbeiner et al., 2009c). Subsequently, using conserved sequence primers, HuAstV-VA2, -VA3, and -MLB2 were reported from patients from Vellore, India and St. Louis, USA (Finkbeiner et al., 2009a). At the same time, another group discovered HuAstV-HMO-A, -B, and -C in various stool samples from Nigeria, Pakistan, and Nepal (Kapoor et al., 2009). Phylogenetically, these viruses clustered together with sheep and mink astroviruses and thus were designated "human, mink, and ovine (HMO)-like." In the following years, other human viruses belonging to these genetic clusters were reported: HuAstV-MLB3, as well as HuAstV-VA4 and -VA5 (Jiang et al., 2013; Meyer et al., 2015; Phan et al., 2014). On the genetic level, HuAstV-VA and HuAstV-HMO strains are very similar and referred to as HuAstV-VA1/HMO-C, HuAstV-VA2/HMO-A, and HuAstV-VA3/HMO-B, respectively, whereas HuAstV-MLB strains are part of a distinct group (Bosch et al., 2014). Taken together, human astrovirus strains currently form three distinct genotype clusters: classic human astroviruses with serotypes 1–8 (*MAstV 1*), HuAstV-MLB (*MAstV 6*), and HuAstV-VA/HMO (*MAstV 8* and *9*) (Fig. 3).

7. ASTROVIRUS DISCOVERIES IN MAMMALIAN ANIMALS

7.1 Bovine Astroviruses

Using conserved primer pairs and rapid amplification of cDNA ends (RACE), Tse and colleagues determined the first full-length genome sequences of a series of bovine astrovirus (BoAstV) strains in fecal samples of cattle (Tse et al., 2011). Phylogenetically, these strains cluster in four distinct branches and build the basis for the definition of the genotype species *MAstV 28*, *29*, *30*, and *33* (Fig. 3). Subsequent studies on cattle fecal samples from Scotland (Sharp et al., 2015), Brazil (Candido et al., 2015) and Korea

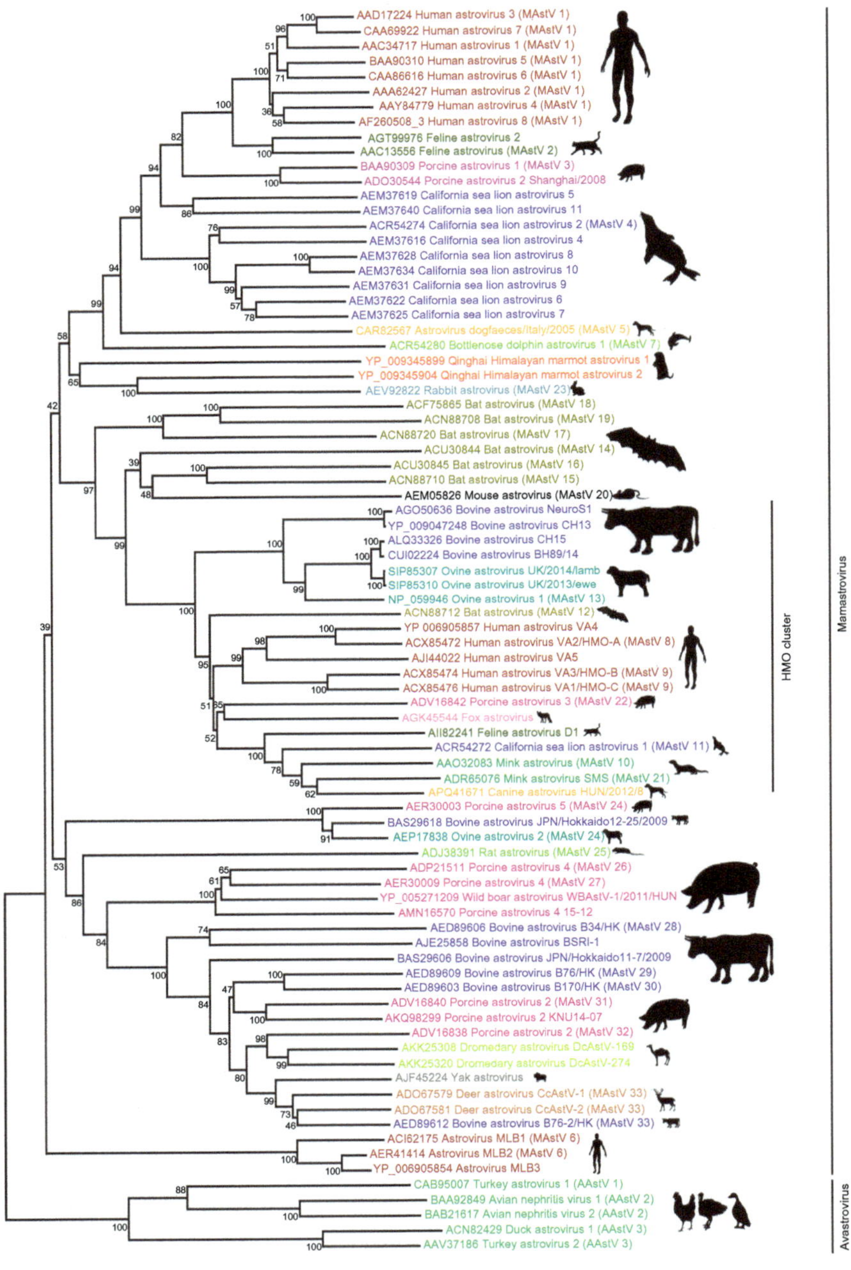

Fig. 3 Spectrum and phylogeny of astrovirus strains as known in the year 2017. The tree was constructed as described for Fig. 2 and includes defined genotype species of the genera Mamastrovirus (MAstV) and Avastrovirus (AAstV), as indicated, as well as so far unclassified strains. Genbank accession numbers are provided.

(Oem and An, 2014) identified additional strains apparently belonging to these four genotypes, based on the comparison of partial amino acid sequences of the RdRp. Different, thus far unclassified BoAstV strains have been reported recently from fecal samples of cattle from Japan (e.g., BoAstV/JPN/Hokkaido 11-7/2009, BoAstV/JPN/Hokkaido 12-25/2009) (Nagai et al., 2015), and another distinct astrovirus was found to be statistically associated with diarrhea in yak calves (*Bos grunniens*) in Tibet (Chen et al., 2015) (Fig. 3). A growing field of interest concerns extraenteric manifestation of astrovirus infections in cattle, in particular, those affected by encephalitis and neurological disease, and these will be presented in Section 9.

7.2 Porcine Astroviruses

Even more diverse than BoAstV are porcine astroviruses (PoAstV), and many studies have addressed their prevalence and genetic characterization in pig holdings worldwide (Lan et al., 2011; Laurin et al., 2011; Luo et al., 2011; Reuter et al., 2012b; Shan et al., 2011; Ulloa and Gutierrez, 2010). Conventionally, PoAstV strains had been attributed to five genetic lineages, which were later classified into seven genotype species including *MAstV 3, 22, 24, 26, 27, 31,* and *32* by the ICTV. Prevalence in feces samples in industrial pig holdings can be as high as 90%, but infections are usually asymptomatic (Brnic et al., 2013; Luo et al., 2011), and so far, there is no clear association of any of the porcine genotype species with pathology or disease. Divergent PoAstV strains were reported from feces samples from individual pigs in China (PoAstV 2 Shanghai/2008) (Shan et al., 2012), as well as Korea (PoAstV 2 KNU 14-07) (Lee et al., 2015), yet similarly their clinical significance remains to be investigated.

7.3 Sheep Astroviruses

Compared to BoAstV and PoAstV, even less is known about the diversity and prevalence of ovine astrovirus (OvAstV) strains. This is surprising, since OvAstV-1 (*MAstV 13*) represented the first report of an astrovirus in an animal species (Snodgrass and Gray, 1977), and its genome sequence was already determined in 2003 by conventional sequencing methods (Jonassen et al., 2003). A second OvAstV genotype strain (OvAstV-2) has been identified by viral metagenomics in a fecal sample of a single sheep in Hungary (Reuter et al., 2012b). A peculiarity is that this strain belongs to the genotype species *MAstV 24* which also contains strains of PoAstV, suggesting that these strains had a recent common ancestor during virus evolution.

7.4 Mink Astroviruses

In minks, mink astrovirus (MiAstV, *MAstV 10)* includes fecal strains and was associated with the so-called preweaning diarrhea syndrome in Denmark and Sweden (Mittelholzer et al., 2003a,b). One of the first astroviruses identified by viral metagenomics in domestic animals was found in brain samples of farmed minks that suffered from a certain neurological disease (commonly named shaking mink syndrome) in Denmark, and the virus was designated as mink astrovirus SMS (MiAstV-SMS, *MAstV 21*) (Blomstrom et al., 2010). Mink astroviruses are phylogenetically grouped together with genotype species found in humans (*MAstV 8* and *9*), pigs (*MAstV 22*), sheep (*MAstV 13*), cattle, bats (*MAstV 12*), and California sea lions (*MAstV 11*), forming the above-mentioned HMO cluster. Strikingly, most strains that have been associated with neurological disease and encephalitis in humans and animals belong to this cluster.

7.5 Astroviruses in Companion Animals

In dogs, there have been increasing reports on canine astroviruses (CaAstV, *MAstV 5*) in feces samples from puppies with diarrhea (Martella et al., 2012; Toffan et al., 2009; Zhu et al., 2011). However, studies on the association of virus detection and disease are contradictory (Grellet et al., 2012). The diversity of canine virus strains within the *MAstV 5* genotype is relatively high, with predicted amino acid sequences of the capsid protein differing by up to 30%. Very recently, a highly divergent canine astrovirus (CaAstV HUN/2012/8) was detected in a puppy with diarrhea in Hungary, of which the capsid protein sequence mapped to the HMO cluster with highest similarity (~65%), especially to *MAstV 21* and *10* (Mihalov-Kovacs et al., 2017) (Fig. 3). A very similar situation has been observed for feline astroviruses (FeAstV) in cats. While most strains cluster in the relatively diverse genotype species *MAstV 2* (Cho et al., 2014; Lau et al., 2013; Ng et al., 2014), one strain from a study of the fecal virome in cats from an animal shelter in California (FeAstV D1) was different and closely related to a fox astrovirus within the HMO clade (Zhang et al., 2014) (Fig. 3). These findings suggest that the astroviruses infecting dogs and cats may be more diverse than currently supposed.

7.6 Astroviruses in Other Terrestrial Mammalian Animals

A large spectrum of astroviruses was found in feces samples of many other mammalian species, further expanding both the spectrum of host species and of virus genotypes: in dromedaries (Woo et al., 2015), foxes (Bodewes et al., 2013), marmots (Ao et al., 2017), rats (Chu et al., 2010), mice (Ng et al., 2013; Phan et al., 2011; Yokoyama et al., 2012), rabbits (Martella et al., 2011b; Stenglein et al., 2012), wild boar (Reuter et al., 2012a), and roe deer (Smits et al., 2010). While most of these strains were genetically distinct from those of other host species, strains of roe deer clustered with *MAstV 33*, which also includes BoAstV strains, and those of wild boar were closely related to PoAstV 4 (*MAstV 27*) (Fig. 3).

Because bats are considered a major reservoir and a niche for emerging viruses with zoonotic potential, many efforts have been put forward to investigate their virome, not only in fecal samples but also in urine and oropharyngeal swabs. Results indicate that astrovirus infection in bats is very frequent and that genetic virus diversity is high. Currently, more than 300 individual sequences of bat astroviruses are available on Genbank, and seven genotype species (*MAstV 12, 14–19*) have been recognized by the ICTV so far. For more detailed information on bat astroviruses, we refer the reader to the recent literature (Fischer et al., 2017).

7.7 Astroviruses in Marine Mammalian Animals

Two studies have been conducted on feces samples of marine mammal host species. Rivera and colleagues described genetically distinct astroviruses in California sea lions (CslAstV), in a Bottlenose dolphin (BdAstV-1), and in a Steller sea lion (SslAstV-1) by EM, conserved primer RT-PCR and Sanger sequencing. Strikingly, CslAstV-2 (*MAstV 4*) revealed the highest similarity of its capsid protein to *MAstV 1* (human), *MAstV2* (feline), and *MAstV 3* (porcine), while CslAstV-1 (*MAstV 11*) mapped to the HMO clade (Rivera et al., 2010). In the second study, Li et al. analyzed fecal samples of 47 California sea lions by viral metagenomics, and ~50% of them were astrovirus positive, with seven additional genetically distinct *MAstV 4*-like, but so far unclassified strains (Li et al., 2011) (Fig. 3).

8. ASSOCIATION WITH DISEASE

Classic human astrovirus strains (*MAstV 1*) are considered to be one of the most frequent causes of viral gastroenteritis in humans, especially in infants, young children, and immunocompromised people

(De Benedictis et al., 2011). Conversely, the association of novel human astroviruses of the HMO clade and of the MLB genotype species with gastro-intestinal disease is currently still unclear. With age, the presence of antibodies against classic human astrovirus (*MAstV 1*) increases in the general population (Bosch et al., 2014), and as healthy adults usually do not develop clinical disease, it seems that a first contact with a specific astrovirus serotype confers protection against new symptomatic infections. Interestingly, astroviruses in mink also induce diarrhea (Englund et al., 2002), whereas for many other mammalian species (including cattle, sheep, cats, dogs, pigs, and bats), a causal link between astrovirus infection and gastroenteritis could not be ascertained, as experimental infections did not induce significant clinical disease or the viruses were found in diarrheic animals but also commonly in apparently healthy ones (De Benedictis et al., 2011; Fischer et al., 2017).

In birds, AAstV are known to be associated with various kinds of diseases. Two genetically different AAstVs cause fatal hepatitis in ducks. In turkeys, two astrovirus species have been reported and linked to growth retardation and enteritis in these animals. In chickens, avian nephritis virus and chicken astrovirus have both been isolated in animals suffering from kidney disease or growth retardation. Again, the lone presence of the virus does not seem to be enough to induce disease, and other factors appear to play a role, such as the age of an animal when infected, the presence of maternal antibodies, the infectious dose, variability in the virulence of specific strains and coinfections with other pathogens (Smyth, 2017).

9. ASTROVIRUSES IN NEUROLOGICAL DISEASES

The impressive power of viral metagenomics in astrovirus discovery was exemplified with the detection of novel astroviruses in conjunction with neurological disease and encephalitis, because members of the *Astroviridae* family were certainly not expected to invade the central nervous system (CNS) at that time, and only the unbiased nature of sequence detection by NGS led to their identification. In 2010, an astrovirus was detected in brain tissues and in association with encephalitis in a boy with X-linked agammaglobulinemia in the United States (Quan et al., 2010). Designated as human astrovirus Puget-Sound (HuAstV-PS), it was genetically closely related to HAstV-VA1/HMO-C. Since then, very similar strains were repeatedly detected in patients afflicted with encephalitis (Brown et al., 2015; Fremond et al., 2015; Lum et al., 2016; Naccache et al., 2015), and HAstV-MLB1 (Sato et al., 2016) as well as HuAstV-MLB2 (Cordey et al., 2016b) were also reported in association with encephalitis cases. There was also evidence for disseminated HAstV-4 (*MAstV 1*) infection in two immunocompromised infants, one of which with fatal severe multiorgan dysfunction and meningoencephalitis (Wunderli et al., 2011).

Using a similar approach, novel astroviruses were subsequently also detected in brain samples of neurologically diseased cattle with nonsuppurative encephalitis. Two groups identified almost identical novel astrovirus strains, termed BoAstV-NeuroS1 and BoAstV-CH13 in the United States and in Switzerland, respectively, which represent members of one proposed new genotype species within the HMO clade (Bouzalas et al., 2014; Li et al., 2013) (Fig. 3). In retrospective studies, BoAstV-CH13/NeuroS1 was found in up to one-third of so far unresolved encephalitis cases in cattle in Switzerland, with such cases dating back to the 1950s (Selimovic-Hamza et al., 2016, 2017). Shortly afterward, two other neurotropic bovine strains, which were again very similar to each other but distinct from BoAstV-CH13/NeuroS1, were found in Switzerland and Germany and designated as BoAstV-CH15 and BoAstV-BH89/14 (Schlottau et al., 2016; Seuberlich et al., 2016). Very recently, two astroviruses with ~99% amino acid similarity to the capsid protein of BoAstV-CH15 and -BH89/14 were detected in nervous tissues of two sheep with fatal encephalitis in Scotland (OvAstV-UK/2013/ewe and -UK/2014/lamb). Both animals succumbed to neurological disease in the same flock, yet 9 months apart (Pfaff et al., 2017). These findings suggest that this type of astrovirus causes neurological disease in cattle and in sheep, and may be transmitted between both species.

The determinants of tissue tropism of astroviruses are mostly unknown. In humans, astroviruses associated with encephalitis had previously all been detected in stool samples. Apart from one case of astrovirus-associated meningitis in an obviously immunocompetent woman (Cordey et al., 2016b), all reports of astroviruses detected in the CNS were in immunocompromised patients. Immunosuppression, therefore, seems to play an important role in the systemic spreading of astroviruses in humans. In animals, the situation is different: in minks, the astrovirus causing diarrhea is quite closely related, but not identical, to its neurotropic counterpart (Blomstrom et al., 2010). In cattle, on the other hand, the astroviruses found in association with encephalitis are genetically distinct from those originating from feces. Interestingly, astrovirus encephalitis in cattle has been described only in individual animals (i.e., never in the frame of an epidemic), which also indicates a role of host-related factors in disease development, as seems to be the case in humans.

10. ASTROVIRUS INFECTION IN NONENTERIC AND NONNERVOUS TISSUES

In contrast to AAstV, which can be found in various organs of their hosts, MAstV infections were long thought to be exclusively enteric. Apart from nervous tissues, MAstV have also been found in a series of other nonenteric and nonnervous tissues. In a human patient, in cattle, and in pigs, MAstV were identified in association with respiratory disease in their airways by metagenomics: (i) HuAstV-VA1 in the nasal cavity of a child with inexplicable respiratory disease (Cordey et al., 2016a), (ii) a putative novel PoAstV (strains 15–12) in nasal swabs of pigs with acute respiratory disease (Padmanabhan and Hause, 2016), and (iii) a divergent BoAstV (BoAstV-BSRI1) in calves affected by the so-called bovine respiratory disease complex (Ng et al., 2015). The significance of these findings, however, remains unknown, and it cannot be excluded that these viruses were only transiently present in the respiratory tract of these individuals. Finally, HAstV-MLB2 has also been identified in the plasma of a febrile child (Holtz et al., 2011), but again this finding alone did not allow the drawing of any further conclusion. In a very recent large RT-PCR-based screening study using tissue samples derived from diarrheic and healthy pigs in China, PoAstV was also detected in lungs, spleens, and kidneys of these animals, with a considerably high rate of up to 50% in nondiarrheic pigs (Xiao et al., 2017). In another study, a viral RNA seroprevalence of 3.89% was determined in healthy slaughtered pigs in Croatia (Brnic et al., 2013). Taken together, all these findings suggest that PoAstV could have a wide tissue tropism, not limited only to the intestine.

11. ASTROVIRUSES AND CAUSALITY OF DISEASE

Case–control studies are a first step toward the assessment of causality between infection and disease. The association of a pathogen with a disease has already been set as the first condition in Koch's postulates. At this stage, most novel astroviruses already fail to show a clear association, as they can also be found in apparently healthy hosts, and therefore may be present as a component of the physiological intestinal virome of these hosts. The second condition of Koch's postulates is the isolation of the pathogen in pure culture and the experimental induction of disease in inoculated animals. As outlined earlier, there have been enormous difficulties in culturing most of the recently discovered astroviruses, although for many of them, it was not reported to which extend this was attempted. Still, in virology, it is never certain how pure such cultures derived from clinical specimen would be, and whether unrecognized agents that are capable of replicating in the same system would bias the results.

Criteria proposed recently by Lipkin may be better suited when trying to assess the level of confidence in a proposed causal relationship between astrovirus infection and disease (Lipkin, 2010, 2013). Even if culture and isolation still remain important in this approach, other factors are easier to apply to astroviruses. Demonstrating the presence of nucleic acids, proteins, or morphological structures of the virus in an affected organism, or of antibodies against it, already speaks in favor of a possible causal relationship. On the next level, a probable causal relationship is supported by showing that the virus can be found in multiple diseased individuals, that the viral burden there is high, that pathological lesions correlate with viral presence, or that the immune response of the host appears to be specific against the virus. Ultimately, alternatively to Koch's requirements, demonstrating that disease can be prevented by the use of specific antiviral therapeutics or vaccines can also confirm this relationship of causality between the virus and disease. Still, astroviruses prove themselves challenging in such ventures, and most of them do not reach the level of "probable causal relationship."

Moreover, all findings which rely solely on the detection of nucleic acid sequences require very careful interpretation. Indeed, the fact that nucleic acid sequences of a microorganism are noticed in a sample does not automatically speak for infection, let alone for its association with a disease. In particular in the case of stool samples, the presence of such sequences may result only from the host's dietary habits, and the virus may simply have been ingested without replication in the host. The development of powerful NGS tools for the detection of new microorganisms is definitely a huge step forward and a source of exciting discoveries; nevertheless, when causality for a disease is claimed, additional criteria need to be fulfilled.

In the case of MAstV associated with encephalitis, even though the presence of astrovirus nucleic sequences in the brain of patients obviously results less from a coincidence than in the gastrointestinal tract or the airways, more thorough investigation to prove their association with disease as recommended by Lipkin remains crucial when trying to establish a link of causality between virus and disease. For BoAstV-CH13/NeuroS1, there was a clear association between the presence of the virus and diseased brains, while it was absent in the brains of healthy cattle. Moreover, different groups demonstrated virus replication by in situ hybridization (Bouzalas et al., 2014; Li et al., 2013) and immunohistochemistry (Boujon et al., 2017) in neurons and in association with the pathological lesions (Fig. 4), which are strong arguments for a probable causal relationship.

Demonstrating the ability of MAstV to cause encephalitis in humans is of course not possible, and as there is also no suitable in vivo model to this end, aiming to prove a probable causal association will remain limited to in vitro studies. On the other hand, in cattle and recently in sheep, once the corresponding astroviruses have been isolated and characterized in vitro,

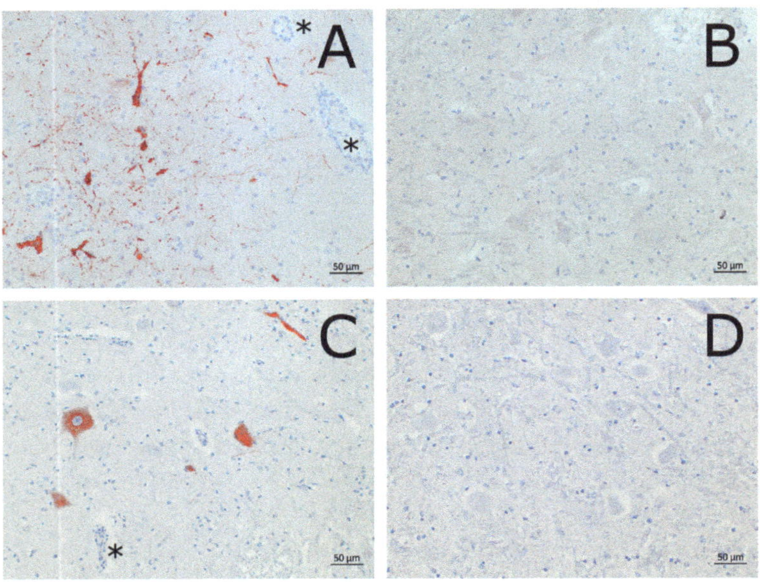

Fig. 4 Immunohistochemistry (IHC) against bovine astrovirus CH13/NeuroS1 (BoAstV-CH13/NeuroS1) and CH15 (BoAstV-CH15) in neurologically diseased cows. Cattle are one of the four species (with humans, minks, and sheep) in which astroviruses have been found to be associated with encephalitis. IHC is a method allowing in situ detection of astrovirus, here in nervous tissue. (A) Cow positive for BoAstV-CH13/NeuroS1, hippocampus, IHC with antibodies directed against the conserved region of the putative capsid protein of BoAstV-CH13/NeuroS1. Positive staining is recognizable by *dark red coloring*. (B) Negative control, obex, IHC with antibodies directed against the conserved region of the putative capsid protein of BoAstV-CH13/NeuroS1. (C) Cow positive for BoAstV-CH15, obex, IHC with antibodies directed against the variable region of the putative capsid protein of BoAstV-CH15. Positive staining is recognizable by *dark red coloring*. (D) Negative control, obex, IHC with antibodies directed against the variable region of the putative capsid protein of BoAstV-CH15. The *asterisks* in (A) and (C) indicate perivascular cuffing.

in vivo experiments could be ethically defendable, especially in view of findings also being relevant for human medicine.

12. HOST SPECIFICITY, INTRASPECIES TRANSMISSION, AND EVOLUTIONARY DYNAMICS

The transmission of astroviruses is assumed to occur via the fecal–oral route. There is no acknowledged vector implicated in the dissemination of astroviruses, the only exception being chicken astroviruses, for which an arthropod is thought to be an actor in their spreading (Smyth, 2017). On the other hand, astroviruses are rather resistant in the environment, and fomites may play an important role in their epidemiology. As astroviruses seem to be stable in water (Abad et al., 1997; Bosch et al., 1997) and different astroviruses, including novel human astroviruses and canine astroviruses, have been found regularly in sewage water and treated wastewater (Hata et al., 2015; Lizasoain et al., 2015a,b; Pinto et al., 2001; Prevost et al., 2015; Qiu et al., 2015; Zhou et al., 2016), consumption of contaminated drinks can result in infections. For humans, contaminated food, such as shellfish (Le Guyader et al., 2000), greens and soft fruit (Bosch et al., 2014), is also a source of astrovirus infections.

For AAstV, infection of different avian species by the same virus strain appears to be the rule rather than the exception. For example, chicken- and turkey-derived astroviruses were found in ducks (Bidin et al., 2012a), and ANV is known to infect chicken, turkeys, ducks, geese, and pigeons (Bidin et al., 2011, 2012b; Pantin-Jackwood et al., 2006; Shirai et al., 1991; Zhao et al., 2011). This situation is in sharp contrast to MAstV, which are thought to be restricted to one host species. However, some reports (i.e., the identification of astrovirus genotype species in unexpected hosts) do challenge this assumption. For instance, human astroviruses were noticed in stool samples from nonhuman primates that lived close to people (Karlsson et al., 2015). Even the genera barrier between AAstV and MAstV appears to be less rigid than initially presumed: a MAstV-like nucleic acid sequence was detected in a carnivorous bird in Hungary (Pankovics et al., 2015), and a sequence closer to those of AAstV than MAstV was found in nonhuman primates feces (Karlsson et al., 2015). One study also reported chicken- and turkey-like astroviruses in fecal samples from farmed minks (Sun et al., 2014). Still, again the simple detection of nucleic acid sequences does not prove infection and could be due to oral intake of astrovirus-contaminated food. On a different note, antibodies against a turkey astrovirus have been noticed in humans, and interestingly with greater frequency in those people with an occupational exposure to turkeys. On the other hand, these data do not allow the drawing of any conclusion as regards replication of this virus or associated disease in humans, and may simply reflect exposure to the respective antigen (Meliopoulos et al., 2014). Despite these reports, based on current knowledge MAstVs generally seem to exhibit relatively strong host specificity.

With the increasing insight on MAstV diversity, the phylogeny of astroviruses progressively reveals its complexity, and the genetic dynamics of their evolution seem to be strong. As RNA viruses, their mutation rate is high and evidence for recombination events within or even across astrovirus species have been increasingly reported. Noteworthy is the fact that recombination events require infection of the same host by distinct viruses at the same time. Such coinfections of a host with different astroviruses have been described in several instances in pigs (Amimo et al., 2014; Ito et al., 2017; Xiao et al., 2013). In humans, findings supporting intra- and intergenotypic recombination are numerous (Babkin et al., 2014; Martella et al., 2013, 2014; Medici et al., 2015; Walter et al., 2001; Wolfaardt et al., 2011; Zhou et al., 2016). In pigs, there are also several studies indicating such events, within and between genotypes but also possibly with HuAstV (Ito et al., 2017; Ulloa and Gutierrez, 2010). These animals, known to be "mixing vessels" for viruses, also for some with zoonotic potential (the most prominent example being influenza; Ma et al., 2008), could therefore have an important role in the emergence of new astroviruses. Interestingly, recombination between a sea lion and a human astrovirus has also been postulated (Rivera et al., 2010). On the other hand, it appears that bats, which can be important reservoirs for human viruses, did not reveal a crucial role in astrovirus epidemiology for other species so far, as MAstV found in this order are genetically mostly different from astroviruses of other mammals (Fischer et al., 2017).

The HMO clade seems to reflect the evolutionary dynamics of astroviruses particularly well, as it is composed of related genotype species affecting humans and at least 10 animal species, which suggests past interspecies transmission events during virus evolution. Some of the members of the HMO clade possess an apparent tropism for nervous tissues, and there is good evidence that these viruses cause encephalitis and neurological disease.

Finally, even if genetically distant, based on the fact that the capsid proteins of astroviruses present several similarities with that of hepatitis E virus (Dong et al., 2011; Dryden et al., 2012; Toh et al., 2016; York et al., 2015), another virus associated with gastroenteric and brain infections, a common ancestor for these viruses has been postulated (Arias and DuBois, 2017).

13. CONCLUSIONS AND PERSPECTIVES

The application of genomics tools in astrovirus research has massively changed our view on the significance of astrovirus infections in humans and animals. Over the past 10 years, the knowledge about the spectrum of astrovirus genotypes and of affected host species has increased immensely. While many virus discoveries could not be clearly associated to disease, so far unresolved fatal pathologies turned out to be likely caused by astrovirus infections. Yet there are still major gaps in the understanding of astrovirus biology, pathology, and epidemiology. Basic mechanisms of viral replication and virus–host interactions still remain unknown, and disease pathogenesis is thus far not well understood. Further studies in this direction are needed to finally develop therapeutic interventions and prophylactic strategies. For this purpose, research tools and in vivo models need to be established. These tools will also assist in the investigation of virus and host factors that determine host specificity and tissue tropism. The huge diversity of astrovirus genotypes and the large spectrum of affected host species point to the risk of interspecies transmissions, and the emergence of new virus variants that could potentially jeopardize human and animal health. All these facts highlight the need for continuous efforts in astrovirus research: the fundamentals are established, and the door is wide open for future activities.

ACKNOWLEDGMENTS

This work was funded by the Swiss Food Safety and Veterinary Office and the Swiss National Science Foundation. The authors thank Michael Garrood (Zürich, Switzerland) for proofreading the manuscript.

REFERENCES

Abad, F.X., Pinto, R.M., Villena, C., Gajardo, R., Bosch, A., 1997. Astrovirus survival in drinking water. Appl. Environ. Microbiol. 63, 3119.

Amimo, J.O., Okoth, E., Junga, J.O., Ogara, W.O., Njahira, M.N., Wang, Q., Vlasova, A.N., Saif, L.J., Djikeng, A., 2014. Molecular detection and genetic characterization of kobuviruses and astroviruses in asymptomatic local pigs in East Africa. Arch. Virol. 159, 1313.

Ao, Y., Yu, J., Li, L., Cao, J., Deng, H., Xin, Y., Liu, M., Lin, L., Lu, S., Xu, J., Duan, Z., 2017. Diverse novel astroviruses identified in wild Himalayan marmots. J. Gen. Virol. 98, 612–623.

Appleton, H., Higgins, P.G., 1975. Letter: viruses and gastroenteritis in infants. Lancet 1, 1297.

Arias, C.F., DuBois, R.M., 2017. The astrovirus capsid: a review. Viruses 9, 15.

Aroonprasert, D., Fagerland, J.A., Kelso, N.E., Zheng, S., Woode, G.N., 1989. Cultivation and partial characterization of bovine astrovirus. Vet. Microbiol. 19, 113.

Babkin, I.V., Tikunov, A.Y., Sedelnikova, D.A., Zhirakovskaia, E.V., Tikunova, N.V., 2014. Recombination analysis based on the HAstV-2 and HAstV-4 complete genomes. Infect. Genet. Evol. 22, 94.

Bass, D.M., Qiu, S., 2000. Proteolytic processing of the astrovirus capsid. J. Virol. 74, 1810.

Belliot, G., Laveran, H., Monroe, S.S., 1997. Detection and genetic differentiation of human astroviruses: phylogenetic grouping varies by coding region. Arch. Virol. 142, 1323.

Bidin, M., Lojkic, I., Bidin, Z., Tisljar, M., Majnaric, D., Mikec, M., 2011. Detection and characterization of avian nephritis virus in ducklings. Avian Pathol. 40, 173.

Bidin, M., Bidin, Z., Majnaric, D., Tisljar, M., Lojkic, I., 2012a. Circulation and phylogenetic relationship of chicken and turkey-origin astroviruses detected in domestic ducks (Anas Platyrhynchos Domesticus). Avian Pathol. 41, 555.

Bidin, M., Lojkic, I., Tisljar, M., Bidin, Z., Majnaric, D., 2012b. Astroviruses associated with stunting and pre-hatching mortality in duck and goose embryos. Avian Pathol. 41, 91.

Blomstrom, A.L., Widen, F., Hammer, A.S., Belak, S., Berg, M., 2010. Detection of a novel astrovirus in brain tissue of mink suffering from shaking mink syndrome by use of viral metagenomics. J. Clin. Microbiol. 48, 4392.

Bodewes, R., van der Giessen, J., Haagmans, B.L., Osterhaus, A.D., Smits, S.L., 2013. Identification of multiple novel viruses, including a parvovirus and a hepevirus, in feces of red foxes. J. Virol. 87, 7758.

Bosch, A., Pinto, R.M., Villena, C., Abad, F.X., 1997. Persistence of human astrovirus in fresh and marine water. Water Sci. Technol. 35, 243.

Bosch, A., Pinto, R.M., Guix, S., 2014. Human astroviruses. Clin. Microbiol. Rev. 27, 1048.

Boujon, C.L., Selimovic-Hamza, S., Bouzalas, I., Seuberlich, T., 2017. Development and validation of an immunohistochemistry procedure for the detection of a neurotropic bovine astrovirus. J. Virol. Methods. 239, 26–33. https://doi.org/10.1016/j.jviromet.2016.10.013. Epub 2016 Oct 29.

Bouzalas, I.G., Wuthrich, D., Walland, J., Drogemuller, C., Zurbriggen, A., Vandevelde, M., Oevermann, A., Bruggmann, R., Seuberlich, T., 2014. Neurotropic astrovirus in cattle with nonsuppurative encephalitis in Europe. J. Clin. Microbiol. 52, 3318.

Bridger, J.C., 1980. Detection by electron microscopy of caliciviruses, astroviruses and rotavirus-like particles in the faeces of piglets with diarrhoea. Vet. Rec. 107, 532.

Brnic, D., Prpic, J., Keros, T., Roic, B., Staresina, V., Jemersic, L., 2013. Porcine astrovirus viremia and high genetic variability in pigs on large holdings in Croatia. Infect. Genet. Evol. 14, 258.

Brown, J.R., Morfopoulou, S., Hubb, J., Emmett, W.A., Ip, W., Shah, D., Brooks, T., Paine, S.M., Anderson, G., Virasami, A., Tong, C.Y., Clark, D.A., Plagnol, V., Jacques, T.S., Qasim, W., Hubank, M., Breuer, J., 2015. Astrovirus VA1/HMO-C: an increasingly recognized neurotropic pathogen in immunocompromised patients. Clin. Infect. Dis. 60, 881.

Candido, M., Alencar, A.L., Almeida-Queiroz, S.R., Buzinaro, M.D., Munin, F.S., de Godoy, S.H., Livonesi, M.C., Fernandes, A.M., de Sousa, R.L., 2015. Molecular detection and phylogenetic analysis of bovine astrovirus in Brazil. Arch. Virol. 160, 1519–1525.

Chapellier, B., Tange, S., Tasaki, H., Yoshida, K., Zhou, Y., Sakon, N., Katayama, K., Nakanishi, A., 2015. Examination of a plasmid-based reverse genetics system for human astrovirus. Microbiol. Immunol. 59, 586.

Chen, X., Zhang, B., Yue, H., Wang, Y., Zhou, F., Zhang, Q., Tang, C., 2015. A novel astrovirus species in the gut of yaks with diarrhea in the Qinghai Tibetan plateau, 2013. J. Gen. Virol. 96, 3672–3680.

Cho, Y.Y., Lim, S.I., Kim, Y.K., Song, J.Y., Lee, J.B., An, D.J., 2014. Molecular characterisation and phylogenetic analysis of feline astrovirus in Korean cats. J. Feline Med. Surg. 16, 679.

Chu, D.K., Poon, L.L., Guan, Y., Peiris, J.S., 2008. Novel astroviruses in insectivorous bats. J. Virol. 82, 9107.

Chu, D.K., Chin, A.W., Smith, G.J., Chan, K.H., Guan, Y., Peiris, J.S., Poon, L.L., 2010. Detection of novel astroviruses in urban brown rats and previously known astroviruses in humans. J. Gen. Virol. 91, 2457.

Cordey, S., Brito, F., Vu, D.L., Turin, L., Kilowoko, M., Kyungu, E., Genton, B., Zdobnov, E.M., D'Acremont, V., Kaiser, L., 2016a. Astrovirus VA1 identified by next-generation sequencing in a nasopharyngeal specimen of a febrile Tanzanian child with acute respiratory disease of unknown etiology. Emerg. Microbes. Infect. 5, e67.

Cordey, S., Vu, D.L., Schibler, M., L'Huillier, A.G., Brito, F., Docquier, M., Posfay-Barbe, K.M., Petty, T.J., Turin, L., Zdobnov, E.M., Kaiser, L., 2016b. Astrovirus MLB2, a new gastroenteric virus associated with meningitis and disseminated infection. Emerg. Infect. Dis. 22, 846.

De Benedictis, P., Schultz-Cherry, S., Burnham, A., Cattoli, G., 2011. Astrovirus infections in humans and animals—molecular biology, genetic diversity, and interspecies transmissions. Infect. Genet. Evol. 11, 1529.

Dong, J., Dong, L., Méndez, E., Tao, Y., 2011. Crystal structure of the human astrovirus capsid spike. Proc. Natl. Acad. Sci. U.S.A. 108, 12681.

Dryden, K.A., Tihova, M., Nowotny, N., Matsui, S.M., Méndez, E., Yeager, M., 2012. Immature and mature human astrovirus: structure, conformational changes, and similarities to hepatitis E virus. J. Mol. Biol. 422, 650.

Edgar, R.C., 2004. MUSCLE: multiple sequence alignment with high accuracy and high throughput. Nucleic Acids Res. 32, 1792.

Englund, L., Chriel, M., Dietz, H.H., Hedlund, K.O., 2002. Astrovirus epidemiologically linked to pre-weaning diarrhoea in mink. Vet. Microbiol. 85, 1.

Finkbeiner, S.R., Allred, A.F., Tarr, P.I., Klein, E.J., Kirkwood, C.D., Wang, D., 2008a. Metagenomic analysis of human diarrhea: viral detection and discovery. PLoS Pathog. 4, e1000011.

Finkbeiner, S.R., Kirkwood, C.D., Wang, D., 2008b. Complete genome sequence of a highly divergent astrovirus isolated from a child with acute diarrhea. Virol. J. 5, 117.

Finkbeiner, S.R., Holtz, L.R., Jiang, Y., Rajendran, P., Franz, C.J., Zhao, G., Kang, G., Wang, D., 2009a. Human stool contains a previously unrecognized diversity of novel astroviruses. Virol. J. 6, 161.

Finkbeiner, S.R., Le, B.M., Holtz, L.R., Storch, G.A., Wang, D., 2009b. Detection of newly described astrovirus MLB1 in stool samples from children. Emerg. Infect. Dis. 15, 441.

Finkbeiner, S.R., Li, Y., Ruone, S., Conrardy, C., Gregoricus, N., Toney, D., Virgin, H.W., Anderson, L.J., Vinje, J., Wang, D., Tong, S., 2009c. Identification of a novel astrovirus (astrovirus VA1) associated with an outbreak of acute gastroenteritis. J. Virol. 83, 10836.

Fischer, K., Pinho Dos Reis, V., Balkema-Buschmann, A., 2017. Bat astroviruses: towards understanding the transmission dynamics of a neglected virus family. Viruses 9, 34.

Fremond, M.L., Perot, P., Muth, E., Cros, G., Dumarest, M., Mahlaoui, N., Seilhean, D., Desguerre, I., Hebert, C., Corre-Catelin, N., Neven, B., Lecuit, M., Blanche, S., Picard, C., Eloit, M., 2015. Next-generation sequencing for diagnosis and tailored therapy: a case report of astrovirus-associated progressive encephalitis. J. Pediatric. Infect. Dis. Soc. 4, e53.

Geigenmüller, U., Ginzton, N.H., Matsui, S.M., 1997. Construction of a genome-length cDNA clone for human astrovirus serotype 1 and synthesis of infectious RNA transcripts. J. Virol. 71, 1713.

Gough, R.E., Collins, M.S., Borland, E., Keymer, L.F., 1984. Astrovirus-like particles associated with hepatitis in ducklings. Vet. Rec. 114, 279.

Grellet, A., De Battisti, C., Feugier, A., Pantile, M., Marciano, S., Grandjean, D., Cattoli, G., 2012. Prevalence and risk factors of astrovirus infection in puppies from French breeding kennels. Vet. Microbiol. 157, 214.

Harbour, D.A., Ashley, C.R., Williams, P.D., Gruffydd-Jones, T.J., 1987. Natural and experimental astrovirus infection of cats. Vet. Rec. 120, 555.

Hata, A., Katayama, H., Kitajima, M., Furumai, H., 2015. Wastewater analysis indicates that genetically diverse astroviruses, including strains belonging to novel clades MLB and VA, are circulating within Japanese populations. Appl. Environ. Microbiol. 81, 4932.

Holtz, L.R., Wylie, K.M., Sodergren, E., Jiang, Y., Franz, C.J., Weinstock, G.M., Storch, G.A., Wang, D., 2011. Astrovirus MLB2 viremia in febrile child. Emerg. Infect. Dis. 17, 2050.

Höper, D., Mettenleiter, T.C., Beer, M., 2016. Metagenomic approaches to identifying infectious agents. Rev. Sci. Tech. 35, 83.

Hoshino, Y., Zimmer, J.F., Moise, N.S., Scott, F.W., 1981. Detection of astroviruses in feces of a cat with diarrhea. Brief report. Arch. Virol. 70, 373.

Ito, M., Kuroda, M., Masuda, T., Akagami, M., Haga, K., Tsuchiaka, S., Kishimoto, M., Naoi, Y., Sano, K., Omatsu, T., Katayama, Y., Oba, M., Aoki, H., Ichimaru, T., Mukono, I., Ouchi, Y., Yamasato, H., Shirai, J., Katayama, K., Mizutani, T., Nagai, M., 2017. Whole genome analysis of porcine astroviruses detected in Japanese pigs reveals genetic diversity and possible intra-genotypic recombination. Infect. Genet. Evol. 50, 38.

Jiang, B., Monroe, S.S., Koonin, E.V., Stine, S.E., Glass, R.I., 1993. RNA sequence of astrovirus: distinctive genomic organization and a putative retrovirus-like ribosomal frameshifting signal that directs the viral replicase synthesis. Proc. Natl. Acad. Sci. U.S.A. 90, 10539.

Jiang, H., Holtz, L.R., Bauer, I., Franz, C.J., Zhao, G., Bodhidatta, L., Shrestha, S.K., Kang, G., Wang, D., 2013. Comparison of novel MLB-clade, VA-clade and classic human astroviruses highlights constrained evolution of the classic human astrovirus nonstructural genes. Virology 436, 8.

Jonassen, C.M., Jonassen, T.T., Sveen, T.M., Grinde, B., 2003. Complete genomic sequences of astroviruses from sheep and turkey: comparison with related viruses. Virus Res. 91, 195.

Kapoor, A., Li, L., Victoria, J., Oderinde, B., Mason, C., Pandey, P., Zaidi, S.Z., Delwart, E., 2009. Multiple novel astrovirus species in human stool. J. Gen. Virol. 90, 2965.

Karlsson, E.A., Small, C.T., Freiden, P., Feeroz, M.M., Matsen, F.A.t., San, S., Hasan, M.K., Wang, D., Jones-Engel, L., Schultz-Cherry, S., 2015. Non-human primates Harbor diverse mammalian and avian astroviruses including those associated with human infections. PLoS Pathog. 11, e1005225.

Kjeldsberg, E., Hem, A., 1985. Detection of astroviruses in gut contents of nude and normal mice. Brief report. Arch. Virol. 84, 135.

Koci, M.D., Schultz-Cherry, S., 2002. Avian astroviruses. Avian Pathol. 31, 213.

Lan, D., Ji, W., Shan, T., Cui, L., Yang, Z., Yuan, C., Hua, X., 2011. Molecular characterization of a porcine astrovirus strain in China. Arch. Virol. 156, 1869.

Lau, S.K., Woo, P.C., Yip, C.C., Bai, R., Wu, Y., Tse, H., Yuen, K.Y., 2013. Complete genome sequence of a novel feline astrovirus from a domestic cat in Hong Kong. Genome Announc. 1, e00708–e00713.

Laurin, M.A., Dastor, M., L'Homme, Y., 2011. Detection and genetic characterization of a novel pig astrovirus: relationship to other astroviruses. Arch. Virol. 156, 2095.
Le Guyader, F., Haugarreau, L., Miossec, L., Dubois, E., Pommepuy, M., 2000. Three-year study to assess human enteric viruses in shellfish. Appl. Environ. Microbiol. 66, 3241.
Lee, T.W., Kurtz, J.B., 1981. Serial propagation of astrovirus in tissue culture with the aid of trypsin. J. Gen. Virol. 57, 421.
Lee, T.W., Kurtz, J.B., 1994. Prevalence of human astrovirus serotypes in the Oxford region 1976-92, with evidence for two new serotypes. Epidemiol. Infect. 112, 187.
Lee, S., Jang, G., Lee, C., 2015. Complete genome sequence of a porcine astrovirus from South Korea. Arch. Virol. 160, 1819.
Lewis, T.L., Greenberg, H.B., Herrmann, J.E., Smith, L.S., Matsui, S.M., 1994. Analysis of astrovirus serotype 1 RNA, identification of the viral RNA-dependent RNA polymerase motif, and expression of a viral structural protein. J. Virol. 68, 77.
Li, L., Shan, T., Wang, C., Cote, C., Kolman, J., Onions, D., Gulland, F.M., Delwart, E., 2011. The fecal viral flora of California sea lions. J. Virol. 85, 9909.
Li, L., Diab, S., McGraw, S., Barr, B., Traslavina, R., Higgins, R., Talbot, T., Blanchard, P., Rimoldi, G., Fahsbender, E., Page, B., Phan, T.G., Wang, C., Deng, X., Pesavento, P., Delwart, E., 2013. Divergent astrovirus associated with neurologic disease in cattle. Emerg. Infect. Dis. 19, 1385.
Lipkin, W.I., 2010. Microbe hunting. Microbiol. Mol. Biol. Rev. 74, 363.
Lipkin, W.I., 2013. The changing face of pathogen discovery and surveillance. Nat. Rev. Microbiol. 11, 133.
Lizasoain, A., Tort, L.F., Garcia, M., Gomez, M.M., Leite, J.P., Miagostovich, M.P., Cristina, J., Berois, M., Colina, R., Victoria, M., 2015a. Sewage surveillance reveals the presence of canine GVII norovirus and canine astrovirus in Uruguay. Arch. Virol. 160, 2839.
Lizasoain, A., Tort, L.F., Garcia, M., Gomez, M.M., Leite, J.P., Miagostovich, M.P., Cristina, J., Colina, R., Victoria, M., 2015b. Environmental assessment reveals the presence of MLB-1 human astrovirus in Uruguay. J. Appl. Microbiol. 119, 859.
Lum, S.H., Turner, A., Guiver, M., Bonney, D., Martland, T., Davies, E., Newbould, M., Brown, J., Morfopoulou, S., Breuer, J., Wynn, R., 2016. An emerging opportunistic infection: fatal astrovirus (VA1/HMO-C) encephalitis in a pediatric stem cell transplant recipient. Transpl. Infect. Dis. 18, 960–964.
Luo, Z., Roi, S., Dastor, M., Gallice, E., Laurin, M.A., L'Homme, Y., 2011. Multiple novel and prevalent astroviruses in pigs. Vet. Microbiol. 149, 316.
Ma, W., Kahn, R.E., Richt, J.A., 2008. The pig as a mixing vessel for influenza viruses: human and veterinary implications. J. Mol. Genet. Med. 3, 158.
Madeley, C.R., 1979. Comparison of the features of astroviruses and caliciviruses seen in samples of feces by electron microscopy. J. Infect. Dis. 139, 519.
Madeley, C.R., Cosgrove, B.P., 1975a. Letter: 28 nm particles in faeces in infantile gastroenteritis. Lancet 2, 451.
Madeley, C.R., Cosgrove, B.P., 1975b. Letter: viruses in infantile gastroenteritis. Lancet 2, 124.
Martella, V., Moschidou, P., Lorusso, E., Mari, V., Camero, M., Bellacicco, A., Losurdo, M., Pinto, P., Desario, C., Banyai, K., Elia, G., Decaro, N., Buonavoglia, C., 2011a. Detection and characterization of canine astroviruses. J. Gen. Virol. 92, 1880.
Martella, V., Moschidou, P., Pinto, P., Catella, C., Desario, C., Larocca, V., Circella, E., Banyai, K., Lavazza, A., Magistrali, C., Decaro, N., Buonavoglia, C., 2011b. Astroviruses in rabbits. Emerg. Infect. Dis. 17, 2287.
Martella, V., Moschidou, P., Catella, C., Larocca, V., Pinto, P., Losurdo, M., Corrente, M., Lorusso, E., Banyai, K., Decaro, N., Lavazza, A., Buonavoglia, C., 2012. Enteric disease in dogs naturally infected by a novel canine astrovirus. J. Clin. Microbiol. 50, 1066.

Martella, V., Medici, M.C., Terio, V., Catella, C., Bozzo, G., Tummolo, F., Calderaro, A., Bonura, F., Di Franco, M., Banyai, K., Giammanco, G.M., De Grazia, S., 2013. Lineage diversification and recombination in type-4 human astroviruses. Infect. Genet. Evol. 20, 330.

Martella, V., Pinto, P., Tummolo, F., De Grazia, S., Giammanco, G.M., Medici, M.C., Ganesh, B., L'Homme, Y., Farkas, T., Jakab, F., Banyai, K., 2014. Analysis of the ORF2 of human astroviruses reveals lineage diversification, recombination and rearrangement and provides the basis for a novel sub-classification system. Arch. Virol. 159, 3185.

Medici, M.C., Tummolo, F., Martella, V., Banyai, K., Bonerba, E., Chezzi, C., Arcangeletti, M.C., De Conto, F., Calderaro, A., 2015. Genetic heterogeneity and recombination in type-3 human astroviruses. Infect. Genet. Evol. 32, 156–160.

Meliopoulos, V.A., Kayali, G., Burnham, A., Oshansky, C.M., Thomas, P.G., Gray, G.C., Beck, M.A., Schultz-Cherry, S., 2014. Detection of antibodies against Turkey astrovirus in humans. PLoS One 9, e96934.

Méndez, E., Salas-Ocampo, E., Arias, C.F., 2004. Caspases mediate processing of the capsid precursor and cell release of human astroviruses. J. Virol. 78, 8601.

Méndez, E., Murillo, A., Velázquez, R., Burnham, A., Arias, C.F., 2013. Replication cycle of astroviruses. In: Schultz-Cherry, S. (Ed.), Astrovirus Research, p. Springer-Verlag, New York, pp. 19–45.

Meyer, C.T., Bauer, I.K., Antonio, M., Adeyemi, M., Saha, D., Oundo, J.O., Ochieng, J.B., Omore, R., Stine, O.C., Wang, D., Holtz, L.R., 2015. Prevalence of classic, MLB-clade and VA-clade astroviruses in Kenya and the Gambia. Virol. J. 12, 78.

Mihalov-Kovacs, E., Martella, V., Lanave, G., Bodnar, L., Feher, E., Marton, S., Kemenesi, G., Jakab, F., Banyai, K., 2017. Genome analysis of canine astroviruses reveals genetic heterogeneity and suggests possible inter-species transmission. Virus Res. 232, 162.

Mittelholzer, C., Englund, L., Hedlund, K.O., Dietz, H.H., Svensson, L., 2003a. Detection and sequence analysis of Danish and Swedish strains of mink astrovirus. J. Clin. Microbiol. 41, 5192.

Mittelholzer, C., Hedlund, K.O., Englund, L., Dietz, H.H., Svensson, L., 2003b. Molecular characterization of a novel astrovirus associated with disease in mink. J. Gen. Virol. 84, 3087.

Naccache, S.N., Peggs, K.S., Mattes, F.M., Phadke, R., Garson, J.A., Grant, P., Samayoa, E., Federman, S., Miller, S., Lunn, M.P., Gant, V., Chiu, C.Y., 2015. Diagnosis of neuroinvasive astrovirus infection in an immunocompromised adult with encephalitis by unbiased next-generation sequencing. Clin. Infect. Dis. 60, 919.

Nagai, M., Omatsu, T., Aoki, H., Otomaru, K., Uto, T., Koizumi, M., Minami-Fukuda, F., Takai, H., Murakami, T., Masuda, T., Yamasato, H., Shiokawa, M., Tsuchiaka, S., Naoi, Y., Sano, K., Okazaki, S., Katayama, Y., Oba, M., Furuya, T., Shirai, J., Mizutani, T., 2015. Full genome analysis of bovine astrovirus from fecal samples of cattle in Japan: identification of possible interspecies transmission of bovine astrovirus. Arch. Virol. 160, 2491–2501.

Ng, T.F., Kondov, N.O., Hayashimoto, N., Uchida, R., Cha, Y., Beyer, A.I., Wong, W., Pesavento, P.A., Suemizu, H., Muench, M.O., Delwart, E., 2013. Identification of an astrovirus commonly infecting laboratory mice in the US and Japan. PLoS One 8, e66937.

Ng, T.F., Mesquita, J.R., Nascimento, M.S., Kondov, N.O., Wong, W., Reuter, G., Knowles, N.J., Vega, E., Esona, M.D., Deng, X., Vinje, J., Delwart, E., 2014. Feline fecal virome reveals novel and prevalent enteric viruses. Vet. Microbiol. 171, 102.

Ng, T.F., Kondov, N.O., Deng, X., Van Eenennaam, A., Neibergs, H.L., Delwart, E., 2015. A metagenomics and case-control study to identify viruses associated with bovine respiratory disease. J. Virol. 89, 5340.

Noel, J.S., Lee, T.W., Kurtz, J.B., Glass, R.I., Monroe, S.S., 1995. Typing of human astroviruses from clinical isolates by enzyme immunoassay and nucleotide sequencing. J. Clin. Microbiol. 33, 797.

Oem, J.K., An, D.J., 2014. Phylogenetic analysis of bovine astrovirus in Korean cattle. Virus Genes 48, 372.

Padmanabhan, A., Hause, B.M., 2016. Detection and characterization of a novel genotype of porcine astrovirus 4 from nasal swabs from pigs with acute respiratory disease. Arch. Virol. 161, 2575.

Pankovics, P., Boros, A., Kiss, T., Delwart, E., Reuter, G., 2015. Detection of a mammalian-like astrovirus in bird, European roller (Coracias garrulus). Infect. Genet. Evol. 34, 114–121.

Pantin-Jackwood, M.J., Spackman, E., Woolcock, P.R., 2006. Molecular characterization and typing of chicken and turkey astroviruses circulating in the United States: implications for diagnostics. Avian Dis. 50, 397.

Perot, P., Lecuit, M., Eloit, M., 2017. Astrovirus diagnostics. Viruses 9, 1–14.

Pfaff, F., Schlottau, K., Scholes, S., Courtenay, A., Hoffmann, B., Hoper, D., Beer, M., 2017. A novel astrovirus associated with encephalitis and ganglionitis in domestic sheep. Transbound. Emerg. Dis. 64, 677–682.

Phan, T.G., Kapusinszky, B., Wang, C., Rose, R.K., Lipton, H.L., Delwart, E.L., 2011. The fecal viral flora of wild rodents. PLoS Pathog. 7, e1002218.

Phan, T.G., Nordgren, J., Ouermi, D., Simpore, J., Nitiema, L.W., Deng, X., Delwart, E., 2014. New astrovirus in human feces from Burkina Faso. J. Clin. Virol. 60, 161.

Pinto, R.M., Villena, C., Le Guyader, F., Guix, S., Caballero, S., Pommepuy, M., Bosch, A., 2001. Astrovirus detection in wastewater samples. Water Sci. Technol. 43, 73.

Prevost, B., Lucas, F.S., Ambert-Balay, K., Pothier, P., Moulin, L., Wurtzer, S., 2015. Deciphering the diversities of astroviruses and noroviruses in wastewater treatment plant effluents by a high-throughput sequencing method. Appl. Environ. Microbiol. 81, 7215.

Qiu, Y., Lee, B.E., Neumann, N., Ashbolt, N., Craik, S., Maal-Bared, R., Pang, X.L., 2015. Assessment of human virus removal during municipal wastewater treatment in Edmonton, Canada. J. Appl. Microbiol. 119, 1729.

Quan, P.L., Wagner, T.A., Briese, T., Torgerson, T.R., Hornig, M., Tashmukhamedova, A., Firth, C., Palacios, G., Baisre-De-Leon, A., Paddock, C.D., Hutchison, S.K., Egholm, M., Zaki, S.R., Goldman, J.E., Ochs, H.D., Lipkin, W.I., 2010. Astrovirus encephalitis in boy with X-linked agammaglobulinemia. Emerg. Infect. Dis. 16, 918.

Reuter, G., Nemes, C., Boros, A., Kapusinszky, B., Delwart, E., Pankovics, P., 2012a. Astrovirus in wild boars (Sus scrofa) in Hungary. Arch. Virol. 157, 1143.

Reuter, G., Pankovics, P., Delwart, E., Boros, A., 2012b. Identification of a novel astrovirus in domestic sheep in Hungary. Arch. Virol. 157, 323.

Rivera, R., Nollens, H.H., Venn-Watson, S., Gulland, F.M., Wellehan Jr., J.F., 2010. Characterization of phylogenetically diverse astroviruses of marine mammals. J. Gen. Virol. 91, 166.

Sato, M., Kuroda, M., Kasai, M., Matsui, H., Fukuyama, T., Katano, H., Tanaka-Taya, K., 2016. Acute encephalopathy in an immunocompromised boy with astrovirus-MLB1 infection detected by next generation sequencing. J. Clin. Virol. 78, 66.

Schlottau, K., Schulze, C., Bilk, S., Hanke, D., Hoper, D., Beer, M., Hoffmann, B., 2016. Detection of a novel bovine astrovirus in a cow with encephalitis. Transbound. Emerg. Dis. 63, 253.

Selimovic-Hamza, S., Bouzalas, I.G., Vandevelde, M., Oevermann, A., Seuberlich, T., 2016. Detection of astrovirus in historical cases of European sporadic bovine encephalitis, Switzerland 1958-1976. Front. Vet. Sci. 3, 91.

Selimovic-Hamza, S., Boujon, C.L., Hilbe, M., Oevermann, A., Seuberlich, T., 2017. Frequency and pathological phenotype of bovine astrovirus CH13/NeuroS1 infection in neurologically-diseased cattle: towards assessment of causality. Viruses 9, 12.

Seuberlich, T., Wuthrich, D., Selimovic-Hamza, S., Drogemuller, C., Oevermann, A., Bruggmann, R., Bouzalas, I., 2016. Identification of a second encephalitis-associated astrovirus in cattle. Emerg. Microbes. Infect. 5, e5.

Shan, T., Li, L., Simmonds, P., Wang, C., Moeser, A., Delwart, E., 2011. The fecal virome of pigs on a high-density farm. J. Virol. 85, 11697.

Shan, T., Wang, C., Tong, W., Zheng, H., Hua, X., Yang, S., Guo, Y., Zhang, W., Tong, G., 2012. Complete genome of a novel porcine astrovirus. J. Virol. 86, 13820.

Sharp, C.P., Gregory, W.F., Mason, C., Bronsvoort, B.M., Beard, P.M., 2015. High prevalence and diversity of bovine astroviruses in the faeces of healthy and diarrhoeic calves in south West Scotland. Vet. Microbiol. 178, 70.

Shimizu, M., Shirai, J., Narita, M., Yamane, T., 1990. Cytopathic astrovirus isolated from porcine acute gastroenteritis in an established cell line derived from porcine embryonic kidney. J. Clin. Microbiol. 28, 201.

Shirai, J., Nakamura, K., Shinohara, K., Kawamura, H., 1991. Pathogenicity and antigenicity of avian nephritis isolates. Avian Dis. 35, 49.

Smits, S.L., van Leeuwen, M., Kuiken, T., Hammer, A.S., Simon, J.H., Osterhaus, A.D., 2010. Identification and characterization of deer astroviruses. J. Gen. Virol. 91, 2719.

Smyth, V.J., 2017. A review of the strain diversity and pathogenesis of chicken astrovirus. Viruses 9.

Snodgrass, D.R., Gray, E.W., 1977. Detection and transmission of 30 nm virus particles (astroviruses) in faeces of lambs with diarrhoea. Arch. Virol. 55, 287.

Stenglein, M.D., Velazquez, E., Greenacre, C., Wilkes, R.P., Ruby, J.G., Lankton, J.S., Ganem, D., Kennedy, M.A., DeRisi, J.L., 2012. Complete genome sequence of an astrovirus identified in a domestic rabbit (Oryctolagus cuniculus) with gastroenteritis. Virol. J. 9, 216.

Sun, N., Yang, Y., Wang, G.S., Shao, X.Q., Zhang, S.Q., Wang, F.X., Tan, B., Tian, F.L., Cheng, S.P., Wen, Y.J., 2014. Detection and characterization of avastrovirus associated with diarrhea isolated from minks in China. Food Environ. Virol. 6, 169.

Tamura, K., Stecher, G., Peterson, D., Filipski, A., Kumar, S., 2013. MEGA6: molecular evolutionary genetics analysis version 6.0. Mol. Biol. Evol. 30, 2725.

Toffan, A., Jonassen, C.M., De Battisti, C., Schiavon, E., Kofstad, T., Capua, I., Cattoli, G., 2009. Genetic characterization of a new astrovirus detected in dogs suffering from diarrhoea. Vet. Microbiol. 139, 147.

Toh, Y., Harper, J., Dryden, K.A., Yeager, M., Arias, C.F., Méndez, E., Tao, Y.J., 2016. Crystal structure of the human astrovirus capsid protein. J. Virol. 90, 9008.

Tse, H., Chan, W.M., Tsoi, H.W., Fan, R.Y., Lau, C.C., Lau, S.K., Woo, P.C., Yuen, K.Y., 2011. Rediscovery and genomic characterization of bovine astroviruses. J. Gen. Virol. 92, 1888.

Tzipori, S., Menzies, J.D., Gray, E.W., 1981. Detection of astrovirus in the faeces of red deer. Vet. Rec. 108, 286.

Ulloa, J.C., Gutierrez, M.F., 2010. Genomic analysis of two ORF2 segments of new porcine astrovirus isolates and their close relationship with human astroviruses. Can. J. Microbiol. 56, 569.

Walter, J.E., Briggs, J., Guerrero, M.L., Matson, D.O., Pickering, L.K., Ruiz-Palacios, G., Berke, T., Mitchell, D.K., 2001. Molecular characterization of a novel recombinant strain of human astrovirus associated with gastroenteritis in children. Arch. Virol. 146, 2357.

Willcocks, M.M., Brown, T.D., Madeley, C.R., Carter, M.J., 1994. The complete sequence of a human astrovirus. J. Gen. Virol. 75 (Pt. 7), 1785.

Williams Jr., F.P., 1980. Astrovirus-like, coronavirus-like, and parvovirus-like particles detected in the diarrheal stools of beagle pups. Arch. Virol. 66, 215.

Wolfaardt, M., Kiulia, N.M., Mwenda, J.M., Taylor, M.B., 2011. Evidence of a recombinant wild-type human astrovirus strain from a Kenyan child with gastroenteritis. J. Clin. Microbiol. 49, 728.

Woo, P.C., Lau, S.K., Teng, J.L., Tsang, A.K., Joseph, S., Xie, J., Jose, S., Fan, R.Y., Wernery, U., Yuen, K.Y., 2015. A novel astrovirus from dromedaries in the Middle East. J. Gen. Virol. 96, 2697–2707.

Woode, G.N., Bridger, J.C., 1978. Isolation of small viruses resembling astroviruses and caliciviruses from acute enteritis of calves. J. Med. Microbiol. 11, 441.

Wunderli, W., Meerbach, A., Gungor, T., Berger, C., Greiner, O., Caduff, R., Trkola, A., Bossart, W., Gerlach, D., Schibler, M., Cordey, S., McKee, T.A., Van Belle, S., Kaiser, L., Tapparel, C., 2011. Astrovirus infection in hospitalized infants with severe combined immunodeficiency after allogeneic hematopoietic stem cell transplantation. PLoS One 6, e27483.

Xiao, C.T., Gimenez-Lirola, L.G., Gerber, P.F., Jiang, Y.H., Halbur, P.G., Opriessnig, T., 2013. Identification and characterization of novel porcine astroviruses (PAstVs) with high prevalence and frequent co-infection of individual pigs with multiple PAstV types. J. Gen. Virol. 94, 570.

Xiao, C.T., Luo, Z., Lv, S.L., Opriessnig, T., Li, R.C., Yu, X.L., 2017. Identification and characterization of multiple porcine astrovirus genotypes in Hunan province, China. Arch. Virol. 162 (4), 943–952. https://doi.org/10.1007/s00705-016-3185-0. Epub 2016 Dec 18.

Yokoyama, C.C., Loh, J., Zhao, G., Stappenbeck, T.S., Wang, D., Huang, H.V., Virgin, H.W., Thackray, L.B., 2012. Adaptive immunity restricts replication of novel murine astroviruses. J. Virol. 86, 12262.

York, R.L., Yousefi, P.A., Bogdanoff, W., Haile, S., Tripathi, S., DuBois, R.M., 2015. Structural, mechanistic, and antigenic characterization of the human astrovirus capsid. J. Virol. 90, 2254–2263.

Zhang, W., Li, L., Deng, X., Kapusinszky, B., Pesavento, P.A., Delwart, E., 2014. Faecal virome of cats in an animal shelter. J. Gen. Virol. 95, 2553.

Zhao, W., Zhu, A.L., Yuan, C.L., Yu, Y., Zhu, C.X., Lan, D.L., Yang, Z.B., Cui, L., Hua, X.G., 2011. Detection of astrovirus infection in pigeons (Columbia livia) during an outbreak of diarrhoea. Avian Pathol. 40, 361.

Zhou, N., Lin, X., Wang, S., Wang, H., Bi, Z., Wang, P., Chen, P., Tao, Z., Song, L., Song, Y., Xu, A., 2016. Molecular characterization of classic human astrovirus in eastern China, as revealed by environmental sewage surveillance. J. Appl. Microbiol. 120, 1436.

Zhu, A.L., Zhao, W., Yin, H., Shan, T.L., Zhu, C.X., Yang, X., Hua, X.G., Cui, L., 2011. Isolation and characterization of canine astrovirus in China. Arch. Virol. 156, 1671.

CHAPTER SIX

New Leaves in the Growing Tree of Pestiviruses

Sandra Blome[1], Martin Beer, Kerstin Wernike
Institute of Diagnostic Virology, Friedrich-Loeffler-Institut, Greifswald-Insel Riems, Germany
[1]Corresponding author: e-mail address: sandra.blome@fli.de

Contents

Abstract

Pestiviruses are a group of viruses of veterinary importance infecting livestock animals like pigs, cattle, and sheep, and also wildlife animals like wild boar and different deer species. While for decades only four classical species (Classical swine fever virus, Bovine viral diarrhea virus types 1 and 2, Border disease virus), and a few so-called atypical pestiviruses were known (e.g., Giraffe virus, Pronghorn virus, HoBi virus), a series of novel pestiviruses was identified in the last years (Bungowannah virus, Bat pestivirus, Norway rat pestivirus, Atypical porcine pestivirus, LINDA virus). The Australian Bungowannah virus could be isolated and further characterized by classical sequencing, but all the other latest novel pestiviruses were identified by metagenomics using next-generation sequencing technologies. Here, we describe these new viruses and their discovery and characterization. Differentiation is made between the occurrence of classical pestiviruses in new species and novel viruses or virus types.

Advances in Virus Research, Volume 99
ISSN 0065-3527
https://doi.org/10.1016/bs.aivir.2017.07.003

1. INTRODUCTION

Pestiviruses are small, enveloped viruses with a single-stranded, positive-sense RNA genome and belong to the *Flaviviridae* family (Simmonds et al., 2017). The genomic organization of all recognized pestiviruses is very similar: the single positive-stranded genome codes for 12 (13) proteins in one large open reading frame that is flanked by 5′- and 3′-nontranslated regions. The translated polyprotein is subsequently processed by viral and cellular proteases, and the resulting proteins are N^{pro}, C (capsid), the envelope proteins E^{rns}, E1, and E2, p7, and nonstructural proteins NS2, NS3, (NS2–3), NS4A, NS4B, NS5A, and NS5B. The nonstructural autoprotease N^{pro} and also the envelope protein E^{rns} are unique to pestiviruses and can be found in all members of the genus but not in other *Flaviviridae* (Lindenbach et al., 2013). Furthermore, members of the *Pestivirus* genus infect a wide range of wild and domestic cloven-hoofed animals with different main target species.

At present, the International Committee on Taxonomy of Viruses has recognized only four pestivirus species. These “classical” members of the genus *Pestivirus* are Classical swine fever virus (CSFV), Bovine viral diarrhea virus (BVDV) types 1 and 2, and Border disease virus (BDV). Especially CSFV and BVDV are very important pathogens of livestock.

Classical swine fever (CSF) is a severe multisystemic disease that is considered a major threat to profitable pig production worldwide. Natural hosts are members of the family *Suidae*. After implementation of strict control measures, several countries in the European Union and Northern America succeeded in eradicating CSF. Nevertheless, in most parts of the world with significant pig production, CSFV is at least sporadically present, and both Asian and South and Central American countries suffer endemic infections. The clinical picture of CSF depends on virus and host factors and is highly variable. It can range from an almost inapparent infection to a hemorrhagic fever-like illness with high mortality. Due to its impact on animal health and pig industry as well as the potential to cause transboundary outbreaks, the disease is notifiable to the World Organization for Animal Health (OIE).

BVDV, the causative agent of Bovine viral diarrhea (BVD), is endemic in cattle populations worldwide. It exists in two different species (BVDV-1 and BVDV-2) which are further subdivided into numerous subtypes based on genetic analysis (Vilcek et al., 2001, 2004). According to their growth in cell culture, BVD viruses are classified into the two distinct biotypes: cytopathic (cp) and noncytopathic (ncp) (Ridpath, 2010). As with CSF, clinical signs can vary considerably, from inapparent infections to unspecific symptoms such as diarrhea, fever, hemorrhagic lesions, pneumonia, or to the inevitably fatal mucosal disease (MD). The outcome of prenatal infections is dependent on the stage of gestation. Fetal infections with ncp BVDV may result in abortion, stillbirth, teratogenic effects, or the birth of immunotolerant, persistently infected (PI), viremic calves (Baker, 1995). The BVDV-specific illness of PI calves, the so-called MD, is associated with the appearance of a cp biotype arising from mutations of the ncp BVDV which is already circulating in this PI animal or with the superinfection of the animal with a cp strain that is very similar to the persisting ncp strain (Lanyon et al., 2014). BVD is one of the most prevalent infectious diseases of cattle causing major economic losses (Houe, 1999, 2003).

BDV infections in sheep have also economic impact (Westaway et al., 1985). The infection of pregnant animals between the 17th and 80th day of gestation can lead to stillbirth, malformations, or the birth of so-called hairy shaker lambs. The latter are characterized by hairy fleece and congenital tremor (Nettleton et al., 1992). These lambs but also their apparently healthy mates can be PI (Terpstra, 1980). Postnatal infection of immunocompetent sheep usually results in only mild disease or no overt signs. BDVs were reclassified into genotypes 1 to 3 (Becher et al., 2003), including the isolates from reindeer that were previously reported as a putative new genotype.

With the advent of modern next-generation sequencing techniques, new pestiviruses have been discovered, and the behavior and dynamics of pestiviral infections have been explored. Especially over the last decade, a growing number of novel pestiviruses has been discovered in various domestic and wild ruminant species as well as pigs (Postel et al., 2015). Based on the phylogeny that is based on a molecular dataset combining the 5′-NTR, N^{pro}, and E2 gene regions, Liu et al. (2009b) proposed nine pestivirus species: BVDV-1, BVDV-2, BVDV-3 (HoBi-like), BDV, Tunisian sheep pestiviruses (TSV), CSFV, Antelope (Pronghorn), Giraffe, and Bungowannah virus (Table 1). However, meanwhile further members of the genus were identified and the tree of pestiviruses is further growing. Here, we describe the new viruses and their discovery and characterization. Differentiation is made between the occurrence of classical pestiviruses in new species and novel viruses or virus types.

Table 1 Pestivirus Diversity and Classification

Species	Virus Name	Abbreviation	Main Target Host	Distribution
Border disease virus	Border disease virus	BDV	Ruminants	All continents
Bovine viral diarrhea virus 1	Bovine viral diarrhea virus 1	BVDV-1	All even-toed ungulates	All continents
Bovine viral diarrhea virus 2	Bovine viral diarrhea virus 2	BVDV-2	All even-toed ungulates	All continents
Bovine viral diarrhea virus 3	HoBi-like virus	BVDV-3	Cattle and water buffalo	America, Europe, Asia
Classical swine fever virus	Classical swine fever virus	CSFV	Swine	All continents
Unassigned	Atypical porcine pestivirus	APPV	Swine	America, Europe, Asia
Unassigned	Bungowannah virus	BuPV	Swine	Australia
Unassigned	Giraffe pestivirus	GiPV	Giraffe	Africa
Unassigned	Norway rat pestivirus	NrPV	Rat	America
Unassigned	Pronghorn antelope pestivirus	PAPV	Pronghorn antelope	America
Unassigned	*Rhinolophus affinis* pestivirus	RaPestV	Bat (*Rhinolophus affinis*)	Asia
Unassigned	Sheep pestivirus Aydin/04/TR	ShPV-Aydin	Small ruminants	Eurasia
Unassigned	Lateral-Shaking Inducing NeuroDegenerative Agent	LINDA virus	Swine	Austria

2. CLASSICAL PESTIVIRUSES IN NEW SPECIES

The classical pestiviruses were originally differentiated according to the host species from which they were isolated and the clinical signs they induce. However, members of the genus *Pestivirus* are able to cross the species barriers as has been repeatedly demonstrated.

As an example, BDV has caused an outbreak of severe disease in free-ranging chamois in the Pyrenees (Arnal et al., 2004; Hurtado et al., 2004; Marco et al., 2007, 2008, 2009, 2011; Pioz et al., 2007; Vilcek et al., 2010). Apparently, the virus had circulated for some time without negative effects before a change in virulence occurred. The clinical signs included general weakness, gait problems, and skin lesions (alopecia and hyper-pigmentation). Overall, the chamois population was significantly reduced due to this outbreak episode. In some herds, also endemic infections were established, while others were able to clear the virus (Fernandez-Sirera et al., 2012).

Despite the fact that new world camelids were rarely reported to be sero-positive for pestiviruses, Goyal et al. (2002) reported the isolation of BVDV from an aborted alpaca fetus, and in 2005 Carman et al. (2005) confirmed persistent BVDV infection in an alpaca cria. The latter was found after BVDV symptoms in an alpaca herd in the United States. This outbreak was linked to other cases which confirmed the risk of BVDV infections for alpacas (Ridpath, 2015).

It has also been demonstrated that BDV is able to induce persistent infection in cattle and can be transmitted from PI cattle to naïve heifers in early pregnancy (Braun et al., 2015). In some cases, these transmissions resulted in additional PI animals. While these transmissions may be rather rare and inefficient events, BDV should also be considered in the framework of BVDV eradication programs.

3. ATYPICAL PESTIVIRUSES

3.1 Giraffe Virus

The Giraffe pestivirus was associated with an outbreak of MD-like symptoms in giraffes in the Nanyuki District of Kenya in 1967 (Dekker et al., 1995; Harasawa et al., 2000). The virus was shown to be divergent from classical pestiviruses (Becher et al., 1997, 2003). A closely related virus, referred to as PG-2, was obtained from a bovine cell culture in the 1990th (Becher et al., 2003) suggesting the presence of further, similar viruses in Africa. The knowledge about the natural host range, virulence, or geographic distribution is scarce until now; however, the transmission of such newly emerging pestiviruses to domestic ruminants could pose a risk for the livestock industry, especially in countries like Switzerland, Denmark, Sweden, or Germany with eradication programs and more and more naïve cattle (Moennig and Becher, 2015; Moennig et al., 2005; Stahl and Alenius, 2012).

3.2 Pronghorn Virus

In 2005, a new and divergent pestivirus was found in a pronghorn antelope, *Antilocapra americana* (Vilcek et al., 2005). The affected animal was a young antelope that was found wandering blind. Routine diagnostic methods revealed a virus that showed some cross-neutralization with pestiviruses. Based on partial sequencing (first phylogenetic analyses were carried out in the 5′-NTR, the entire N^{pro} and E2 region), it was demonstrated that the respective virus was quite divergent from the classical pestiviruses (see Fig. 1), but shared the pestiviral characteristics of conserved 5′-NTR motifs and a N^{pro}-encoding region of the predicted length. The virus was later on fully sequenced using next-generation sequencing on an Illumina MiSeq platform and an Ion Torrent machine (Neill et al., 2014). At present, this virus would be considered the member of a novel pestivirus species, and one of the more divergent lineages (Liu et al., 2009b).

3.3 Tunesian Sheep Pestiviruses

Recent genetic and antigenetic analyses of ovine samples demonstrated the presence of a new pestivirus group referred to as TSV. The first member of this group has been isolated from contaminated sheep pox vaccines following outbreaks of BD-like diseases in immunized animals. Both, sheep pox vaccine-derived pestiviruses and later on isolated Tunisian field strains,

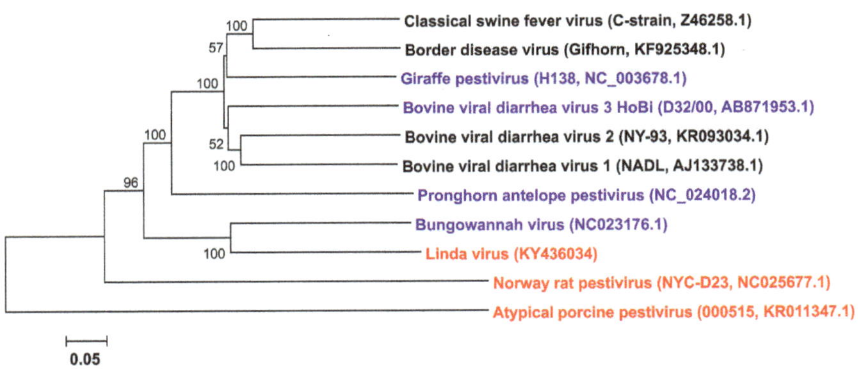

Fig. 1 Neighbor-joining tree based on complete genome sequences of representative pestivirus strains, isolate name and accession number are given in parentheses. The classical pestiviruses are printed in *black*, atypical pestiviruses detected prior to 2010 are highlighted in *blue*, and viruses isolated since 2010 are shown in *red*. Statistical support for nodes was obtained by bootstrapping (1000 replicates); only values ≥50% are shown. The *scale bar* indicates nucleotide substitutions per site.

induce mild clinical sign such as fever and leucopenia for a few days (Thabti et al., 2002). Based on the 5′-NTR and N^{pro}–E2 coding sequences, the Tunisian isolates form a separate branch within the genus *Pestivirus* with sequence divergences between the Tunisian viruses and BDV and CSFV of 36.5%–40% and 32.8%–35.1%, respectively (Thabti et al., 2005). Antigenically the Tunisian isolates are also closer related to BDV reference strains than to CSFV or BVDV, and limited cross-reactivity was observed between BDV and the Tunisian viruses (Thabti et al., 2005). Apart from Northern Africa, Tunisian-like pestivirus has been found in small ruminants in Italy and France (Ciulli et al.,2017). Martin et al. (2015) demonstrated their widespread distribution among domestic and wild ruminants in the French Southern Alps.

3.4 Ruminant Pestiviruses With Close Relation to CSFV (Aydin/Burdur)

In Turkey, two ruminant pestiviruses were found after routine surveillance in a sheep and a goat that showed considerable cross-reactions with routine CSFV diagnostics. The affected animals, however, showed signs of border disease (Oguzoglu et al., 2009). Follow-up of strains Aydin/04 and Burdur/05, including next-generation sequencing of the full genomes, revealed a close relation to CSFV both genetically and antigenically (Becher et al., 2012; Postel et al., 2015). These strains can cause serious diagnostic problems in established CSFV serology and form a distinct group located between CSFV and BDV in phylogenetic trees (Postel et al., 2015). In contrast to the above-mentioned Tunesian sheep pestiviruses that are antigenically closer to BDV, the Turkish viruses are much closer related to CSFV. Given this close relationship to CSFV, an experimental infection of pigs was carried out. Nevertheless, while the animals seroconverted with strong reactions in neutralization assays, no clinical signs were observed.

3.5 HoBi-Like Pestiviruses

The so-called HoBi-like viruses were first described in 2004 as a contaminant in a batch of fetal bovine serum originating from Brazil, and the name was derived from the initial isolate "HoBi_D32/00" (Schirrmeier et al., 2004). After this initial description, several similar viruses have been detected (Bauermann and Ridpath, 2015; Peletto et al., 2012; Stalder et al., 2005; Xia et al., 2011, 2013) and were named HoBi-like, atypical pestivirus, or

BVDV-3 (Liu et al., 2009b). The suggested high prevalence of HoBi-like viruses in fetal bovine sera batches, especially from South America, indicates that these viruses are probably widespread in South America (Bauermann and Ridpath, 2015). Apart from South American sera, materials from Australia, Canada, Mexico, and the United States were found to be contaminated with HoBi-like viruses (Xia et al., 2011). Apart from these detections as contaminants, natural infections have been reported from Southeast Asia (Haider et al., 2014; Mishra et al., 2014; Stahl et al., 2007), Italy (Decaro et al., 2011, 2016), and Brazil (Cortez et al., 2006; Stalder et al., 2005).

The origin of the HoBi-like viruses is unknown. However, water buffalos are discussed as a potential source (Bauermann and Ridpath, 2015). In fact, the HoBi-like virus "BrazBuf9" was isolated in the late 1990 from a disease outbreak in Brazilian water buffalos (Stalder et al., 2005).

HoBi-like viruses were shown to be related to BVDV at both the genetic and the antigenic level with high identities among different isolates. The isolate from Thailand (Th/04_KhonKaen) is the most distantly related HoBi-like virus (Bauermann and Ridpath, 2015).

The disease caused by HoBi-like viruses resembles clinical signs that were historically associated with BVDV infection including growth retardation, reduced milk production, respiratory disease, fertility problems, and increased mortality in young animals (Bauermann and Ridpath, 2015). The first evidence of natural infection is the above-mentioned outbreak in water buffalos in Brazil. Thereafter, Brazilian HoBi-like viruses were associated with abortions, reproductive failure, and blind calves that were born after insemination with virus-positive semen (Bauermann and Ridpath, 2015; Cortez et al., 2006). An additional association was found with a fatal enteric disease in Midwestern Brazil (reviewed by Bauermann and Ridpath, 2015). Recently, MD-like symptoms were reported from young animals at a farm in Brazil (Weber et al., 2016). In Thailand, HoBi-like viruses were found as part of Bovine herpesvirus and BVDV monitoring in multiple herds without a link to overt illness (Kampa et al., 2009). The isolate "Th/04_KhonKaen" was obtained through RNA transfection into bovine turbinate cells (Liu et al., 2009a) and was thereafter sequenced (see earlier). Further isolates were obtained in India and Bangladesh (Haider et al., 2014; Mishra et al., 2014). In Italy, HoBi virus infections were associated with respiratory signs of different severity and reproductive failure (Decaro et al., 2011, 2012a). Interestingly, reproductive failure was observed upon reoccurrence of the virus in the herd previously affected by respiratory signs. Also in the same herd, the first cytopathogenic HoBi virus isolate was found in an animal with respiratory disease (Decaro et al., 2012c).

Evidence exists that HoBi-like viruses circulated in Italy already in 2007. The respective viruses were found in archival samples from animals with respiratory disease, and the sequences are more closely related to South American HoBi-like viruses (Decaro et al., 2013). Experimental infections using "HoBi_D32/00," "Italy-1/10-1," or "Th/04_KhonKaen" resulted in mild clinical signs including nasal and ocular discharge, conjunctivitis, and slightly elevated body temperature (Decaro et al., 2012b; Larska et al., 2012; Schirrmeier et al., 2004). Experimental infection of pregnant animals led to reproductive failure and persistent infections (Decaro et al., 2015).

It is important to know that some BVDV diagnostic tests may fail to detect HoBi-like viruses or may not differentiate between BVDV and HoBi-like viruses. However, diagnostic tests have been adapted and tailored that overcome these problems. Moreover, cross-protection against HoBi-like viruses conferred by current BVDV vaccines cannot be expected (Bauermann and Ridpath, 2015).

Concluding, HoBi-like viruses show an almost worldwide distribution with significant prevalence in Brazil (Bauermann and Ridpath, 2015; Decaro et al., 2011, 2016; Giammarioli et al., 2015; Haider et al., 2014; Mao et al., 2012; Maya et al., 2016; Mishra et al., 2014; Shi et al., 2016; Silveira et al., 2017; Weber et al., 2016). The economic impact is not completely known, but under certain conditions, HoBi-like viruses can show considerable effects on the reproductive performance (Decaro et al., 2016). It is likely that the HoBi-like pestiviruses will be classified as the new pestiviral species BVDV-3.

3.6 Bungowannah Virus

In June 2003, an outbreak of "porcine myocarditis syndrome" of late term and neonatal piglets was reported in the largest integrated pig farm in New South Wales, Australia. The infectious disease of unknown cause spread through the farm, and the pigs that survived developed immunity. Consequently, the problem declined and soon, only sporadic cases were reported. However, for a period of app. 16–18 weeks, preweaning mortalities rose to 35% in affected litters, and stillbirths were around 20%–30%.

Affected piglets often died during routine handling with some of them showing cyanosis and hyperventilation before death. A constant histopathological finding was an acute to subacute multifocal, nonsuppurative myocarditis with myonecrosis (McOrist et al., 2004). Electron microscopy of fixed heart tissues revealed small virus-like particles that underlined the assumption of a viral etiology. However, a wide range of common and exotic pathogens was excluded.

The "mystery piglet deaths" prompted an extensive animal health investigation that finally led to the discovery of Bungowannah virus, named after the location of its first occurrence (Kirkland et al., 2007; McOrist et al., 2004; Thornley, 2004). While virus isolation was unsuccessful in the beginning, transmission of an infectious agent was proven by inoculation of clinical material into porcine fetuses (Finlaison et al., 2010).

Ultimately, the novel virus was discovered in serum pools of experimentally infected fetuses (collected at 17–21 days postinoculation) using a modified sequence-independent single primer amplification (SISPA) method followed by sequencing of derived clones. Based on the obtained sequences, molecular methods (tailored nested RT-PCR and rapid amplification of cDNA ends) were utilized that allowed both confirmation of SISPA products and closure of sequence gaps based on the investigation of clinical and experimental specimens. Testing of an extensive sample set from the outbreak farms confirmed Bungowannah virus as the cause of the disease (Finlaison et al., 2009). Moreover, virus isolation on porcine kidney cells and indirect immunoperoxidase staining (the latter based on a polyclonal pig serum) were successfully implemented (Kirkland et al., 2007).

Based on the discovery of conserved motifs within the 5′-nontranslated region and a putative N^{pro}-encoding region, the virus was placed into the *Pestivirus* genus with which it shares also a very limited antigenic cross-reactivity (Kirkland et al., 2007). After completion of the genome, it was shown that the virus has all sequence information to encode the known structural and nonstructural proteins typical for a pestivirus (Kirkland et al., 2015b), especially the unique proteins N^{pro} and E^{rns}.

At the time of its first description, Bungowannah virus was the most divergent pestivirus with the Pronghorn virus being its closest relative with app. 63% nucleotide identity only, followed by the Giraffe virus with roughly 56% identity. None of the tested "Panpesti" monoclonal antibodies reacted with Bungowannah virus; however, more recently, monoclonal antibodies specific to Bungowannah virus have been developed and extensive in vitro studies have been carried out (Kirkland et al., 2015a; Richter et al., 2011, 2014). It was, e.g., shown that Bungowannah virus shows a much broader host range in vitro than all other known pestiviruses with several cell lines supporting viral replication. Susceptible cell lines included cells from humans, African green monkey, bats, mice, and several farm animals (Kirkland et al., 2015a). Molecular studies using the BVDV type 1 "CP7" infectious cDNA clone to place Bungowannah virus genes encoding key structural and nonstructural proteins into the BVDV backbone showed that the chimeric viruses remain fully functional, including the type-I interferon inhibition by N^{pro} (Richter et al., 2011, 2014). Yet, the abovementioned broad host range was not retained when all structural proteins of the BVDV backbone were replaced by Bungowannah virus. The latter led to the conclusion that also the replication machinery of Bungowannah virus is essential for its capacity to infect this wide range of cell cultures (Richter et al., 2011).

To better understand pathogenesis and disease dynamics, experimental infections have been carried out. Following infection of pigs at day 35 of gestation, stillbirths, and mummifications prevailed, while infection at days 55 and 75 led to much lesser reduction in viable piglets. Piglets born alive from sows infected at 35 days of pregnancy showed persistent infection and were stunted and weak. Piglets after infection of their mothers at day 55 of pregnancy were chronically infected but cleared the virus after several months. Acute myocarditis was associated with infection in late gestation (day 90) (Finlaison et al., 2010). Experimental infection of weaner pigs resulted in a transient infection with moderate viral shedding and seroconversion from about 10 days postinoculation (Finlaison et al., 2012). No apparent clinical signs were observed, and transmission to naïve pen-mates did not occur (which did occur when naïve animals were cohoused with persistently or chronically infected animals).

A study with nonpregnant sheep and cattle showed infection after intranasal challenge with Bungowannah virus but no disease (reviewed by Kirkland et al., 2015b), and a calf got infected when cohoused with a PI piglet. Inoculation of seven cows at 53–65 days of pregnancy resulted in three calves that showed high neutralizing antibody titers in precolostral sera and Bungowannah virus genome in skin biopsy samples. However, only one of these calves showed low-level viremia, and no virus could be detected in any other tissues of this calf or any of the others (Kirkland et al., 2015b).

So far, Bungowannah virus was not reported from any other country or region, and origin and reservoir remain unclear. A published surveillance action of Bungowannah virus in Upper Midwestern United States by specific RT-qPCRs of tissues from undiagnosed cases of abortion or respiratory disease did not give any positive results (Abrahante et al., 2014).

3.7 Pestivirus Infections Outside the *Artiodactyla* Order

Only recently, a pestiviral sequence has been detected in a member of the order *Chiroptera*. Analysis of the virome from *Rhinolophus affinis* bats in China revealed a 5-kb contig of a putative pestivirus (Wu et al., 2012) named *R. affinis* pestivirus (RaPV). The obtained partial sequence of RaPV showed only 46.5%–47.3% nucleotide sequence identities to any other known pestivirus sequence in the databases, indicating that this putative novel species occurred evolutionarily very distant from the known pestiviruses (see Fig. 2).

Furthermore, metagenomic sequencing of Norway rats in New York City that were carried out to give an unbiased insight into the infectious agents present in the commensal rat population also identified a complete pestivirus genome, named Norway rat pestivirus, NrPV (Firth et al., 2014). The viral sequence showed a standard 5′-NTR including formation of an internal ribosomal entry site. The polyprotein sequence of 3993 amino acids shared a maximum identity of 60% with other pestivirus polyproteins and was predicted to encode all known standard pestivirus proteins. Firth et al. (2014) described that the N^{pro}-encoding region of NrPV was substantially longer than that of other pestiviruses (273 amino acids vs 168) but showed the conserved catalytic residues (though present at other positions).

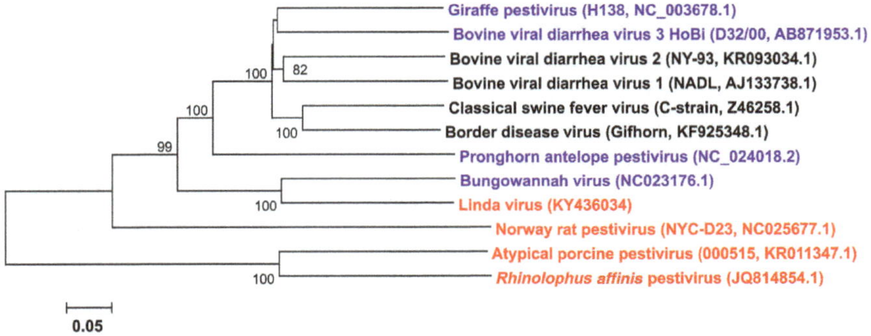

Fig. 2 Neighbor-joining tree based on NS2–3 sequences of representative pestivirus strains, isolate name and accession number are given in parentheses. The classical pestiviruses are printed in *black*, atypical pestiviruses detected prior to 2010 are highlighted in *blue*, and viruses isolated since 2010 are shown in *red*. Statistical support for nodes was obtained by bootstrapping (1000 replicates); only values ≥50% are shown. The *scale bar* indicates nucleotide substitutions per site.

3.8 Atypical Porcine Pestiviruses

In 2015, a putative new porcine pestivirus was found in the United States through metagenomic analyses of porcine samples that were subjected to NGS as part of a porcine reproductive and respiratory syndrome virus metagenomic sequencing project (Hause et al., 2015). Six contigs gave hits with pestiviruses and subsequent resequencing coupled with PCRs to close the gaps in the assembly resulted in a single 11,276 bp contig encoding a predicted 3635 amino acid polyprotein with 68% identity to RaPV and app. 40% identity to complete BVDV, BDV, and CSFV amino acid sequences. The 5′-NTR seemed considerably shorter than that of other pestiviruses, but confirmation by random amplification of cDNA ends failed. The putative catalytic and cleavage sites of N^{pro} where identified but beyond that no significant similarity was found (e.g., only 9%–18% pairwise amino acid identity to other N^{pro} sequences). Likewise, a RNase T2 domain was identified in the E^{rns} despite low identity with other pestivirus E^{rns} proteins (32.9%–39%). Low identities were also revealed for E1 and E2. Furthermore, both APPV and RaPV seem to have a significantly shorter E2 protein (Hause et al., 2015). Protease domains of NS2 were identified that differed from that of other pestiviruses, and the same was true for active sites of other nonstructural proteins. Subsequent studies by tailored RT-PCRs and an E^{rns} antibody ELISA revealed that that virus appeared widespread in the United States with over 90% seroprevalence. The virus was named atypical porcine pestivirus (APPV).

Based on next-generation sequencing of samples from piglets with congenital tremor on an Illumina MiSeq platform, APPV was later shown to correlate with this condition in newborn piglets, and an infection model was established with intranasal/intravenous inoculation of pregnant sows at either 45 or 62 days of pregnancy and simultaneous direct inoculation of fetal amniotic vesicles (Arruda et al., 2016). The inoculum consisted of serum from a viraemic animal. This virus showed 88% and 94.6% nucleotide and amino acid identity with the above-mentioned virus, respectively. It was demonstrated that 57%–100% of the inoculated fetuses showed congenital tremor after farrowing, while control animals were negative. From the inoculated litters, both unaffected and congenital tremor piglets were found positive by RT-PCR. The virus was found in experimental samples from the brain stem, spinal cord, cerebellum, cerebrum, mesenteric and tracheobronchial lymph nodes, feces, serum, nasal swabs, kidney, thymus, hear, spleen, and whole blood. Tailored RT-qPCR assays targeting the NS3 region of the genome demonstrated the virus in central nervous tissues from animals with congenital tremor but not in central nervous tissues from unaffected piglets. However, the virus was also found in 6% of routine diagnostic submission from animals with varied clinical signs (21 out of 362 samples) (Arruda et al., 2016).

Following the description of the virus in the United States, APPV was also found in European countries including Germany (Beer et al., 2016; Postel et al., 2016), Italy (Postel et al., 2017), Austria (Schwarz et al., 2017), the Netherlands (de Groof et al., 2016), Sweden (Blomström et al., 2016), and Great Britain (Williamson, 2017). Indications exist that the virus is also present in China. Several studies confirmed the link to congenital tremor (see details later).

Postel et al. (2016) showed the existence of APPV in Germany with an estimated individual prevalence of 2.4% and 10% at farm level. Viral RNA was found by RT-qPCR targeting the NS3- and NS4B-encoding regions in apparently healthy adult pigs but also newborn piglets affected by congenital tremor. Retrospective analyses demonstrated the virus in samples from piglets with congenital tremor of unknown etiology back in 2007. With regard to tissue tropism, viral RNA was especially found in the glands of the *Arcus palatoglossus*, and the mandibular lymph nodes. Moreover, cerebellum (inner granular cell layer unlike with CSFV), trigeminal and spinal ganglia but also peripheral nerves were shown to contain viral genome and antigen. Other tissues that show high viral loads in CSFV infection, like spleen or kidney, show much less APPV genome. APPV sequences obtained in this study (full genome obtained on an Illumina HiSeq) showed high variability across Germany, and an 88.2% identity on nucleotide level to the United States APPV.

High variability of German strains was confirmed by Beer et al. (2016). In the respective study that comprised tonsil and serum samples from different parts of the pig-producing sector, high prevalences were detected in both tonsils (9%) and sera (up to 22%), but in this sample set, no link to a clinical syndrome was observed. Phylogenetic studies were done with 21 partial NS5B sequences and revealed high variability with local clusters. Another full-length genome (obtained with an Illumina MiSeq platform) showed 87.5% identity to the whole genome reported by Hause et al. (2015). The large deletion in the N-terminal region of the APPV E2 that leads to truncation of the envelope protein and affects a known fusion peptide region as well as a putative disulfide isomerase motif was confirmed. Also, the linear B cell antigenic "TAVSPTTLR" epitope (Liu et al., 2006) is missing in the APPV E2. This kind of epitope or a similar sequence is usually present in classical pestiviruses. However, another virus missing this epitope is NrPV.

In a study reported by de Groof et al. (2016), several strains of APPV were found in Dutch (and Spanish) piglets with clinical signs of congenital tremor, while piglets on a farm with no history of congenital tremor were negative. Clinical cases since 2012 were found APPV positive. The viruses showed quite high variability but clustered in different groups. The authors were able to prove vertical transmission after intramuscular inoculation of pregnant sows at day 32 of pregnancy. In contrast to the study reported by Arruda et al., sows were shown to be viraemic upon inoculation. Long-term follow-up of affected animals showed a declining viral load in serum after 5 month but continued shedding via feces which could explain persistence on the farm level.

Follow-up of outbreaks of congenital tremor in several Austrian piglet-producing farms showed the presence of APPV also in Austria both recently and retrospectively in 2013 collected samples (Schwarz et al., 2017). A full-length genome and several partial sequences were determined showing a 90% identity to the United States APPV sequences and 92% to the above-mentioned German sequences. The genomes were identified in different body fluids and tissues including the central nervous systems of diseased piglets which is in line with previously reported experience (see earlier). Interestingly, the recovering piglets showed declining, probably maternally derived antibodies but persistent viral loads in saliva but also semen. Both of the latter would indicate a persistent infection. One farm was affected both in 2013 and 2016 with almost the same virus. Here, viral persistence in the herd is highly likely.

Concluding, APPV is the latest and most distantly related novel pestivirus, and seems to be at least one cause of congenital tremor type A-II that has long been regarded as a transmissible disease of unknown viral etiology. While the presence of APPV genomes in newborn piglets clearly correlates with disease, viremia in adult animals is not necessarily linked to disease. Generally, litters from gilts are more often affected than litters from sows. It can be speculated that the infection of the central nervous system is transient as the piglets with congenital tremor usually recover. The detection of APPV genomes in clinically healthy animals with rather high prevalence could indicate long-term persistence in postnatally infected animals without clinical impact. Sows seem to mount immunity that protects subsequent litters from congenital tremor.

Very recently, it was shown that APPV does—as expected—not interfere with CSF diagnosis (Postel et al., 2017). Neither the well-established RT-qPCR systems nor routine serological assays gave false-positive CSF results when testing APPV-positive materials.

Further studies to elucidate pathogenesis, possible control options, and detection are necessary for a better understanding of the biology of APPV.

3.9 LINDA Virus

Very recently, another novel porcine pestivirus has been described in Austria (Lamp et al., 2017). It was discovered in a piglet-producing farm with severe congenital tremor cases and was provisionally termed LINDA (Lateral-Shaking Inducing NeuroDegenerative Agent) virus. The affected animals showed lateral shaking and inability to suckle. Severe hypomyelination was observed in the white matter of the spinal cord, and viral antigen was demonstrated in the brain of affected piglets. LINDA virus (see Fig. 2) shows only 60% identity (<54% amino acid identity) with classical pestiviruses or APPV, but 68% with Bungowannah virus (69% amino acid identity). LINDA virus infected SK6 or MDBK cells could be visualized using a BVDV E2-specific antibody. Thus, unlike the above-mentioned APPV (that was ruled out for these cases), LINDA virus may have an impact on CSFV surveillance.

4. CONCLUSIONS AND IMPLICATIONS

The tree of members of the genus *Pestivirus* is constantly growing, and the area of NGS-based metagenomics has paved the way for the discovery of several completely new and unexpected pestiviruses. The description of those novel pestiviruses and variants will also impact on pestivirus classification, which has to be completely updated and adapted to the number of available novel whole genomes. The proposal by Liu et al. from 2009 suggests nine different pestivirus species. However, already now, new pestiviruses emerged that may need proper classification, and it can be expected that metagenomics will result in further members of the genus *Pestivirus*, may be also in new animal hosts.

REFERENCES

Abrahante, J.E., Zhang, J.W., Rossow, K., Zimmerman, J.J., Murtaugh, M.P., 2014. Surveillance of Bungowannah pestivirus in the upper Midwestern USA. Transbound. Emerg. Dis. 61, 375.

Arnal, M., Fernandez-de-Luco, D., Riba, L., Maley, M., Gilray, J., Willoughby, K., Vilcek, S., Nettleton, P.F., 2004. A novel pestivirus associated with deaths in Pyrenean chamois (Rupicapra pyrenaica pyrenaica). J. Gen. Virol. 85, 3653.

Arruda, B.L., Arruda, P.H., Magstadt, D.R., Schwartz, K.J., Dohlman, T., Schleining, J.A., Patterson, A.R., Visek, C.A., Victoria, J.G., 2016. Identification of a divergent lineage porcine pestivirus in nursing piglets with congenital tremors and reproduction of disease following experimental inoculation. PLoS One 11, e0150104.

Baker, J.C., 1995. The clinical manifestations of bovine viral diarrhea infection. Vet. Clin. North Am. Food Anim. Pract. 11, 425.

Bauermann, F.V., Ridpath, J.F., 2015. HoBi-like viruses—the typical 'atypical bovine pestivirus'. Anim. Health Res. Rev. 16, 64.

Becher, P., Orlich, M., Shannon, A.D., Horner, G., Konig, M., Thiel, H.J., 1997. Phylogenetic analysis of pestiviruses from domestic and wild ruminants. J. Gen. Virol. 78 (Pt. 6), 1357.

Becher, P., Avalos Ramirez, R., Orlich, M., Cedillo Rosales, S., König, M., Schweizer, M., Stalder, H., Schirrmeier, H., Thiel, H.J., 2003. Genetic and antigenic characterization of novel pestivirus genotypes: implications for classification. Virology 311, 96.

Becher, P., Schmeiser, S., Oguzoglu, T.C., Postel, A., 2012. Complete genome sequence of a novel pestivirus from sheep. J. Virol. 86, 11412.

Beer, M., Wernike, K., Dräger, C., Höper, D., Pohlmann, A., Bergermann, C., Schröder, C., Klinkhammer, S., Blome, S., Hoffmann, B., 2016. High prevalence of highly variable atypical porcine pestiviruses found in Germany. Transbound. Emerg. Dis. https://doi.org/10.1111/tbed.12532.

Blomström, A.L., Fossum, C., Wallgren, P., Berg, M., 2016. Viral metagenomic analysis displays the co-infection situation in healthy and PMWS affected pigs. PLoS One 11, e0166863.

Braun, U., Hilbe, M., Janett, F., Hassig, M., Zanoni, R., Frei, S., Schweizer, M., 2015. Transmission of border disease virus from a persistently infected calf to seronegative heifers in early pregnancy. BMC Vet. Res. 11, 43.

Carman, S., Carr, N., DeLay, J., Baxi, M., Deregt, D., Hazlett, M., 2005. Bovine viral diarrhea virus in alpaca: abortion and persistent infection. J. Vet. Diagn. Investig. 17, 589.

Ciulli, S., Purpari, G., Agnello, S., Di Marco, P., Di Bella, S., Volpe, E., Mira, F., de Aguiar Saldanha Pinheiro, A.C., Vullo, S., Guercio, A., 2017. Evidence for Tunisian-like pestiviruses presence in small ruminants in Italy since 2007. Transbound. Emerg. Dis. 64 (4), 1243–1253. https://doi.org/10.1111/tbed.12498. Epub 2016 Apr 3.

Cortez, A., Heinemann, M.B., de Castro, A.M.M., Soares, R.M., Pinto, A.M.V., Alfieri, A.A., Flores, E.F., Leite, R.C., Richtzenhain, L.J., 2006. Genetic characterization of Brazilian bovine viral diarrhea virus isolates by partial nucleotide sequencing of the 5′-UTR region. Pesqui. Vet. Bras. 26, 211.

Decaro, N., Lucente, M.S., Mari, V., Cirone, F., Cordioli, P., Camero, M., Sciarretta, R., Losurdo, M., Lorusso, E., Buonavoglia, C., 2011. Atypical pestivirus and severe respiratory disease in calves, Europe. Emerg. Infect. Dis. 17, 1549.

Decaro, N., Lucente, M.S., Mari, V., Sciarretta, R., Pinto, P., Buonavoglia, D., Martella, V., Buonavoglia, C., 2012a. Hobi-like pestivirus in aborted bovine fetuses. J. Clin. Microbiol. 50, 509.

Decaro, N., Mari, V., Lucente, M.S., Sciarretta, R., Moreno, A., Armenise, C., Losurdo, M., Camero, M., Lorusso, E., Cordioli, P., Buonavoglia, C., 2012b. Experimental infection of cattle, sheep and pigs with 'Hobi'-like pestivirus. Vet. Microbiol. 155, 165.

Decaro, N., Mari, V., Pinto, P., Lucente, M.S., Sciarretta, R., Cirone, F., Colaianni, M.L., Elia, G., Thiel, H.J., Buonavoglia, C., 2012c. Hobi-like pestivirus: both biotypes isolated from a diseased animal. J. Gen. Virol. 93, 1976.

Decaro, N., Mari, V., Lucente, M.S., Sciarretta, R., Elia, G., Ridpath, J.F., Buonavoglia, C., 2013. Detection of a Hobi-like virus in archival samples suggests circulation of this emerging pestivirus species in Europe prior to 2007. Vet. Microbiol. 167, 307.

Decaro, N., Losurdo, M., Larocca, V., Lucente, M.S., Mari, V., Varello, K., Patruno, G., Camero, M., Sciarra, M., Occhiogrosso, L., Tempesta, M., Iulini, B., Buonavoglia, C., 2015. HoBi-like pestivirus experimental infection in pregnant ewes: reproductive disorders and generation of persistently infected lambs. Vet. Microbiol. 178, 173.
Decaro, N., Lucente, M.S., Losurdo, M., Larocca, V., Elia, G., Occhiogrosso, L., Marino, P.A., Cirone, F., Buonavoglia, C., 2016. HoBi-like pestivirus and its impact on cattle productivity. Transbound. Emerg. Dis. 63, 469.
de Groof, A., Deijs, M., Guelen, L., van Grinsven, L., van Os-Galdos, L., Vogels, W., Derks, C., Cruijsen, T., Geurts, V., Vrijenhoek, M., Suijskens, J., van Doorn, P., van Leengoed, L., Schrier, C., van der Hoek, L., 2016. Atypical porcine pestivirus: a possible cause of congenital tremor type A-II in newborn piglets. Viruses 8, 271.
Dekker, A., Wensvoort, G., Terpstra, C., 1995. Six antigenic groups within the genus pestivirus as identified by cross neutralization assays. Vet. Microbiol. 47, 317.
Fernandez-Sirera, L., Cabezon, O., Allepuz, A., Rosell, R., Riquelme, C., Serrano, E., Lavin, S., Marco, I., 2012. Two different epidemiological scenarios of border disease in the populations of Pyrenean chamois (Rupicapra p. pyrenaica) after the first disease outbreaks. PLoS One 7, e51031.
Finlaison, D.S., King, K.R., Frost, M.J., Kirkland, P.D., 2009. Field and laboratory evidence that Bungowannah virus, a recently recognised pestivirus, is the causative agent of the porcine myocarditis syndrome (PMC). Vet. Microbiol. 136, 259.
Finlaison, D.S., Cook, R.W., Srivastava, M., Frost, M.J., King, K.R., Kirkland, P.D., 2010. Experimental infections of the porcine foetus with Bungowannah virus, a novel pestivirus. Vet. Microbiol. 144, 32.
Finlaison, D.S., King, K.R., Gabor, M., Kirkland, P.D., 2012. An experimental study of Bungowannah virus infection in weaner aged pigs. Vet. Microbiol. 160, 245.
Firth, C., Bhat, M., Firth, M.A., Williams, S.H., Frye, M.J., Simmonds, P., Conte, J.M., Ng, J., Garcia, J., Bhuva, N.P., Lee, B., Che, X., Quan, P.L., Lipkin, W.I., 2014. Detection of zoonotic pathogens and characterization of novel viruses carried by commensal Rattus norvegicus in New York City. MBio 5, e01933.
Giammarioli, M., Ridpath, J.F., Rossi, E., Bazzucchi, M., Casciari, C., De Mia, G.M., 2015. Genetic detection and characterization of emerging HoBi-like viruses in archival foetal bovine serum batches. Biologicals 43, 220.
Goyal, S.M., Bouljihad, M., Haugerud, S., Ridpath, J.F., 2002. Isolation of bovine viral diarrhea virus from an alpaca. J. Vet. Diagn. Investig. 14, 523.
Haider, N., Rahman, M.S., Khan, S.U., Mikolon, A., Gurley, E.S., Osmani, M.G., Shanta, I.S., Paul, S.K., Macfarlane-Berry, L., Islam, A., Desmond, J., Epstein, J.H., Daszak, P., Azim, T., Luby, S.P., Zeidner, N., Rahman, M.Z., 2014. Identification and epidemiology of a rare HoBi-like pestivirus strain in Bangladesh. Transbound. Emerg. Dis. 61, 193.
Harasawa, R., Giangaspero, M., Ibata, G., Paton, D.J., 2000. Giraffe strain of pestivirus: its taxonomic status based on the 5′-untranslated region. Microbiol. Immunol. 44, 915.
Hause, B.M., Collin, E.A., Peddireddi, L., Yuan, F., Chen, Z., Hesse, R.A., Gauger, P.C., Clement, T., Fang, Y., Anderson, G., 2015. Discovery of a novel putative atypical porcine pestivirus in pigs in the USA. J. Gen. Virol. 96, 2994.
Houe, H., 1999. Epidemiological features and economical importance of bovine virus diarrhoea virus (BVDV) infections. Vet. Microbiol. 64, 89.
Houe, H., 2003. Economic impact of BVDV infection in dairies. Biologicals 31, 137.
Hurtado, A., Aduriz, G., Gomez, N., Oporto, B., Juste, R.A., Lavin, S., Lopez-Olvera, J.R., Marco, I., 2004. Molecular identification of a new pestivirus associated with increased mortality in the Pyrenean Chamois (Rupicapra pyrenaica pyrenaica) in Spain. J. Wildl. Dis. 40, 796.

Kampa, J., Alenius, S., Emanuelson, U., Chanlun, A., Aiumlamai, S., 2009. Bovine herpesvirus type 1 (BHV-1) and bovine viral diarrhoea virus (BVDV) infections in dairy herds: self clearance and the detection of seroconversions against a new atypical pestivirus. Vet. J. 182, 223.
Kirkland, P.D., Frost, M.J., Finlaison, D.S., King, K.R., Ridpath, J.F., Gu, X., 2007. Identification of a novel virus in pigs—Bungowannah virus: a possible new species of pestivirus. Virus Res. 129, 26.
Kirkland, P.D., Frost, M.J., King, K.R., Finlaison, D.S., Hornitzky, C.L., Gu, X., Richter, M., Reimann, I., Dauber, M., Schirrmeier, H., Beer, M., Ridpath, J.F., 2015a. Genetic and antigenic characterization of Bungowannah virus, a novel pestivirus. Vet. Microbiol. 178, 252.
Kirkland, P.D., Read, A.J., Frost, M.J., Finlaison, D.S., 2015b. Bungowannah virus—a probable new species of pestivirus—what have we found in the last 10 years? Anim. Health Res. Rev. 16, 60.
Lamp, B., Schwarz, L., Hogler, S., Riedel, C., Sinn, L., Rebel-Bauder, B., Weissenbock, H., Ladinig, A., Rumenapf, T., 2017. Novel pestivirus species in pigs, Austria, 2015. Emerg. Infect. Dis. 23, 1176.
Lanyon, S.R., Hill, F.I., Reichel, M.P., Brownlie, J., 2014. Bovine viral diarrhoea: pathogenesis and diagnosis. Vet. J. 199, 201.
Larska, M., Polak, M.P., Riitho, V., Strong, R., Belak, S., Alenius, S., Uttenthal, A., Liu, L., 2012. Kinetics of single and dual infection of calves with an Asian atypical bovine pestivirus and a highly virulent strain of bovine viral diarrhoea virus 1. Comp. Immunol. Microbiol. Infect. Dis. 35, 381.
Lindenbach, B.D., Murray, C.L., Thiel, H.J., Rice, C.M., 2013. Flaviviridae. In: Knipe, D.M., Howley, P.M. (Eds.), Fields Virology. Lippincott Williams and Wilkins, Philadelphia, PA, p. 712.
Liu, S., Tu, C., Wang, C., Yu, X., Wu, J., Guo, S., Shao, M., Gong, Q., Zhu, Q., Kong, X., 2006. The protective immune response induced by B cell epitope of classical swine fever virus glycoprotein E2. J. Virol. Methods 134, 125.
Liu, L., Kampa, J., Belak, S., Baule, C., 2009a. Virus recovery and full-length sequence analysis of atypical bovine pestivirus Th/04_KhonKaen. Vet. Microbiol. 138, 62.
Liu, L., Xia, H., Wahlberg, N., Belak, S., Baule, C., 2009b. Phylogeny, classification and evolutionary insights into pestiviruses. Virology 385, 351.
Mao, L., Li, W., Zhang, W., Yang, L., Jiang, J., 2012. Genome sequence of a novel Hobi-like pestivirus in China. J. Virol. 86, 12444.
Marco, I., Lopez-Olvera, J.R., Rosell, R., Vidal, E., Hurtado, A., Juste, R., Pumarola, M., Lavin, S., 2007. Severe outbreak of disease in the southern chamois (Rupicapra pyrenaica) associated with border disease virus infection. Vet. Microbiol. 120, 33.
Marco, I., Rosell, R., Cabezon, O., Mentaberre, G., Casas, E., Velarde, R., Lopez-Olvera, J.R., Hurtado, A., Lavin, S., 2008. Epidemiological study of border disease virus infection in Southern chamois (Rupicapra pyrenaica) after an outbreak of disease in the Pyrenees (NE Spain). Vet. Microbiol. 127, 29.
Marco, I., Rosell, R., Cabezon, O., Mentaberre, G., Casas, E., Velarde, R., Lavin, S., 2009. Border disease virus among chamois, Spain. Emerg. Infect. Dis. 15, 448.
Marco, I., Cabezon, O., Rosell, R., Fernandez-Sirera, L., Allepuz, A., Lavin, S., 2011. Retrospective study of pestivirus infection in Pyrenean chamois (Rupicapra pyrenaica) and other ungulates in the Pyrenees (NE Spain). Vet. Microbiol. 149, 17.
Martin, C., Duquesne, V., Adam, G., Belleau, E., Gauthier, D., Champion, J.L., Saegerman, C., Thiery, R., Dubois, E., 2015. Pestiviruses infections at the wild and domestic ruminants interface in the French Southern Alps. Vet. Microbiol. 175, 341.

Maya, L., Puentes, R., Reolon, E., Acuna, P., Riet, F., Rivero, R., Cristina, J., Colina, R., 2016. Molecular diversity of bovine viral diarrhea virus in uruguay. Arch. Virol. 161, 529.

McOrist, S., Thornton, E., Peake, A., Walker, R., Robson, S., Finlaison, D., Kirkland, P., Reece, R., Ross, A., Walker, K., Hyatt, A., Morrissy, C., 2004. An infectious myocarditis syndrome affecting late-term and neonatal piglets. Aust. Vet. J. 82, 509.

Mishra, N., Rajukumar, K., Pateriya, A., Kumar, M., Dubey, P., Behera, S.P., Verma, A., Bhardwaj, P., Kulkarni, D.D., Vijaykrishna, D., Reddy, N.D., 2014. Identification and molecular characterization of novel and divergent HoBi-like pestiviruses from naturally infected cattle in India. Vet. Microbiol. 174, 239.

Moennig, V., Becher, P., 2015. Pestivirus control programs: how far have we come and where are we going? Anim. Health Res. Rev. 16, 83.

Moennig, V., Houe, H., Lindberg, A., 2005. BVD control in Europe: current status and perspectives. Anim. Health Res. Rev. 6, 63.

Neill, J.D., Ridpath, J.F., Fischer, N., Grundhoff, A., Postel, A., Becher, P., 2014. Complete genome sequence of pronghorn virus, a pestivirus. Genome Announc. 2 (3). pii: e00575-14. https://doi.org/10.1128/genomeA.00575-14.

Nettleton, P.F., Gilmour, J.S., Herring, J.A., Sinclair, J.A., 1992. The production and survival of lambs persistently infected with border disease virus. Comp. Immunol. Microbiol. Infect. Dis. 15, 179.

Oguzoglu, T.C., Tan, M.T., Toplu, N., Demir, A.B., Bilge-Dagalp, S., Karaoglu, T., Ozkul, A., Alkan, F., Burgu, I., Haas, L., Greiser-Wilke, I., 2009. Border disease virus (BDV) infections of small ruminants in Turkey: a new BDV subgroup? Vet. Microbiol. 135, 374.

Peletto, S., Zuccon, F., Pitti, M., Gobbi, E., Marco, L.D., Caramelli, M., Masoero, L., Acutis, P.L., 2012. Detection and Phylogenetic Analysis of an Atypical Pestivirus, Strain IZSPLV_To. Res. Vet. Sci. 92, 147–150.

Pioz, M., Loison, A., Gibert, P., Dubray, D., Menaut, P., Le Tallec, B., Artois, M., Gilot-Fromont, E., 2007. Transmission of a pestivirus infection in a population of Pyrenean chamois. Vet. Microbiol. 119, 19.

Postel, A., Schmeiser, S., Oguzoglu, T.C., Indenbirken, D., Alawi, M., Fischer, N., Grundhoff, A., Becher, P., 2015. Close relationship of ruminant pestiviruses and classical swine fever virus. Emerg. Infect. Dis. 21, 668.

Postel, A., Hansmann, F., Baechlein, C., Fischer, N., Alawi, M., Grundhoff, A., Derking, S., Tenhundfeld, J., Pfankuche, V.M., Herder, V., Baumgartner, W., Wendt, M., Becher, P., 2016. Presence of atypical porcine pestivirus (APPV) genomes in newborn piglets correlates with congenital tremor. Sci. Rep. 6, 27735.

Postel, A., Meyer, D., Petrov, A., Becher, P., 2017. Recent emergence of a novel porcine pestivirus: interference with classical swine fever diagnosis? Emerg. Microbes Infect. 6, e19.

Richter, M., Reimann, I., Wegelt, A., Kirkland, P.D., Beer, M., 2011. Complementation studies with the novel "Bungowannah" virus provide new insights in the compatibility of pestivirus proteins. Virology 418, 113.

Richter, M., Reimann, I., Schirrmeier, H., Kirkland, P.D., Beer, M., 2014. The viral envelope is not sufficient to transfer the unique broad cell tropism of Bungowannah virus to a related pestivirus. J. Gen. Virol. 95, 2216.

Ridpath, J.F., 2010. Bovine viral diarrhea virus: global status. Vet. Clin. North Am. Food Anim. Pract. 26, 105.

Ridpath, J.F., 2015. Emerging pestiviruses infecting domestic and wildlife hosts. Anim. Health Res. Rev. 16, 55.

Schirrmeier, H., Strebelow, G., Depner, K., Hoffmann, B., Beer, M., 2004. Genetic and antigenic characterization of an atypical pestivirus isolate, a putative member of a novel pestivirus species. J. Gen. Virol. 85, 3647.

Schwarz, L., Riedel, C., Hogler, S., Sinn, L.J., Voglmayr, T., Wochtl, B., Dinhopl, N., Rebel-Bauder, B., Weissenbock, H., Ladinig, A., Rümenapf, T., Lamp, B., 2017. Congenital infection with atypical porcine pestivirus (APPV) is associated with disease and viral persistence. Vet. Res. 48, 1.
Shi, H., Kan, Y., Yao, L., Leng, C., Tang, Q., Ji, J., Sun, S., 2016. Identification of natural infections in sheep/goats with HoBi-like pestiviruses in China. Transbound. Emerg. Dis. 63, 480.
Silveira, S., Weber, M.N., Mosena, A.C., da Silva, M.S., Streck, A.F., Pescador, C.A., Flores, E.F., Weiblen, R., Driemeier, D., Ridpath, J.F., Canal, C.W., 2017. Genetic diversity of Brazilian bovine pestiviruses detected between 1995 and 2014. Transbound. Emerg. Dis. 64, 613.
Simmonds, P., Becher, P., Bukh, J., Gould, E.A., Meyers, G., Monath, T., Muerhoff, S., Pletnev, A., Rico-Hesse, R., Smith, D.B., Stapleton, J.T., Ictv Report, C., 2017. ICTV virus taxonomy profile: Flaviviridae. J. Gen. Virol. 98, 2.
Stahl, K., Alenius, S., 2012. BVDV control and eradication in Europe—an update. Jpn. J. Vet. Res. 60 (Suppl), S31.
Stahl, K., Kampa, J., Alenius, S., Persson Wadman, A., Baule, C., Aiumlamai, S., Belak, S., 2007. Natural infection of cattle with an atypical 'HoBi'-like pestivirus—implications for BVD control and for the safety of biological products. Vet. Res. 38, 517.
Stalder, H.P., Meier, P., Pfaffen, G., Wageck-Canal, C., Rufenacht, J., Schaller, P., Bachofen, C., Marti, S., Vogt, H.R., Peterhans, E., 2005. Genetic heterogeneity of pestiviruses of ruminants in Switzerland. Prev. Vet. Med. 72, 37.
Terpstra, C., 1980. Border disease: a persistent virus infection in sheep (author's transl). Tijdschr. Diergeneeskd. 105, 650.
Thabti, F., Fronzaroli, L., Dlissi, E., Guibert, J.M., Hammami, S., Pepin, M., Russo, P., 2002. Experimental model of Border Disease Virus infection in lambs: comparative pathogenicity of pestiviruses isolated in France and Tunisia. Vet. Res. 33, 35.
Thabti, F., Letellier, C., Hammami, S., Pepin, M., Ribiere, M., Mesplede, A., Kerkhofs, P., Russo, P., 2005. Detection of a novel border disease virus subgroup in Tunisian sheep. Arch. Virol. 150, 215.
Thornley, M., 2004. Mystery piglet deaths prompt animal health investigation. Aust. Vet. J. 82, 448.
Vilcek, S., Paton, D.J., Durkovic, B., Strojny, L., Ibata, G., Moussa, A., Loitsch, A., Rossmanith, W., Vega, S., Scicluna, M.T., Paifi, V., 2001. Bovine viral diarrhoea virus genotype 1 can be separated into at least eleven genetic groups. Arch. Virol. 146, 99.
Vilcek, S., Durkovic, B., Kolesarova, M., Greiser-Wilke, I., Paton, D., 2004. Genetic diversity of international bovine viral diarrhoea virus (BVDV) isolates: identification of a new BVDV-1 genetic group. Vet. Res. 35, 609.
Vilcek, S., Ridpath, J.F., Van Campen, H., Cavender, J.L., Warg, J., 2005. Characterization of a novel pestivirus originating from a pronghorn antelope. Virus Res. 108, 187.
Vilcek, S., Willoughby, K., Nettleton, P., Becher, P., 2010. Complete genomic sequence of a border disease virus isolated from Pyrenean chamois. Virus Res. 152, 164.
Weber, M.N., Mosena, A.C., Simoes, S.V., Almeida, L.L., Pessoa, C.R., Budaszewski, R.F., Silva, T.R., Ridpath, J.F., Riet-Correa, F., Driemeier, D., Canal, C.W., 2016. Clinical presentation resembling mucosal disease associated with 'HoBi'-like pestivirus in a field outbreak. Transbound. Emerg. Dis. 63, 92.
Westaway, E.G., Brinton, M.A., Gaidamovich, S., Horzinek, M.C., Igarashi, A., Kaariainen, L., Lvov, D.K., Porterfield, J.S., Russell, P.K., Trent, D.W., 1985. Togaviridae. Intervirology 24, 125.
Williamson, S., 2017. Congenital tremor associated with atypical porcine pestivirus. Vet. Rec. 180, 42.

Wu, Z., Ren, X., Yang, L., Hu, Y., Yang, J., He, G., Zhang, J., Dong, J., Sun, L., Du, J., Liu, L., Xue, Y., Wang, J., Yang, F., Zhang, S., Jin, Q., 2012. Virome analysis for identification of novel mammalian viruses in bat species from Chinese provinces. J. Virol. 86, 10999.

Xia, H., Vijayaraghavan, B., Belak, S., Liu, L., 2011. Detection and identification of the atypical bovine pestiviruses in commercial foetal bovine serum batches. PLoS One 6, e28553.

Xia, H., Larska, M., Giammarioli, M., De Mia, G.M., Cardeti, G., Zhou, W., Alenius, S., Belak, S., Liu, L., 2013. Genetic detection and characterization of atypical bovine pestiviruses in foetal bovine sera claimed to be of Australian origin. Transbound. Emerg. Dis. 60, 284.

CHAPTER SEVEN

Bluetongue Virus: From BTV-1 to BTV-27

Guillaume Belbis[1,2], Stéphan Zientara[1,2], Emmanuel Bréard, Corinne Sailleau, Grégory Caignard, Damien Vitour, Houssam Attoui
UMR1161 Virologie, ANSES, INRA, Ecole Nationale Vétérinaire d'Alfort, Université Paris-Est, Maisons-Alfort, France
[1]Corresponding authors: e-mail address: guillaume.belbis@vet-alfort.fr; stephan.zientara@vet-alfort.fr

Contents

Abstract

Bluetongue virus (BTV) is the type species of genus *Orbivirus* within family *Reoviridae*. Bluetongue virus is transmitted between its ruminant hosts by the bite of *Culicoides* spp. midges. Severe BT cases are characterized by symptoms including hemorrhagic fever, particularly in sheep, loss of productivity, and death. To date, 27 BTV serotypes have been documented. These include novel isolates of atypical BTV, which have been almost fully characterized using deep sequencing technologies and do not rely on *Culicoides* vectors for their transmission among hosts. Due to its high economic impact, BT is an Office International des Epizooties (OIE) listed disease that is strictly controlled in international commercial exchanges. During the 20th century, BTV has been endemic in

[2] Both authors contributed equally to this work.

Advances in Virus Research, Volume 99
ISSN 0065-3527
https://doi.org/10.1016/bs.aivir.2017.08.003

subtropical regions. In the last 15 years, novel strains of nine "typical" BTV serotypes (1, 2, 4, 6, 8, 9, 11, 14, and 16) invaded Europe, some of which caused disease in naive sheep and unexpectedly in bovine herds (particularly serotype 8). Over the past few years, three novel "atypical" serotypes (25–27) were characterized during sequencing studies of animal samples from Switzerland, Kuwait, and France, respectively. Classical serotype-specific inactivated vaccines, although expensive, were very successful in controlling outbreaks as shown with the northern European BTV-8 outbreak which started in the summer of 2006. Technological jumps in deep sequencing methodologies made rapid full characterizations of BTV genome from isolates/tissues feasible. Next-generation sequencing (NGS) approaches are powerful tools to study the variability of BTV genomes on a fine scale. This paper provides information on how NGS impacted our knowledge of the BTV genome.

1. INTRODUCTION

Bluetongue (BT) is an infectious noncontagious disease caused by bluetongue virus (BTV). *Bluetongue virus* is the type species of genus *Orbivirus* within the family *Reoviridae* (Mertens et al., 2005). BTV has a wide host range and affects domestic and wild ruminants. BTV infection causes a hemorrhagic disease in sheep, which are the most clinically affected species, while clinical signs have not been observed in cattle until the introduction of serotypes 1 and 8 into northern Europe. BTV is an arbovirus that is transmitted between its mammalian hosts by competent species of *Culicoides* biting midges (Dijkstra et al., 2008). BTV transmission can also occur through the semen of infected animals, and specific strains/serotypes have been shown to cross the placental barrier. Transplacental transmission has been first described with attenuated vaccine strains belonging to serotypes 10, 11, 13, and 17 (MacLachlan et al., 2000). During the BTV-8 outbreak in northern Europe between 2006 and 2010, it became evident that the circulating wild-type BTV-8 strain can cross the placental barrier (Backx et al., 2009; Belbis et al., 2013; Coetzee et al., 2013; De Clercq et al., 2008; Desmecht et al., 2008; Menzies et al., 2008; van der Sluijs et al., 2011; Worwa et al., 2009; Zanella et al., 2012). Until 2008, 24 serotypes of BTV were documented. The discovery of further novel serotypes added to the existing serotype complexity. The serotype in BTV is defined by the interaction of neutralizing antibodies with the VP2 outer capsid protein. With VP2 exhibiting high levels of amino acid sequence variability between serotypes (up to 72.9% aa divergence; Maan et al., 2007a), it is not surprising that low levels of cross-protection exist between serotypes, which complicates vaccination strategies (Bréard et al., 2011).

2. THE BLUETONGUE VIRUS

Bluetongue virus is nonenveloped, with a genome of approximately 19,200 base pairs (bp) composed of 10 linear segments of double-stranded RNA (dsRNA). The genome codes for seven structural (VP1–VP7) and four nonstructural (NS) proteins (NS1–NS4) (Belhouchet et al., 2011; Ratinier et al., 2016; Schwartz-Cornil et al., 2008). Genome segment 10 is considered to harbor an additional overlapping open reading frame which is believed to encode a fifth nonstructural protein, although the existence of this protein has not been proven in infected cells (Sealfon et al., 2015; Stewart et al., 2015).

The largest dsRNA segment for BTV is 3954 bp (BTV-10) and encodes VP1, while the smallest encodes NS3/NS3A and has a length of 822 bp (BTV-10) (http://www.reoviridae.org/dsRNA_virus_proteins/BTV.htm). Noncoding termini of BTV have conserved hexanucleotides at both 5′ and 3′ ends, which are 5′-GUUAAA-3′ and 5′-ACUUAC-3′, respectively, on the sequence of all 10 positive strands of the dsRNA genome (Mertens and Sangar, 1985).

The genome is packaged within a triple layered icosahedral protein capsid (approximately 90 nm). Structural studies using cryoelectronmicroscopy, X-ray crystallography, and radiolabelling shaped our knowledge about the structural organization of the BTV particle (Grimes et al., 1998; Nason et al., 2004; Stuart et al., 1998). Two structural proteins form the outer coat of the particle. These are VP2 (~111 kDa), arranged as 60 trimers and VP5 (~59 kDa) arranged as 120 timers. VP2, the largest outer capsid and serotype determining protein, is the least conserved among all BTV proteins (Maan et al., 2007a; Mertens et al., 1987; Roy, 1989). The core of the particle is composed of two structural proteins (Schwartz-Cornil et al., 2008). The VP7 (~38 kDa), the outer layer of the core, forms the T13 layer of the icosahedral particle and is organized as 260 trimers (780 copies of VP7). The inner layer of the core (T2 layer) is made of 60 dimers of VP3 (~100 kDa). These dimers associate to form 12 decamers. The core contains a single copy of each dsRNA genome segment. Three other proteins are contained within the T2 layer and located as an enzymatic complex at each of the 12 vertices of the icosahedron. Each complex contains a monomer of VP1 polymerase, a dimer of the VP4 capping enzyme (with both guanylyltransferase and methyltransferase activities) and a hexamer of the VP6 helicase (Boyce et al., 2004; Matsuo and Roy, 2009, 2011; Stuart et al., 1998; Sutton et al., 2007).

VP2 is the main antigen for neutralization and interacts with serotype-specific antibodies. VP5 helps determining the serotype, possibly by imposing structural constraints on the VP2 (Schwartz-Cornil et al., 2008). Furthermore, nonstructural proteins play important roles in BTV replication (Kar et al., 2007), maturation and export from infected cells (Forzan et al., 2004; Han and Harty, 2004; Hyatt et al., 1993; Owens et al., 2004; Wirblich et al., 2006).

The evolutionary dynamics of BTV and/or other orbiviruses has been further studied, and evidence of coevolution of orbiviruses and their arthropod vectors has been proposed (Mohd Jaafar et al., 2014). It has been shown that molecular evolutionary rates of the genomes of insect-borne orbiviruses were in the range of 2.3×10^{-4} to 4.52×10^{-4} substitutions/site/year (Mohd Jaafar et al., 2014). Molecular evolutionary rates specific to BTV were found in the range of 0.5×10^{-4} and 7×10^{-4} substitutions/site/year (Carpi et al., 2010), a rate that is considered lower than that of vector-borne positive-sense single-strand RNA viral genomes. In addition reassortment is an important driver that shapes evolution and epidemiology/pathogenicity of BTV in the field (Nomikou et al., 2015).

3. BTV PATHOGENESIS

3.1 Clinical Impact

BTV infection has been associated with clinical sings mainly in sheep and particular wild ruminants (Backx et al., 2007). Our knowledge of the symptomatology associated to BTV infection evolved with the arrival of BTV-8 into northern Europe in the summer of 2006. Although susceptible to BTV infection, cattle and goats did not develop overt clinical signs in the past, when infected with BTV, but have been considered as reservoirs for the virus (Anderson et al., 1987). The BTV-8 strain which circulated in northern Europe between 2006 and 2010 was a "game changer," and was associated with an unusual host range. A large number of infected cattle clearly developed clinical signs (Elbers et al., 2008). Goats and very rarely wild ruminants also developed clinical signs with this strain (Backx et al., 2007; Drolet et al., 2013). Similar observations have been made with BTV-1 in Europe, in terms of host range and clinical signs developed in wild ruminants (Fernández-Pacheco et al., 2008; Meyer et al., 2009).

Clinical symptomatology in BTV-infected animals of the same species ranges from subclinical infection to mild disease, to acute or even fatal disease (Backx et al., 2007; Elbers et al., 2008; Worwa et al., 2010). In sheep, clinical signs typically observed include pyrexia, nasal discharge, tachypnea, lethargy, buccal ulcers, and edema (of the lips, face, or tongue) (Schwartz-Cornil et al., 2008). In severe cases, tongue cyanosis had been observed, hence the name of the disease, "bluetongue." Salient pathological observations include generalized edema and hemorrhages (especially in the pulmonary artery, lungs, heart, skeletal muscles, and lymph nodes). Other prominent pathological changes include necrosis of the buccal, nasal, and digestive tract mucosae. Microscopic lesions include endothelial hypertrophy, vascular stasis, and thrombosis with tissue infarction (Schwartz-Cornil et al., 2008).

3.2 Cell Tropism, Dissemination, and Reservoir

BTV is classically transmitted to its ruminant host by intradermal inoculation via *Culicoides* bites. Saliva of *Culicoides* may play a role in viral replication, first by recruiting leucocytes such as gamma delta-T-cells, where the virus can replicate but also by cleaving VP2 and therefore generating infectious subviral particles, thus enhancing particle infectivity (Darpel et al., 2011). After an initial replication in agranular leucocytes at the site of inoculation (Darpel et al., 2012) and in regional lymph nodes (Barratt-Boyes and MacLachlan, 1995), infected lymphocytes, monocytes, dendritic cells, and macrophages disseminate the virus to secondary sites (mainly lungs, spleen, and lymph nodes) where it replicates in endothelial cells and mononuclear phagocytes (Barratt-Boyes and MacLachlan, 1994; MacLachlan, 2004; MacLachlan et al., 1990). Primary replication runs for 2–3 days, while secondary replication has been estimated between 4 and 20 days. BTV can disseminate via the lymph (Barratt-Boyes and MacLachlan, 1994) and/or via blood (Brodie et al., 1998), the virus being physically associated to mononuclear cells, erythrocytes, and thrombocytes. BTV infects monocytes both in vivo and in vitro. In vivo, infectious BTV can be recovered transiently (<1 week) from monocytes (minimum 10^5 cells) (Brodie et al., 1998; Whetter et al., 1989).

Infectious BTV can be detected associated with erythrocytes, in which it does not replicate but persists in invaginations of cell membranes in association with glycophorins (MacLachlan et al., 1994; Whetter et al., 1989). Association of infectious BTV with erythrocytes is detected soon after infection (24 h) and persists throughout the duration of viraemia, protecting the virus from immune clearance. In ruminants a prolonged cell-associated viraemia persists even in the presence of high levels of neutralizing antibodies

(MacLachlan et al., 1990), and animals recovering from infection are immune to reinfection with a homologous BTV serotype. In the presence of neutralizing antibodies, erythrocyte membrane-associated virus particles are likely neutralized and may not serve as a source of reinfection. In addition, bluetongue virus cores have been suspected to directly penetrate cell membranes of ovine erythrocytes. Association of BTV to the cell membrane of erythrocytes or the phenomenon of direct cell penetration does not imply virus replication in these cells and represents a different case scenario from other dsRNA arboviruses, particularly the coltiviruses, which infect hematopoietic progenitor cells and remain sheltered within erythrocytes upon maturation of the progenitors (Attoui et al., 2005).

Infectious virus had been isolated from blood samples collected 39–56 days postinfection (Di Gialleonardo et al., 2011; MacLachlan et al., 1994) but could not be isolated from all RT-PCR positive samples collected after 60 days postinfection (Katz et al., 1994). According to species, age, and serotype, BTV RNA can be detected by RT-PCR until 167 days postinfection in sheep and 213 days in cattle (Zanella et al., 2013). Some other studies indicated that BTV RNA can be detected even up to 222 days (Bonneau et al., 2002) and up to 2 years or even slightly longer concerning BTV-25 in goats (Vögtlin et al., 2013).

3.3 Pathogenicity

Difference of clinical impact after infection is observed between sheep and cattle. This difference could be linked to interspecies differences of sensitivity to infection of endothelial cells but also to differential levels of proinflammatory cytokines production (Coen et al., 1991). BTV infection of endothelial cells leads to production of several inflammatory mediators: infection is associated to an increase of PGF2α production (Osburn, 1994), thromboxane and prostacyclin infection (Schwartz-Cornil et al., 2008) and induces the transcription of IL-1, -6, and -8, cyclo-oxygenase-2, and nitric oxide synthase (DeMaula et al., 2002). Ovine endothelial cells produce higher levels of proinflammatory cytokines than bovine endothelial cells, despite a less important level of BTV replication (DeMaula et al., 2001, 2002; Russell et al., 1996). Moreover significantly higher ratios of prostacyclin/thromboxane are observed in cattle (Schwartz-Cornil et al., 2008): as thromboxane is a strong procoagulant factor and prostacyclin a vasodilation factor and an inhibitor of platelet aggregation, this observation could explain predominance of hemorrhagic syndrome, thrombosis, and edema in ovine species in comparison to other ruminants (DeMaula et al., 2001, 2002).

BTV has been known as a strong inducer of interferon type I in cell cultures (Jameson and Grossberg, 1978; Vitour et al., 2014) and in vivo in sheep (Foster et al., 1991), in cattle (MacLachlan and Thompson, 1985), and also in mice (Jameson et al., 1978). NS3 and NS4 interfere with induction of the IFN response, implying the importance of these proteins in BTV pathogenesis (Chauveau et al., 2013; Ratinier et al., 2016).

3.4 Immune Response

BTV infection in ruminants induce rapid antibodies production to both structural and nonstructural proteins (Adkison et al., 1987; Anderson et al., 1993; Barros et al., 2009; Richards et al., 1988). Fetal immune competence to BTV develops by mid-gestation (Enright and Osburn, 1980; MacLachlan et al., 1984; Osburn et al., 1982), and therefore those animals infected in the late stages of gestation are born without significant malformation in their central nervous system.

In BTV-10, using escape mutants, neutralizing epitopes have been associated with two amino acid regions (aa 199–213 and aa 321–346) (DeMaula et al., 2000). These two regions are least conserved between all BTV serotypes (Maan et al., 2007a), which may explain the low levels of cross-protection among different serotypes. However, the presence of potential common neutralizing epitopes to more closely related serotypes cannot be excluded, as shown by the existence of serological cross-reactions between particular BTV serotypes. Furthermore, lower levels of protection have been observed in animals that have been inoculated/immunized with certain BTV serotypes, then challenged by heterologous serotypes (Bréard et al., 2015; Dungu et al., 2004; Rossitto and MacLachlan, 1992).

Maan and colleagues grouped all 24 known "typical" serotypes within 10 nucleotypes (A–J), based on Seg-2 sequencing and phylogenetic analyses (Maan et al., 2007a). The higher levels of identity between Seg-2 (VP2) of different serotypes grouped within the same nucleotype reflect serological cross-reactions observed between these serotypes. Two additional nucleotypes of Seg-2 (K and L), which are grouped as "atypical" BTVs, have been described (Fig. 1).

BTV-25 and BTV-27 cluster within nucleotype K, while BTV-26 belongs to nucleotype L (Maan et al., 2010). Eight nucleotypes had also been described for Seg-6 (A–H) (Maan et al., 2010, 2011a) with the "typical" BTV-1 to -24 serotypes. BTV-25, -26, and -27 discovery and sequencing led to description of another two novel nucleotypes in Seg-6 (I and H) (Schulz et al., 2016).

Fig. 1 Phylogenetic tree of BTV genome segment 2, depicting 12 nucleotypes designated A–L. The most recently identified types 25, 26, and 27 are grouped within nucleotypes K and L.

Neutralizing antibodies confer a long-lasting protection to reinfection by a homologous BTV serotype, a property which is at the basis of vaccination strategies against BTV (Huismans et al., 1987; Noad and Roy, 2009; Savini et al., 2008). Antibodies to VP7 are group-specific, VP7 being well conserved between different serotypes. Detection of anti-VP7 antibodies is actually used for serological diagnosis of BTV infection using several ELISA tests, which are commonly competition ELISA-based (Gumm and Newman, 1982; Hamblin, 2004; MacLachlan and Mayo, 2013; Vandenbussche et al., 2008; Zhou et al., 2001).

The level of neutralizing antibodies detected by serum neutralization tests (SNT) does not necessarily correlate with the degree of protection (Jeggo et al., 1983). Cellular immunity plays an important role in protection against BTV infection. Its importance had been demonstrated by experiments of passive transfer of cytotoxic T cells between monozygous sheep, which conferred partial protection to infection (Jeggo et al., 1985). Duration of cellular immunity, secondary to natural infection, or vaccination is however not known.

NS1 and VP2 have been shown as main targets to cellular immunity following BTV infection in sheep (Andrew et al., 1995; Calvo-Pinilla et al., 2009). Cytotoxic response induced by NS1 is cross-reactive, while VP2-induced response is serotype specific (Andrew et al., 1995). VP3, VP5, and VP7 also induce cytotoxic response in sheep but appear to be less immune-dominant (Calvo-Pinilla et al., 2009; Janardhana et al., 1999). VP4, VP6, NS2, and NS3 do not induce such cellular responses (NS2 can induce cytotoxic response in mice; Jones et al., 1997).

Finally, "helper" T cell response is observed during a BTV infection. VP2 and to a lesser extent VP5 contain serotype-specific determinants, and the two major core structural proteins, VP3 and VP7, bear determinants that are cross-reactive (Takamatsu et al., 1990).

4. EPIDEMIOLOGICAL SITUATION

4.1 Modes of Transmission

BTV, with the exception of BTV-25 and -26 (and very likely BTV-27), is transmitted between its ruminant hosts through the bite of virus-infected hematophagous *Culicoides* midges. Once infected (following a blood meal from a BTV-infected ruminant and an extrinsic incubation period of approximately 10 days, depending on ambient temperature), *Culicoides* midges are set to transmit the virus for the remainder of their life span. *Culicoides imicola* was known as the main BTV vector in Africa, and in the Mediterranean Basin (Nolan et al., 2008) since 1998. However, the emergence/incursions of BTV-8 in northern Europe in 2006 could not be linked to *C. imicola*, which is absent from the first affected geographical areas. *Culicoides obsoletus* and *Culicoides scoticus* have been associated as the main vectors of BTV in northern and western Europe (Carpenter et al., 2008), but *Culicoides dewulfi* and *Culicoides chiopterus* have also been designated as potential vectors (Dijkstra et al., 2008).

BTV can be transmitted by vector-independent route(s), although their importance in BTV epidemiology remains unclear. The European strain of BTV-8 crosses the placental barrier and infects the fetus. Transplacental transmission of BTV-8 was observed in many experimental (Belbis et al., 2013; Coetzee et al., 2013; van der Sluijs et al., 2011) and field studies (De Clercq et al., 2008; Menzies et al., 2008; Santman-Berends et al., 2010). Transplacental transmission rates ranged from 10% to almost 70% in experimental models (De Clercq et al., 2008; Santman-Berends et al., 2010; van der Sluijs et al., 2011). Transplacental transmission was also observed, under experimental conditions, with field strains of BTV-1 (van der Sluijs et al., 2013) and BTV-2 (Rasmussen et al., 2013). Previously, transplacental transmission had been considered to occur with live-attenuated vaccine strains of BTV (MacLachlan et al., 2000). However, the impact of transplacental transmission on the overwintering of BTV is considered low: if transient viraemia is detected in the neonate by RT-PCR (sometimes up to 5 months of age) (De Clercq et al., 2008; Santman-Berends et al., 2010), there are no data indicating that infectious virus could be isolated from the blood of a young ruminant contaminated by the transplacental route.

Finally, BTV transmission by oral infection has been reported in ruminant livestock, involving only BTV serotype 8, including infection of calves which were fed infected colostrum (Backx et al., 2009; Menzies et al., 2008). Contact-transmission by aerosol, between infected ruminants, has been considered as the main route of spreading for the novel serotypes BTV-25 and BTV-26. While in case of BTV-25, this route had been only suspected (Reber et al., 2012), it has been experimentally confirmed for BTV-26 (Batten et al., 2014, 2013).

4.2 Situation in Europe

4.2.1 From 1990 to 2008

Until the last century, bluetongue was considered as an exotic disease to Europe. Description of BT was originally made in the 19th century, upon importation of Merino sheep into South Africa (Spreull, 1905). During the beginning of the 20th century, BT spread in Africa and other continents. Subsequently and prior to 1998, geographical regions between latitudes 40°S and 53°N in America, Africa, Australia, and Asia have been recognized as endemic for BTV. Only few sporadic incursions of the disease were described in Europe before 1998 (Spain and Portugal between 1956 and 1960; Cyprus in 1977; Greece in 1979) (Sellers, 1975; Sellers et al., 1978, 1979).

Between 1998 and 2006, at least eight distinct BTV strains from five different serotypes (serotypes 1, 2, 4, 9, and 16) emerged in the Mediterranean basin: BTV-1, -4, -9, and -16 coming from middle-east and invaded eastern parts of the Mediterranean basin while strains, coming from North Africa, including BTV-1, -2, -4, and -16 were encountered in the west part of the Mediterranean basin (Corsica, Sardinia, ...) (Purse et al., 2005; Saegerman et al., 2008; Zientara et al., 2002).

In August 2006, BTV-8 was first identified in the Netherlands and spread into Belgium, Germany, and France (Ardennes). In 2007, the virus substantially extended its distribution in these four countries (6661 cases in Belgium, 11,487 cases in Germany, 5798 cases in the Netherlands, 15,257 cases in France) and spread into Luxemburg (1315 cases), Denmark (1 case), Switzerland (5 cases), the United Kingdom (65 cases), and the Czech Republic (1 case) (Schwartz-Cornil et al., 2008). In the same year, BTV-1, detected earlier in 2006 in North Africa, was found in Spain and subsequently in the south west of France (3 cases in 2007) (Corbière et al., 2012).

In 2008, vaccination against BTV-8 was implemented in most of European countries: compulsory vaccination programs were implemented in Germany (Baetza, 2014), France, and Switzerland (Willgert et al., 2011), while a voluntary vaccination program was implemented in Great Britain (Szmaragd et al., 2010; Webb et al., 2011).

The same year, an additional two BTV serotypes were detected in northern Europe: BTV-6 in Germany and the Netherlands (Eschbaumer et al., 2010; Maan et al., 2010), and BTV-11 in Belgium (De Clercq et al., 2009). Fortunately, these two BTV strains did not spread further and finally disappeared from the field.

4.2.2 Between 2008 and 2016: Evolution and Identification of Novel Serotypes

The year 2008 was not only characterized by the occurrence of BTV-6 and -11 in Germany, the Netherlands, and Belgium. In 2008, in Toggenburg (Switzerland), clinically healthy goats were identified as BTV positive using a broad-reactive BTV-specific rRT-PCR assay, although biological samples were negative or weakly positive for BTV using the "then standard" rRT-PCR tests. These "standard" assays were designed to detect BTV reference strains representing all 24 known "typical" serotypes (Chaignat et al., 2009). The virus was designated Toggenburg orbivirus (TOV). The complete coding sequences of TOV Seg-2 and Seg-5–10 were determined (Hofmann et al., 2008). Phylogenetic analyses supported the classification of TOV as a novel BTV serotype, representing therefore serotype 25. TOV is associated with subclinical infections in goats to mild symptoms in sheep (Chaignat et al., 2009). Prolonged detection of BTV-25 genome (for more than 2 years) seems possible (Vögtlin et al., 2013).

In February 2010, small ruminants in Kuwait showing bluetongue-like symptoms were found positive for BTV by serology (Maan et al., 2011a). Only two samples gave positive results by BTV-specific rRT-PCR targeting Seg-10 and virus could be isolated from one blood sample (KUW2010/02). However, RT-qPCRs targeting Seg-1 and Seg-2 gave negative results. Full-length cDNA copies of Seg-2 and Seg-3 of KUW2010/02 were obtained and sequenced using FLAC methodology (Maan et al., 2011b). Sequence analyses of Seg-2/VP2 showed nucleotide/amino acid identity level of 63.9%/61.5% with BTV-25. This molecular data, confirmed by the absence of viral neutralization by any of the reference sera against BTV-1 to BTV-24 and antisera of goats infected by BTV-25, identified KUW2010/02 as the 26th BTV serotype (Maan et al., 2011b).

The last case of BTV in continental France was identified as BTV-1 detected in June 2010. After this date, neither BTV-1 nor BTV-8 were isolated from field surveillance samples. Effective vaccination campaigns across many European countries, as well as natural immunity to infection, permitted controlling the disease. As BTV was no longer identified/isolated in field samples after 2011, a BTV-free status was declared in several European countries, including the United Kingdom, Belgium, the Netherlands, Germany, and Switzerland (Baetza, 2014; Büchi et al., 2014; Zientara et al., 2012). France recovered its BTV-free status in 2012 and that status was maintained until 2015 (Bluetongue outbreaks continue in France, 2015b).

However, a BTV-free status of certain countries in northern and western Europe could not mask the presence of several bluetongue outbreaks in Europe after 2011. BTV-1 was detected in Corsica, Sardinia, and Sicily in 2012 and 2013 (Sailleau et al., 2015b) and central Italy since 2013 (Conte et al., 2016; Lorusso et al., 2014a). BTV-4 outbreaks, first detected in Peloponnese in May 2014, were present in 2014 in Greece (Katsoulos et al., 2016), Bulgaria, Macedonia, Albania, Romania, and Hungary (Hornyák et al., 2015).

More recently, and surprisingly, BTV-8 was detected in August 2015 in a sheep exhibiting respiratory disease in the Allier area, in the centre of France (Sailleau et al., 2015a). Between August 2015 and December

2016, BTV-8 has been detected in France in farms (149 in 2015 and in 2016) through surveillance program (Bluetongue in France, 2015a). Many hypotheses were imagined to explain this (re-)emergence. It was initially suspected that the 2015 strain could correspond to the emergence of a novel strain of BTV-8 in France. It was also considered that the 2015 strains could have persisted somehow in the field (in wild or domestic ruminants) and is indeed the same/derived from the one involved in 2006–10 outbreaks (Bluetongue in France, 2015a).

Full-genome sequencing was performed using next-generation sequencing (NGS) to assess sequence relatedness between strains of 2008 and 2015 in France. The results confirmed the close relatedness between the two strains (Bréard et al., 2016), therefore supporting the hypothesis of a persistence of the BTV-8 strain from the earlier 2006 to 2010 outbreak in France.

5. NGS DRAMATICALLY IMPROVED FEASIBILITY OF BTV GENOME CHARACTERIZATION

Full-length genome sequencing approaches reveal the extent of genetic diversities in RNA virus genomes. BTV genetic diversity is generated by accumulating point mutations (drift) and/or reassortment (shift) of genome segments during coinfection (Batten et al., 2008; Kumar et al., 2016; Maan et al., 2012). Reassortant viruses are genetically distinct from their parental virus strains, generating novel virus phenotypes, with potentially biological consequences, changes in transmissibility, host/vector ranges, virulence, or pathogenicity (Nomikou et al., 2015).

The growing database of BTV sequences helped designing conventional and real-time RT-PCR assays, not only for the identification of BTV but also the serotype involved. Until these tools were developed, identification of a BTV serotype relied principally on SNT, which is a time-consuming assay, requiring up to 2 weeks for results to be validated. Moreover, the use of immunological tools to define BTV strain relatedness underestimates the degree of genetic similarity between viruses (Gibbs and Greiner, 1994). Sequencing of a large number of BTV genomes helped designing Seg-2/VP2-based conventional and real-time RT-PCR assays for the identification of the genotype/serotype. These have become a gold standard in reference laboratories for serotype identification of cell culture isolates/bloods, concomitantly with a Sanger sequencing approach of the resulting PCR product.

Sequence determination of BTV genome coding sequences (open reading frames) using "conserved" primers has been the principal sequencing approach, until BTV-specific sequence-independent approaches were developed. These include the single primer amplification technique or SPAT (Attoui et al., 2000; Lambden et al., 1992), and a derived more effective version, the full-length amplification of complementary DNA or FLAC (Maan et al., 2007b). These methods allow sequence determination of individual full-length genome segments or even full-length genomes, and hence characterization of gene constellations (genotypes) and geographical variants of a given genome segment (topotypes). Phylogenetic analyses became more thorough with the growing databases of full-length genomes, providing important information about the history, epidemiology, and origin of a BTV strain.

Over the past 10 years, major technological jumps in sequencing of dsRNA genomes came with the development of a range of NGS methodologies (Coetzee and Venter, 2015; Dennis et al., 2014; Jamnikar-Ciglenecki et al., 2016; Magagula et al., 2015; Masuda et al., 2014; Minami-Fukuda et al., 2013; Rao et al., 2013; Theuns et al., 2016). The consequences of these continuously evolving methodologies for sequence determination have resulted in an immense expansion of the BTV sequence databases. One can only regret the lack of accurate annotation of a number of these sequences in terms of origin and date of isolation. However, looking at the bright side, the combination of these approaches has been a great success. Since 2008, three novel serotypes (BTV-25, -26, and -27) have been identified in goats in Switzerland (Chaignat et al., 2009), in sheep in Kuwait (Maan et al., 2011b), and in goats in Corsica, respectively. Their full sequences have been characterized using a combination of these approaches.

Structural BTV proteins, including VP1, VP3, VP4, VP6, and nonstructural proteins NS1, NS2, and NS4 are highly conserved among serotypes (Maan et al., 2015). Sequence variations in these proteins correlate with topotypes (Balasuriya et al., 2008; Daniels et al., 2004; Maan et al., 2010; Nomikou et al., 2009). In general, most BTV isolates have been assigned to the major "eastern" and "western" topotypes (Maan et al., 2011b). Some examples are detailed later showing how full-genome characterization by NGS helped determining topotypes and provided crucial information regarding BTV epidemiology, in particular identification of cross-topotype field reassortants (involving wild-type strains and/or live attenuated vaccine strains).

5.1 Molecular Tools Used for Genetic and Epidemiological Studies

Sanger sequencing (Sanger and Coulson, 1975), using labeled dideoxynucleotides, has been the principal sequencing methodology for BTV and is still used nowadays in epidemiological studies. Genomic dsRNAs are subjected to reverse transcription followed by PCR amplifications of either full-length or partial genome sequences. BTV-specific primers are then used for Sanger sequencing of the PCR product, producing single-stranded DNA fragments of varying lengths ending each by an incorporated labeled dideoxynucleotide. This robust technique usually generates high quality reads of up to ~1 kbp bases on average. A large number of coding sequences of BTV genome segments have been characterized by this method using primers derived from the 5′ and 3′ conserved terminal sequences. For longer genome segments, gene walking using several predefined specific primers was the way forward to fully characterize an ORF, excluding therefore the sequence determination of the termini which served as footprints for the amplification/sequencing primers. At that time SPAT and FLAC were developed, they have been considered a technological leap for the full-length sequence determination of a BTV genome segment/genome. However, NGS approaches are increasingly becoming the methods of choice for sequencing RNA virus genomes, including BTV. Several NGS technologies have been used for BTV genome sequence characterizations.

In these approaches, libraries are prepared which require the reverse transcription of the BTV dsRNA segments into DNA using universal, specific primers, or anchor spacer-ligated primers. The DNA is then randomly fragmented enzymatically or by sonication, followed by ligation of custom linkers (or adaptors) which enable annealing of the sequence to a complementary counterpart. A clonal amplification of the library is then performed, followed by sequencing of the amplified library. Both pyrosequencing and the Ion torrent sequencing are "sequencing by synthesis" methodologies. Pyrosequencing detects the release of pyrophosphate when nucleotides are added to the DNA chain, while Ion torrent sequencing involves the synthesis of a new DNA strand, complementary to the target strand, one base at a time. A semiconductor chip detects the hydrogen ions produced during the process of DNA polymerization.

Full-length cDNA copies of individual genome segments or a group of segments of orbivirus genomes are commonly synthesized and amplified by

RT-PCR using SPAT/FLAC (Attoui et al., 2000; Maan et al., 2007b; Potgieter et al., 2009). These methods have been used for cloning/PCR amplification of dsRNA genome segments without the prior knowledge of genome sequence. Individual plasmid clones of genome segments or purified PCR products of individual genome segments (Attoui et al., 2000; Maan et al., 2007b; Potgieter et al., 2002, 2009; Vreede et al., 1998) are then sequenced by using Sanger sequencing, pyrosequencing, or Ion torrent-based methodologies.

Full-genome sequencing provided rapid identification of BTV following the more recent incursions into Europe. The emergence of BTV-8 in northern Europe in 2006 raised questions about the origin of that particular strain. After initial characterization using conventional RT-PCR, the full-genome was rapidly characterized by FLAC. Using established BTV sequence databases (Maan et al., 2008), it became clear that this northern European strain of BTV-8 was of subsaharan African origin and was not derived from the BTV-8 vaccine strain as it may have been originally suspected.

BTV-6 was identified in mid-September 2008 in the eastern Netherlands. This serotype had previously been identified in Africa, Middle East, India, Pakistan, USA, Central and South America but never in Europe (Mo et al., 1994). Again, full-genome sequencing and phylogenetic studies revealed that 8 out of the 10 genome segments exhibit high levels of sequence identity with their homologous segments in the South African BTV-6 live attenuated vaccine. In addition, Seg-10 exhibited high levels of nucleotide identity (98.4%) to the BTV-2 vaccine (RSAcvv2/02) and only 93.6% with the BTV-6 vaccine strain, indicating that this BTV-6 strain (NET2008/05) had been generated by multiple reassortments (Maan et al., 2010). In 2008 in Belgium a BTV serotype 11 was isolated from a cow with a subclinical infection, which gave early birth to a newborn calf that had died soon after delivery. Full-length cDNA copies of all 10 BTV genome segments were synthesized according to the protocol developed by Potgieter et al. (2009) and sequenced using the 454 GS FLX Titanium chemistry (Roche Diagnostics). Full-genome sequencing rapidly pointed that this virus was derived from vaccine strains (De Clercq et al., 2009; Vandenbussche et al., 2015), possibly resulting from the potential illegal use of live attenuated vaccine (Maclachlan, 2010). It remains so far unclear how these three BTV strains/serotypes have been introduced into northern Europe. However, genetic studies pointed out that these strains are of African origin and that at least two of them (serotypes 6 and 11) might result from use and subsequent field transmissions of live attenuated vaccine strains.

Other recent studies, all based on full-genome studies, determined the origin and topotype of BTV serotypes commonly found in different parts of the world. One of the most studied serotype is the BTV-1. Indeed, in the Mediterranean basin, a report on BTV-1 circulating in Italy in 2014 (Lorusso et al., 2014b) showed that this strain is of "western" topotype, like BTV-1 from Algeria, Morocco, and Italy in 2006, and in Portugal and France in 2007 (Cêtre-Sossah et al., 2011). In Argentina, analysis of Seg-3, Seg-7, and Seg-10 of BTV-1 isolates revealed that Argentinian isolates are also of the "western" topotype, indicating the African/European origin of this strain (Legisa et al., 2013). In Asia, BTV-1 isolated recently in India was found to be an "eastern" topotype (Chand et al., 2016), as was the BTV-1 strain (Y863) isolated in China (Zhu et al., 2013). The full-genomes of two BTV-1 field strains isolated in China were also sequenced (Yang et al., 2012). All segments of the BTV-1 strain (Y863 strain), isolated in 1979, are of the "eastern" topotype, while strain SZ97/1, isolated in 1996, was closest related to a South African BTV-1 strain, suggesting an introduction of African lineage strains and reassortment with "local" BTV strains. In an effort to understand the molecular epidemiology of BTV in China, the full-genome sequence of a BTV-7, isolated in China in 2014, was reported (Yang et al., 2016). Results showed that Seg-1, -3, and -4 of this strain are of the "western" topotype and most closely related to BTV strains from Africa, indicating a spread of BTV from Africa to Asia.

Phylogenetic analyses revealed that most of BTV strains circulating in India, cluster within "eastern" BTV topotypes. Indeed, the full-genome sequence of a BTV-16 strain isolated from southern India revealed that strain IND2014/01 was derived mainly from eastern strains (Seg-1, -3, -7, -8, -9, and -10 belonging to eastern topotype) (Kumar et al., 2016). However, Seg-5 was found to be of "western" origin, probably due to the reassortment with one of the other BTV serotypes circulating in India. Similarly, Seg-2 versions of "western" origin were identified in three Indian strains of BTV-2 isolated in 2003 (IND2003/01, IND2003/02, and IND2003/03). These "western" Seg-2 versions exhibited the highest nucleotide identity (99.7%) with a South African BTV-2 vaccine (Maan et al., 2015). Moreover, Seg-2 sequences for three Indian BTV-10 strains isolated between 2003 and 2005 (IND2004/01, IND2003/06, and IND2005/04) are also of "western" origin and are identical (99.9% nt identity) with Seg-2 of the BTV-10 strain used in a live attenuated BTV-10 vaccine (Maan et al., 2015). Two southern isolates of BTV-2 (IND1993/01 and IND1994/01) had a Seg-6 identical to that of an Indian BTV-1 strain. Molecular epidemiology revealed that, after

1982, a "western" topotype Seg-5 (NS1) spread through "eastern" topotype Indian strains of BTV, suggesting that this version of NS1 enhanced the transmission of eastern BTV strains in India (Maan et al., 2015).

Data obtained from Asian BTV strains revealed occurrence of reassortment events between "western" and "eastern" strains and demonstrated spread of BTV from west to east. Nevertheless, it remains unclear how the "western" BTV strains were introduced. Studies, with North American and Australian BTV isolates, also demonstrated reassortments in the field, which can occur in *Culicoides* vector as well as ruminant hosts (Oberst et al., 1987; Samal et al., 1987a,b).

Recently, an important work has been carried out on the full-genome of 116 field isolates collected between 1958 and 2012 from Europe, the Mediterranean region, and African countries (Nomikou et al., 2015). A total of nine BTV field serotypes (1, 2, 3, 4, 8, 9, 14, 15, and 16), as well as four live attenuated BTV vaccine strains (serotypes 1, 2, 9, and 16), were fully sequenced by NGS. Full-length cDNA copies of individual BTV genome segments were synthesized and amplified by RT-PCR using the FLAC as previously described (Maan et al., 2007b; Potgieter et al., 2009). Individual cDNA amplicons were purified and sequenced by 454 pyrosequencing. This study showed that reassortment was a common and widespread phenomenon that occurred after virus introduction into Europe, affecting therefore the genomic composition of European BTV strains. Reassortment frequency appeared to be largely driven by opportunity, suggesting a lack of any fundamental barriers preventing different BTV strains from exchanging genetic material during natural infections (Nomikou et al., 2015). The study also demonstrated that segments involved in reassortment events were not the result of a completely random process. Seg-1, Seg-4, and Seg-9, which code for the viral replication complex, coevolve in order to function optimally. Seg-7 and Seg-10 evolve independently of the rest of the BTV genome, while Seg-2 and Seg-6 encoding the outer coat proteins VP2 and VP5 were strongly associated. The authors clearly confirmed earlier reports involving live attenuated vaccine strains in reassortment occurring in the field, particularly live attenuated vaccine-derived Seg-1 to Seg-7 and Seg-10. These reassortant viruses from the field are a proof that reassortment between live attenuated vaccines and wild-type strains has consequences of augmented fitness over the parental strains.

NGS tools proved to provide an effective response to emergencies in suspected BTV outbreaks. The reemergence of BTV-8 in central France in 2015 was originally detected using RT-qPCR kits allowing group and type identification (Sailleau et al., 2015a). The origin of this strain was puzzling, and it was not known whether it resulted from an introduction of a novel strain or a reemergence of the strain which circulated in France before 2011. The BTV-8 strain was isolated after a single passage on BSR cells, and NGS was performed on RNA extracts from infected cultures (Bréard et al., 2016). Sequence analysis of the 10 dsRNA genome segments showed a close relationship to all 10 segments of BTV-8 isolated in 2008 in France. Sequences of noncoding regions in Seg-1 to Seg-10 were 100% identical to those of the 2008 strain of BTV-8, while only 19 substitutions were identified in 18,444 bases constituting the coding sequences. The consequence of these nucleotide changes was a total of four amino acid changes (1 change in each of Seg-1, 3, 8, and 9) (Table 1). Indeed, these results clearly showed that BTV-8 isolated in 2015 was nearly identical to the BTV-8 strain which circulated in France and in Europe until 2010. Accordingly, the current BTV-8 strain has most likely persisted in the field by an unknown mechanism, with a low prevalence since it was last detected in France in 2009 (Bréard et al., 2016). The 2015 BTV-8 outbreak in France continued in 2016 through 2017 (Fig. 2). Furthermore, NGS analysis performed on

Table 1 Nucleotide and Amino Acid (AA) Substitutions Found in the Full Genomes of BTV-8 Strains Isolated in France in 2008 and 2016 Compared With BTV-8 Strain Isolated in France in 2015

	BTV-8 2008			BTV-8 2016		
	Nucleotide			Nucleotide		
BTV-8 2015	Coding	Noncoding	AA	Coding	Noncoding	AA
Seg 1	3	0	1	7	0	2
Seg 2	3	0	0	3	0	0
Seg 3	4	0	1	3	0	1
Seg 4	1	0	0	1	0	1
Seg 5	3	0	0	2	0	1
Seg 6	0	0	0	1	0	0
Seg 7	1	0	0	1	0	0
Seg 8	1	0	1	0	0	0
Seg 9	2	0	1	2	0	0
Seg 10	1	0	0	1	0	0
Total substitutions	19	0	4	21	0	5

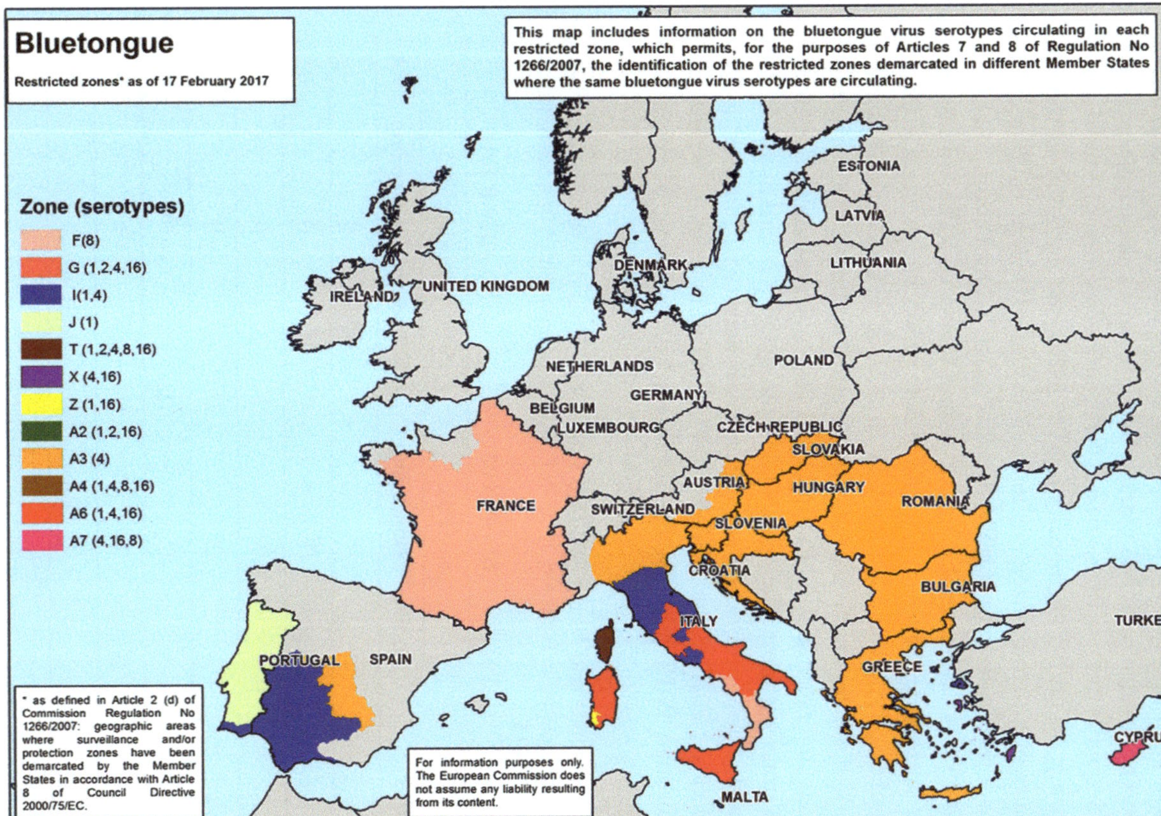

Fig. 2 Current known distribution and restriction zones for circulating BTV serotypes in Europe. The map reflects the epidemiological situation as of the 17th of February 2017 (https://ec.europa.eu/food/sites/food/files/animals/docs/ad_control-measures_bt_restrictedzones-map.jpg).

the first BTV-8 isolate in September 2016 showed the 2016 strain to be identical to the 2015 isolates (Table 1).

In December 2016, abnormal mortalities were detected in a mixed sheep–goat herd in Corsica (France). Nine ewes presented bluetongue-like symptoms, and eight died from the disease. The French National Reference Laboratory for BTV, detected BTV genome by RT-qPCR and isolated BTV-4 from biological samples. In Hungary, a virulent strain of BTV-4 was detected in 2015 (Hornyák et al., 2015) and has spread into western Europe. This BTV-4 strain was detected in Sardinia (Sailleau et al., 2017) and other parts of Italy in 2016. However, serotype 4 was also present in Corsica in 2003 and 2004. The full-genome sequence of the 2016 Italian BTV-4 showed that this strain was identical to the 2015 BTV-4 strain present in the Balkans (C. Sailleau et al., unpublished data). Indeed, to confirm that the 2016 Corsican BTV-4 was similar/identical to the Balkan strain of BTV-4 (and exclude reemergence), full-genome NGS confirmed that all 10 segments were identical to those of the Hungarian BTV-4 (C. Sailleau et al., unpublished data).

These in-depth studies provided crucial information with respect to reassortment events and origin of BTV strains detected in the field, including the ability to distinguish a state reemergence from an introduction/reintroduction. However, several questions remain unanswered. Although reassortment can occur in either vertebrate hosts or insect vectors (Samal et al., 1987a,b), the relative contributions of these potential "mixing vessels" to the observed patterns require further investigations (Bonneau et al., 2001; Caporale et al., 2014). Certainly, the insect vector plays a crucial role in the fitness expression of novel reassortant viruses. However, experimentation with vectors are not accessible to all. Similarly, it remains unclear to which extent reassortment might have facilitated efficient invasion of novel host and vector communities in Europe, a relevant question that requires further examination.

5.2 Full-Genome Sequencing as a Tool to Identify New Serotypes: The Example of BTV-25, -26, and -27

Advances in sequencing technologies facilitated characterization of novel serotypes like BTV-25 and BTV-26 and isolates of BTV-27. Improved and novel cloning/sequencing methodologies not only provide rapidly full-length sequences but also allow a fine analysis of quasi-species of viruses, including BTV.

These approaches helped characterizing the full coding sequences of TOV and its identification as a novel and 25th serotype of BTV (Hofmann et al., 2008). Two years later, a BTV strain isolated from sheep from Kuwait was identified as a novel 26th serotype after sequencing Seg-2 and Seg-3 (Maan et al., 2011a) followed by the full-length genome characterization (Maan et al., 2011b).

A novel 27th bluetongue virus serotype (BTV-27) was discovered in 2014 in Corsica (France) in biological samples collected from asymptomatic goats during field surveillance and a mandatory vaccination program against BTV-1 (Zientara et al., 2014). Detection of Seg-10 of the BTV-27 genome was performed using commercial RT-qPCR kits validated for molecular diagnosis of BTV in France. Amplification by conventional RT-PCR of a Seg-2 partial sequence, using consensus primers (Zientara et al., 2014) allowed to compare isolate's Seg-2 sequence with those of Seg-2 of the existing 26 serotypes. Sequence comparison showed identities of only 73% at the nucleotide level (75% aa) with Seg-2 of BTV-25.

However, attempts to characterize all nine remaining genome segments with primers designed to amplify full coding sequences or partial genome sequences of BTV serotypes 1–24 were unsuccessful. NGS helped characterizing the full coding sequences of the genome. After total RNA extraction from infected cell cultures, a library was prepared which contained double-stranded cDNA derived from all 10 genome segments. The cDNA was fragmented by sonication down to ~300bp sized fragments, which were sequenced using a procedure developed on an Illumina MiSeq Instrument. The complete coding genome of BTV-27 was characterized and published (Jenckel et al., 2015) confirming the status of this virus as the 27th BTV serotype. In 2014, in Corsica, 35 goat samples were tested positive for BTV-27 RNA using a conventional BTV-27 RT-PCR assay. Partial amplicons of Seg-2 were generated. Sequencing and phylogenetic analyses of 35 amplicons revealed the presence of 3 main sequences/clusters of Seg-2, showing up to 8% divergence in their nucleotide sequences (Schulz et al., 2016). Eight strains were isolated in cell culture from the 35 positive goats. Further studies were carried out using three isolates each representing a Seg-2 variant. These variants were designated BTV-27v01, BTV-27v02, and BTV-27v03. BTV-27v01 was the first BTV-27 strain sequenced by Jenckel et al. (2015). BTV-27v02 and BTV-27v03 were later sequenced using Ion torrent sequencing. Although closely related, sequence differences were identified between the three designated variants of BTV-27. Seg-2 sequence identities were similar between all three variants (92%–93% nt/

90%–91%), confirming all three variants belong to serotype 27 [Different isolates of a single BTV serotype show >68.4% nt identity in Seg-2 (72.6% aa identity in VP2) while between distinct serotypes identities of 40.5%–71.5% in Seg-2 are seen (22.1%–77.8% in VP2)] (Maan et al., 2011a). Interestingly, Seg-2 sequences of BTV-27 showed a closer relationship to BTV-25 (73%–74% nt/75% aa identity) than to BTV-26 (62.6%–63.3% nt/59.9%–60.4% aa identity) (Schulz et al., 2016).

Comparisons of the full-length genome segments 1(VP1) and 3(VP3) which correlate with the species, showed aa identities of ~92%–93% (VP1) and ~95% (VP3) with BTV-25, and ~87% (VP1) and ~90% (VP3) with BTV-26, showing a closer relationship to BTV-25. When the sequences of the remaining eight segments were analyzed separately, segments 2, 4, 8, and 9 of all three BTV-27 variants revealed highest sequence homology to their homologous segments in BTV-25. Important differences in relationship with BTV-25 and -26 were observed for the remaining segments, with Seg-5 of BTV-27v01 and v03, Seg-6 of BTV-27v02 and v03, and Seg-10 of BTV-27v01 and v03 being most closely related to homologous segments of BTV-26. BTV-27v01 and v03 are more closely related to each other (93.2% nt/64.8% aa) than to BTV-27v02. Data obtained by full-genome sequencing indicated that BTV-27 is evolving through active circulation in Corsica. Sequence data are suggestive that BTV-27v01 may have originated from an ancestor of BTV-25 and BTV-26, or a BTV-25 that reassorted Seg-5 and Seg-7 with a BTV-26. BTV-27v03 evolved from v01 or an ancestor, followed by reassortment of Seg-6 of a BTV-26 or an ancestor of BTV-26. BTV-27v02 evolved from v03 or an ancestor and was subsequently reassorted with Seg-10 and Seg-9 from a BTV-25 (Schulz et al., 2016).

The detection of three putative novel serotypes in a Capripox vaccine in Israel (BTV-28) (Bumbarov et al., 2016), in an alpaca in South Africa (BTV-29) (Wright, 2014), and in small ruminants in China (strain XJ1407) were recently described.

In South Africa in November 2008, blood samples from an alpaca were tested for BTV. The alpaca has died after developing an acute disease including respiratory distress, suffusions of the buccal cavity, and frothing of the mouth. The presence of BTV RNA was confirmed by a BTV-specific rRT-PCR at Onderstepoort Veterinary Institute. However viral neutralization tests were negative, suggesting a potential novel BTV serotype (Wright, 2014). Sequences of Seg-2 and Seg-5 were first obtained after RT-PCR amplification and Sanger sequencing. NS1 sequences confirmed first that a virus isolated from this alpaca was BTV indeed. Sequence analysis of Seg-2 revealed low identity between this alpaca strain and BTV-15 (73% nt/77% aa identity). Full-genome sequencing was then achieved using the sequencing method described by Potgieter et al. (2009) and revealed high level of identity in Seg-1 (96% nt, 99% aa) and Seg-9 (98% nt/98% aa) with BTV-1, high identity in Seg-3, Seg-4, and Seg-8 with vaccine strains (Seg-3: 90% nt with BTV-4 vaccine strain; Seg-4: 92% nt with BTV-11 vaccine strain; and Seg-8: 97% nt with BTV-2 vaccine strain). A high level of identity was also identified with Seg-5 of BTV-12 (97% nt), Seg-6 and Seg-7 of BTV-15 South African reference strain (91% and 97% nt/98% and 100% aa) and Seg-10 of BTV-3 (96% nt). When compared to criteria proposed by Maan to determine a novel serotype, full-genome sequencing results could not support defining the novel BTV as a new serotype (Maan et al., 2015). In the same way, a BTV strain (XJ1407) was recently isolated in small ruminants in China. This particular BTV strain was not neutralized by BTV-1–24 reactive antisera. Full-genome sequencing was undertaken and showed that Seg-2 of XJ1407 shared higher identity with BTV-27 (75% nt/78% aa) and BTV-25 (73% nt/75% aa), and less than 66% nt/62% aa with others BTV serotypes (Sun et al., 2016). Phylogenetic analyses of the nine other segments revealed they had different origins, most of them belonging to BTV-25 or BTV-27. The status of BTV XJ1407 remains unclear due to close relatedness with BTV-25 and BTV-27 (Sun et al., 2016). SNT might be of help in resolving this issue, by raising monospecific antibodies to VP2 of BTV-25, -27, XJ1407 and putative BTV-28.

6. CONCLUSIONS

The worrying epidemiological map of BTV infections in Europe illustrates the risk to the entire world of emerging pathogens that were previously confined to specific geographic areas, spreading, e.g., due to global warming and increased trading. BTV is an interesting example which illustrate the success of the genomic era, in developing a range of methodologies for cloning/sequencing of the virus genome. NGS is fast becoming the method of choice to generate full-length genome sequences of BTV to study the epidemiology and the incursion and assess threats posed by emerging viruses.

Most current molecular typing tools capture only a small portion of the true genetic variation and consequently lack the discriminatory power to distinguish between closely related or slowly evolving pathogens. A more comprehensive approach based on analysis of the entire genome could reveal important information about the origins, evolution, and characteristics of each viral population (e.g., reassortment, pathogenicity) (Van Borm et al., 2016). The different incursions of BTV serotypes have driven the emergence of reassortant viruses as shown in France between BTV-1 and BTV-8 (Shaw et al., 2013).

Similar to electron microscopy, metagenomics provides a "catch-all" method for detecting a wide range of pathogens. Owing to their broad detection range, all metagenomic approaches are likely to reveal a wide range of microorganisms in both diseased and healthy animals (Mathijs et al., 2016). For example, various isolates of BTV-27 were detected in healthy goats in Corsica (Zientara et al., 2014). NGS approaches would most likely unveil novel serotypes of BTV and will add to our knowledge of the tremendous diversity of BTV in particular and the orbiviruses in general.

Whole-genome sequencing is a useful tool for studying molecular epidemiology. Genomic data can ultimately drive a reconstruction of virus movements on multiple scales ranging from intrahost, to farm, to region, to country or even on a wider global picture. For BTV, for example, whole-genome sequencing studies of many BTV strains have allowed to better understand the origins of the different BTV strains that have emerged in Europe since 1998. It also helped confirming that the 2015 French BTV-8 isolate from the centre of France was basically the same as the BTV-8 strain which has invaded the country in 2006.

Not only for BTV but for many other viral infections, the genomics revolution is changing the way outbreak tracing is approached on fundamental levels and changing how such information is handled and disseminated.

Third-generation sequencing technologies will provide the opportunity to further improve genome assemblies by generating, e.g., longer reads. A new cohort of techniques has been developed using single molecule sequencing and single real-time sequencing, removing the need for clonal amplification. This reduces errors caused by PCR, simplifies library preparation, and, most importantly, gives a much higher read length using higher-throughput platforms. A nanopore-based device provides single-molecule detection and analytical capabilities that are achieved by electrophoretically driving molecules in solution through a nanoscale pore. The nanopore provides a highly confined space within which single nucleic acid polymers can be analyzed at high throughput (Branton et al., 2008). Kilobase length polymers (single-stranded genomic DNA or RNA) or small molecules (e.g., nucleosides) can be identified and characterized without amplification or labeling, a unique analytical capability that makes also rapid RNA sequencing a possibility in the future.

REFERENCES

Adkison, M.A., Stott, J.L., Osburn, B.I., 1987. Identification of bluetongue virus protein-specific antibody responses in sheep by immunoblotting. Am. J. Vet. Res. 48, 1194–1198.

Anderson, G.A., Stott, J.L., Gershwin, L.J., Osburn, B.I., 1987. Identification of bluetongue virus-specific immunoglobulin E in cattle. J. Gen. Virol. 68 (Pt. 9), 2509–2514. https://doi.org/10.1099/0022-1317-68-9-2509.

Anderson, J., Mertens, P.P., Herniman, K.A., 1993. A competitive ELISA for the detection of anti-tubule antibodies using a monoclonal antibody against bluetongue virus non-structural protein NS1. J. Virol. Methods 43, 167–175.

Andrew, M., Whiteley, P., Janardhana, V., Lobato, Z., Gould, A., Coupar, B., 1995. Antigen specificity of the ovine cytotoxic T lymphocyte response to bluetongue virus. Vet. Immunol. Immunopathol. 47, 311–322.

Attoui, H., Billoir, F., Cantaloube, J.F., Biagini, P., de Micco, P., de Lamballerie, X., 2000. Strategies for the sequence determination of viral dsRNA genomes. J. Virol. Methods 89, 147–158.

Attoui, H., Mohd Jaafar, F., de Micco, P., de Lamballerie, X., 2005. Coltiviruses and seadornaviruses in North America, Europe, and Asia. Emerg. Infect. Dis. 11, 1673–1679. https://doi.org/10.3201/eid1111.050868.

Backx, A., Heutink, C.G., van Rooij, E.M.A., van Rijn, P.A., 2007. Clinical signs of bluetongue virus serotype 8 infection in sheep and goats. Vet. Rec. 161, 591–592.

Backx, A., Heutink, R., van Rooij, E., van Rijn, P., 2009. Transplacental and oral transmission of wild-type bluetongue virus serotype 8 in cattle after experimental infection. Vet. Microbiol. 138, 235–243. https://doi.org/10.1016/j.vetmic.2009.04.003.

Baetza, H.-J., 2014. Eradication of bluetongue disease in Germany by vaccination. Vet. Immunol. Immunopathol. 158, 116–119. https://doi.org/10.1016/j.vetimm.2013.09.001.

Balasuriya, U.B.R., Nadler, S.A., Wilson, W.C., Pritchard, L.I., Smythe, A.B., Savini, G., Monaco, F., De Santis, P., Zhang, N., Tabachnick, W.J., Maclachlan, N.J., 2008. The NS3 proteins of global strains of bluetongue virus evolve into regional topotypes through negative (purifying) selection. Vet. Microbiol. 126, 91–100. https://doi.org/10.1016/j.vetmic.2007.07.006.

Barratt-Boyes, S.M., MacLachlan, N.J., 1994. Dynamics of viral spread in bluetongue virus infected calves. Vet. Microbiol. 40, 361–371.

Barratt-Boyes, S.M., MacLachlan, N.J., 1995. Pathogenesis of bluetongue virus infection of cattle. J. Am. Vet. Med. Assoc. 206, 1322–1329.

Barros, S.C., Cruz, B., Luís, T.M., Ramos, F., Fagulha, T., Duarte, M., Henriques, M., Fevereiro, M., 2009. A DIVA system based on the detection of antibodies to non-structural protein 3 (NS3) of bluetongue virus. Vet. Microbiol. 137, 252–259. https://doi.org/10.1016/j.vetmic.2009.01.033.

Batten, C.A., Maan, S., Shaw, A.E., Maan, N.S., Mertens, P.P.C., 2008. A European field strain of bluetongue virus derived from two parental vaccine strains by genome segment reassortment. Virus Res. 137, 56–63. https://doi.org/10.1016/j.virusres.2008.05.016.

Batten, C.A., Henstock, M.R., Steedman, H.M., Waddington, S., Edwards, L., Oura, C.A.L., 2013. Bluetongue virus serotype 26: infection kinetics, pathogenesis and possible contact transmission in goats. Vet. Microbiol. 162, 62–67. https://doi.org/10.1016/j.vetmic.2012.08.014.

Batten, C., Darpel, K., Henstock, M., Fay, P., Veronesi, E., Gubbins, S., Graves, S., Frost, L., Oura, C., 2014. Evidence for transmission of bluetongue virus serotype 26 through direct contact. PLoS One 9, e96049. https://doi.org/10.1371/journal.pone.0096049.

Belbis, G., Bréard, E., Cordonnier, N., Moulin, V., Desprat, A., Sailleau, C., Viarouge, C., Doceul, V., Zientara, S., Millemann, Y., 2013. Evidence of transplacental transmission of bluetongue virus serotype 8 in goats. Vet. Microbiol. 166, 394–404. https://doi.org/10.1016/j.vetmic.2013.06.020.

Belhouchet, M., Mohd Jaafar, F., Firth, A.E., Grimes, J.M., Mertens, P.P.C., Attoui, H., 2011. Detection of a fourth orbivirus non-structural protein. PLoS One 6, e25697. https://doi.org/10.1371/journal.pone.0025697.

Bluetongue in France: a role for wildlife in the latest outbreaks? 2015. Vet. Rec. 177, 455. https://doi.org/10.1136/vr.h5853.

Bluetongue outbreaks continue in France. 2015. Vet. Rec. 177, 353. https://doi.org/10.1136/vr.h5328.

Bonneau, K.R., Mullens, B.A., MacLachlan, N.J., 2001. Occurrence of genetic drift and founder effect during quasispecies evolution of the VP2 and NS3/NS3A genes of bluetongue virus upon passage between sheep, cattle, and Culicoides sonorensis. J. Virol. 75, 8298–8305.

Bonneau, K.R., DeMaula, C.D., Mullens, B.A., MacLachlan, N.J., 2002. Duration of viraemia infectious to Culicoides sonorensis in bluetongue virus-infected cattle and sheep. Vet. Microbiol. 88, 115–125.

Boyce, M., Wehrfritz, J., Noad, R., Roy, P., 2004. Purified recombinant bluetongue virus VP1 exhibits RNA replicase activity. J. Virol. 78, 3994–4002.

Branton, D., Deamer, D.W., Marziali, A., Bayley, H., Benner, S.A., Butler, T., Di Ventra, M., Garaj, S., Hibbs, A., Huang, X., Jovanovich, S.B., Krstic, P.S., Lindsay, S., Ling, X.S., Mastrangelo, C.H., Meller, A., Oliver, J.S., Pershin, Y.V., Ramsey, J.M., Riehn, R., Soni, G.V., Tabard-Cossa, V., Wanunu, M., Wiggin, M., Schloss, J.A., 2008. The potential and challenges of nanopore sequencing. Nat. Biotechnol. 26, 1146–1153. https://doi.org/10.1038/nbt.1495.

Bréard, E., Belbis, G., Hamers, C., Moulin, V., Lilin, T., Moreau, F., Millemann, Y., Montange, C., Sailleau, C., Durand, B., Desprat, A., Viarouge, C., Hoffmann, B., de Smit, H., Goutebroze, S., Hudelet, P., Zientara, S., 2011. Evaluation of humoral response and protective efficacy of two inactivated vaccines against bluetongue virus after vaccination of goats. Vaccine 29, 2495–2502. https://doi.org/10.1016/j.vaccine.2010.12.105.

Bréard, E., Belbis, G., Viarouge, C., Nomikou, K., Haegeman, A., De Clercq, K., Hudelet, P., Lilin, T., Sailleau, C., Zientara, S., 2015. Evaluation of adaptative immune responses and heterologous protection induced by multivalent inactivated vaccines. Vaccine 33, 512–518. https://doi.org/10.1016/j.vaccine.2014.11.053.

Bréard, E., Sailleau, C., Quenault, H., Lucas, P., Viarouge, C., Touzain, F., Fablet, A., Vitour, D., Attoui, H., Zientara, S., Blanchard, Y., 2016. Complete genome sequence of bluetongue virus serotype 8, which reemerged in France in August 2015. Genome Announc. 4. https://doi.org/10.1128/genomeA.00163-16.

Brodie, S.J., Wilson, W.C., O'Hearn, P.M., Muthui, D., Diem, K., Pearson, L.D., 1998. The effects of pharmacological and lentivirus-induced immune suppression on orbivirus pathogenesis: assessment of virus burden in blood monocytes and tissues by reverse transcription-in situ PCR. J. Virol. 72, 5599–5609.

Büchi, M., Abril, C., Vögtlin, A., Schwermer, H., 2014. Prevalence of antibodies against bluetongue virus serotype 8 in bulk-tank milk samples from dairy cattle herds located in risk areas for bluetongue virus transmission after a vaccination programme in Switzerland. Berl. Munch. Tierarztl. Wochenschr. 127, 158–162.

Bumbarov, V., Golender, N., Erster, O., Khinich, Y., 2016. Detection and isolation of bluetongue virus from commercial vaccine batches. Vaccine 34, 3317–3323. https://doi.org/10.1016/j.vaccine.2016.03.097.

Calvo-Pinilla, E., Rodríguez-Calvo, T., Sevilla, N., Ortego, J., 2009. Heterologous prime boost vaccination with DNA and recombinant modified vaccinia virus Ankara protects IFNAR(−/−) mice against lethal bluetongue infection. Vaccine 28, 437–445. https://doi.org/10.1016/j.vaccine.2009.10.027.

Caporale, M., Di Gialleonorado, L., Janowicz, A., Wilkie, G., Shaw, A., Savini, G., Van Rijn, P.A., Mertens, P., Di Ventura, M., Palmarini, M., 2014. Virus and host factors affecting the clinical outcome of bluetongue virus infection. J. Virol. 88, 10399–10411. https://doi.org/10.1128/JVI.01641-14.

Carpenter, S., McArthur, C., Selby, R., Ward, R., Nolan, D.V., Luntz, A.J.M., Dallas, J.F., Tripet, F., Mellor, P.S., 2008. Experimental infection studies of UK Culicoides species midges with bluetongue virus serotypes 8 and 9. Vet. Rec. 163, 589–592.

Carpi, G., Holmes, E.C., Kitchen, A., 2010. The evolutionary dynamics of bluetongue virus. J. Mol. Evol. 70, 583–592. https://doi.org/10.1007/s00239-010-9354-y.

Cêtre-Sossah, C., Madani, H., Sailleau, C., Nomikou, K., Sadaoui, H., Zientara, S., Maan, S., Maan, N., Mertens, P., Albina, E., 2011. Molecular epidemiology of bluetongue virus serotype 1 isolated in 2006 from Algeria. Res. Vet. Sci. 91, 486–497. https://doi.org/10.1016/j.rvsc.2010.10.002.

Chaignat, V., Worwa, G., Scherrer, N., Hilbe, M., Ehrensperger, F., Batten, C., Cortyen, M., Hofmann, M., Thuer, B., 2009. Toggenburg Orbivirus, a new bluetongue virus: initial detection, first observations in field and experimental infection of goats and sheep. Vet. Microbiol. 138, 11–19. https://doi.org/10.1016/j.vetmic.2009.02.003.

Chand, K., Biswas, S.K., Sharma, G., Saxena, A., Tewari, N., Mahajan, S., Pandey, A.B., 2016. Full genome sequencing of the bluetongue virus-1 isolate MKD20/08/Ind from goat in India. Braz. J. Microbiol. 47, 527–528. https://doi.org/10.1016/j.bjm.2016.04.022.

Chauveau, E., Doceul, V., Lara, E., Breard, E., Sailleau, C., Vidalain, P.-O., Meurs, E.F., Dabo, S., Schwartz-Cornil, I., Zientara, S., Vitour, D., 2013. NS3 of bluetongue virus interferes with the induction of type I interferon. J. Virol. 87, 8241–8246. https://doi.org/10.1128/JVI.00678-13.

Coen, M.L., Ellis, J.A., O'Toole, D.T., Wilson, W.C., 1991. Cytokine modulation of the interaction between bluetongue virus and endothelial cells in vitro. Vet. Pathol. 28, 524–532.

Coetzee, P., Venter, E.H., 2015. Recent advances in knowledge of BTV–host–vector interaction. Vet. Ital. 51, 301–311. https://doi.org/10.12834/VetIt.490.2367.1.

Coetzee, P., Stokstad, M., Myrmel, M., Mutowembwa, P., Loken, T., Venter, E.H., Van Vuuren, M., 2013. Transplacental infection in goats experimentally infected with a European strain of bluetongue virus serotype 8. Vet. J. 1997 (197), 335–341. https://doi.org/10.1016/j.tvjl.2013.01.005.

Conte, A., Goffredo, M., Candeloro, L., Calistri, P., Curci, G., Colaiuda, V., Quaglia, M., Mancini, G., Santilli, A., Di Lorenzo, A., Tora, S., Savini, L., Savini, G., 2016. Analysis of climatic factors involved in the BTV-1 incursion in Central Italy in 2014. Vet. Ital. 52, 223–229. https://doi.org/10.12834/VetIt.69.198.1.

Corbière, F., Nussbaum, S., Alzieu, J.-P., Lemaire, M., Meyer, G., Foucras, G., Schelcher, F., 2012. Bluetongue virus serotype 1 in wild ruminants, France, 2008-10. J. Wildl. Dis. 48, 1047–1051. https://doi.org/10.7589/2011-12-359.

Daniels, P.W., Sendow, I., Pritchard, L.I., Sukarsih, Eaton, B.T., 2004. Regional overview of bluetongue viruses in South-East Asia: viruses, vectors and surveillance. Vet. Ital. 40, 94–100.

Darpel, K.E., Langner, K.F.A., Nimtz, M., Anthony, S.J., Brownlie, J., Takamatsu, H.-H., Mellor, P.S., Mertens, P.P.C., 2011. Saliva proteins of vector Culicoides modify structure and infectivity of bluetongue virus particles. PLoS One 6, e17545. https://doi.org/10.1371/journal.pone.0017545.

Darpel, K.E., Monaghan, P., Simpson, J., Anthony, S.J., Veronesi, E., Brooks, H.W., Elliott, H., Brownlie, J., Takamatsu, H.-H., Mellor, P.S., Mertens, P.P., 2012. Involvement of the skin during bluetongue virus infection and replication in the ruminant host. Vet. Res. 43, 40. https://doi.org/10.1186/1297-9716-43-40.

De Clercq, K., De Leeuw, I., Verheyden, B., Vandemeulebroucke, E., Vanbinst, T., Herr, C., Méroc, E., Bertels, G., Steurbaut, N., Miry, C., De Bleecker, K., Maquet, G., Bughin, J., Saulmont, M., Lebrun, M., Sustronck, B., De Deken, R., Hooyberghs, J., Houdart, P., Raemaekers, M., Mintiens, K., Kerkhofs, P., Goris, N., Vandenbussche, F., 2008. Transplacental infection and apparently immunotolerance induced by a wild-type bluetongue virus serotype 8 natural infection. Transbound. Emerg. Dis. 55, 352–359. https://doi.org/10.1111/j.1865-1682.2008.01044.x.

De Clercq, K., Mertens, P., De Leeuw, I., Oura, C., Houdart, P., Potgieter, A.C., Maan, S., Hooyberghs, J., Batten, C., Vandemeulebroucke, E., Wright, I.M., Maan, N., Riocreux, F., Sanders, A., Vanderstede, Y., Nomikou, K., Raemaekers, M., Bin-Tarif, A., Shaw, A., Henstock, M., Bréard, E., Dubois, E., Gastaldi-Thiéry, C., Zientara, S., Verheyden, B., Vandenbussche, F., 2009. Emergence of bluetongue serotypes in Europe, part 2: the occurrence of a BTV-11 strain in Belgium. Transbound. Emerg. Dis. 56, 355–361. https://doi.org/10.1111/j.1865-1682.2009.01092.x.

DeMaula, C.D., Bonneau, K.R., MacLachlan, N.J., 2000. Changes in the outer capsid proteins of bluetongue virus serotype ten that abrogate neutralization by monoclonal antibodies. Virus Res. 67, 59–66.

DeMaula, C.D., Jutila, M.A., Wilson, D.W., MacLachlan, N.J., 2001. Infection kinetics, prostacyclin release and cytokine-mediated modulation of the mechanism of cell death during bluetongue virus infection of cultured ovine and bovine pulmonary artery and lung microvascular endothelial cells. J. Gen. Virol. 82, 787–794.

DeMaula, C.D., Leutenegger, C.M., Bonneau, K.R., MacLachlan, N.J., 2002. The role of endothelial cell-derived inflammatory and vasoactive mediators in the pathogenesis of bluetongue. Virology 296, 330–337. https://doi.org/10.1006/viro.2002.1476.

Dennis, F.E., Fujii, Y., Haga, K., Damanka, S., Lartey, B., Agbemabiese, C.A., Ohta, N., Armah, G.E., Katayama, K., 2014. Identification of novel Ghanaian G8P[6] human-bovine reassortant rotavirus strain by next generation sequencing. PLoS One 9, e100699. https://doi.org/10.1371/journal.pone.0100699.

Desmecht, D., Bergh, R.V., Sartelet, A., Leclerc, M., Mignot, C., Misse, F., Sudraud, C., Berthemin, S., Jolly, S., Mousset, B., Linden, A., Coignoul, F., Cassart, D., 2008. Evidence for transplacental transmission of the current wild-type strain of bluetongue virus serotype 8 in cattle. Vet. Rec. 163, 50–52.

Di Gialleonardo, L., Migliaccio, P., Teodori, L., Savini, G., 2011. The length of BTV-8 viraemia in cattle according to infection doses and diagnostic techniques. Res. Vet. Sci. 91, 316–320. https://doi.org/10.1016/j.rvsc.2010.12.017.

Dijkstra, E., van der Ven, I.J.K., Meiswinkel, R., Hölzel, D.R., Van Rijn, P.A., Meiswinkel, R., 2008. Culicoides chiopterus as a potential vector of bluetongue virus in Europe. Vet. Rec. 162, 422.

Drolet, B.S., Reister, L.M., Rigg, T.D., Nol, P., Podell, B.K., Mecham, J.O., VerCauteren, K.C., van Rijn, P.A., Wilson, W.C., Bowen, R.A., 2013. Experimental infection of white-tailed deer (Odocoileus virginianus) with Northern European bluetongue virus serotype 8. Vet. Microbiol. 166, 347–355. https://doi.org/10.1016/j.vetmic.2013.05.027.

Dungu, B., Gerdes, T., Smit, T., 2004. The use of vaccination in the control of bluetongue in southern Africa. Vet. Ital. 40, 616–622.

Elbers, A.R.W., Backx, A., Mintiens, K., Gerbier, G., Staubach, C., Hendrickx, G., van der Spek, A., 2008. Field observations during the Bluetongue serotype 8 epidemic in 2006. II. Morbidity and mortality rate, case fatality and clinical recovery in sheep and cattle in the Netherlands. Prev. Vet. Med. 87, 31–40. https://doi.org/10.1016/j.prevetmed.2008.06.003.

Enright, F.M., Osburn, B.I., 1980. Ontogeny of host responses in ovine fetuses infected with bluetongue virus. Am. J. Vet. Res. 41, 224–229.

Eschbaumer, M., Hoffmann, B., Moss, A., Savini, G., Leone, A., König, P., Zemke, J., Conraths, F., Beer, M., 2010. Emergence of bluetongue virus serotype 6 in Europe–German field data and experimental infection of cattle. Vet. Microbiol. 143, 189–195. https://doi.org/10.1016/j.vetmic.2009.11.040.

Fernández-Pacheco, P., Fernández-Pinero, J., Agüero, M., Jiménez-Clavero, M.A., 2008. Bluetongue virus serotype 1 in wild mouflons in Spain. Vet. Rec. 162, 659–660.

Forzan, M., Wirblich, C., Roy, P., 2004. A capsid protein of nonenveloped Bluetongue virus exhibits membrane fusion activity. Proc. Natl. Acad. Sci. U.S.A. 101, 2100–2105. https://doi.org/10.1073/pnas.0306448101.

Foster, N.M., Luedke, A.J., Parsonson, I.M., Walton, T.E., 1991. Temporal relationships of viremia, interferon activity, and antibody responses of sheep infected with several bluetongue virus strains. Am. J. Vet. Res. 52, 192–196.

Gibbs, E.P.J., Greiner, E.C., 1994. The epidemiology of bluetongue. Comp. Immunol. Microbiol. Infect. Dis. 17, 207–220. https://doi.org/10.1016/0147-9571(94)90044-2.

Grimes, J.M., Burroughs, J.N., Gouet, P., Diprose, J.M., Malby, R., Ziéntara, S., Mertens, P.P., Stuart, D.I., 1998. The atomic structure of the bluetongue virus core. Nature 395, 470–478. https://doi.org/10.1038/26694.

Gumm, I.D., Newman, J.F., 1982. The preparation of purified bluetongue virus group antigen for use as a diagnostic reagent. Arch. Virol. 72, 83–93.

Hamblin, C., 2004. Bluetongue virus antigen and antibody detection, and the application of laboratory diagnostic techniques. Vet. Ital. 40, 538–545.

Han, Z., Harty, R.N., 2004. The NS3 protein of bluetongue virus exhibits viroporin-like properties. J. Biol. Chem. 279, 43092–43097. https://doi.org/10.1074/jbc.M403663200.

Hofmann, M.A., Renzullo, S., Mader, M., Chaignat, V., Worwa, G., Thuer, B., 2008. Genetic characterization of toggenburg orbivirus, a new bluetongue virus, from goats, Switzerland. Emerg. Infect. Dis. 14, 1855–1861. https://doi.org/10.3201/eid1412.080818.

Hornyák, Á., Malik, P., Marton, S., Dóró, R., Cadar, D., Bányai, K., 2015. Emergence of multireassortant bluetongue virus serotype 4 in Hungary. Infect. Genet. Evol. 33, 6–10. https://doi.org/10.1016/j.meegid.2015.03.036.

Huismans, H., van der Walt, N.T., Cloete, M., Erasmus, B.J., 1987. Isolation of a capsid protein of bluetongue virus that induces a protective immune response in sheep. Virology 157, 172–179.

Hyatt, A.D., Zhao, Y., Roy, P., 1993. Release of bluetongue virus-like particles from insect cells is mediated by BTV nonstructural protein NS3/NS3A. Virology 193, 592–603. https://doi.org/10.1006/viro.1993.1167.

Jameson, P., Grossberg, S.E., 1978. Production of interferon in human cell cultures by a new, potent viral inducer. Adv. Exp. Med. Biol. 110, 37–53.

Jameson, P., Schoenherr, C.K., Grossberg, S.E., 1978. Bluetongue virus, an exceptionally potent interferon inducer in mice. Infect. Immun. 20, 321–323.

Jamnikar-Ciglenecki, U., Kuhar, U., Sturm, S., Kirbis, A., Racki, N., Steyer, A., 2016. The first detection and whole genome characterization of the G6P[15] group A rotavirus strain from roe deer. Vet. Microbiol. 191, 52–59. https://doi.org/10.1016/j.vetmic.2016.05.019.

Janardhana, V., Andrew, M.E., Lobato, Z.I., Coupar, B.E., 1999. The ovine cytotoxic T lymphocyte responses to bluetongue virus. Res. Vet. Sci. 67, 213–221. https://doi.org/10.1053/rvsc.1999.0306.

Jeggo, M.H., Gumm, I.D., Taylor, W.P., 1983. Clinical and serological response of sheep to serial challenge with different bluetongue virus types. Res. Vet. Sci. 34, 205–211.

Jeggo, M.H., Wardley, R.C., Brownlie, J., 1985. Importance of ovine cytotoxic T cells in protection against bluetongue virus infection. Prog. Clin. Biol. Res. 178, 477–487.

Jenckel, M., Bréard, E., Schulz, C., Sailleau, C., Viarouge, C., Hoffmann, B., Höper, D., Beer, M., Zientara, S., 2015. Complete coding genome sequence of putative novel bluetongue virus serotype 27. Genome Announc. 3. https://doi.org/10.1128/genomeA.00016-15.

Jones, L.D., Williams, T., Bishop, D., Roy, P., 1997. Baculovirus-expressed nonstructural protein NS2 of bluetongue virus induces a cytotoxic T-cell response in mice which affords partial protection. Clin. Diagn. Lab. Immunol. 4, 297–301.

Kar, A.K., Bhattacharya, B., Roy, P., 2007. Bluetongue virus RNA binding protein NS2 is a modulator of viral replication and assembly. BMC Mol. Biol. 8, 4. https://doi.org/10.1186/1471-2199-8-4.

Katsoulos, P.-D., Giadinis, N.D., Chaintoutis, S.C., Dovas, C.I., Kiossis, E., Tsousis, G., Psychas, V., Vlemmas, I., Papadopoulos, T., Papadopoulos, O., Zientara, S., Karatzias, H., Boscos, C., 2016. Epidemiological characteristics and clinicopathological features of bluetongue in sheep and cattle, during the 2014 BTV serotype 4 incursion in Greece. Tropl. Anim. Health Prod. 48, 469–477. https://doi.org/10.1007/s11250-015-0974-5.

Katz, J., Alstad, D., Gustafson, G., Evermann, J., 1994. Diagnostic analysis of the prolonged bluetongue virus RNA presence found in the blood of naturally infected cattle and experimentally infected sheep. J. Vet. Diagn. Invest. 6, 139–142.

Kumar, L., Batra, K., Chaudhary, D., Gupta, A.K., Dalal, A., Kalyanaraman, B., Irulappan, G.P., Kumar, V., Maan, S., 2016. Full-genome sequence analysis of a reassortant strain of bluetongue virus serotype 16 from Southern India. Genome Announc. 4. https://doi.org/10.1128/genomeA.00783-16.

Lambden, P.R., Cooke, S.J., Caul, E.O., Clarke, I.N., 1992. Cloning of noncultivatable human rotavirus by single primer amplification. J. Virol. 66, 1817–1822.

Legisa, D., Gonzalez, F., De Stefano, G., Pereda, A., Dus Santos, M.J., 2013. Phylogenetic analysis of bluetongue virus serotype 4 field isolates from Argentina. J. Gen. Virol. 94, 652–662. https://doi.org/10.1099/vir.0.046896-0.

Lorusso, A., Marcacci, M., Ancora, M., Mangone, I., Leone, A., Marini, V., Cammà, C., Savini, G., 2014a. Complete genome sequence of bluetongue virus serotype 1 circulating in Italy, obtained through a fast next-generation sequencing protocol. Genome Announc. 2. https://doi.org/10.1128/genomeA.00093-14.

Lorusso, A., Sghaier, S., Ancora, M., Marcacci, M., Di Gennaro, A., Portanti, O., Mangone, I., Teodori, L., Leone, A., Camma', C., Petrini, A., Hammami, S., Savini, G., 2014b. Molecular epidemiology of bluetongue virus serotype 1 circulating in Italy and its connection with northern Africa. Infect. Genet. Evol. 28, 144–149. https://doi.org/10.1016/j.meegid.2014.09.014.

Maan, S., Maan, N.S., Samuel, A.R., Rao, S., Attoui, H., Mertens, P.P.C., 2007a. Analysis and phylogenetic comparisons of full-length VP2 genes of the 24 bluetongue virus serotypes. J. Gen. Virol. 88, 621–630. https://doi.org/10.1099/vir.0.82456-0.

Maan, S., Rao, S., Maan, N.S., Anthony, S.J., Attoui, H., Samuel, A.R., Mertens, P.P.C., 2007b. Rapid cDNA synthesis and sequencing techniques for the genetic study of bluetongue and other dsRNA viruses. J. Virol. Methods 143, 132–139. https://doi.org/10.1016/j.jviromet.2007.02.016.

Maan, S., Maan, N.S., Ross-smith, N., Batten, C.A., Shaw, A.E., Anthony, S.J., Samuel, A.R., Darpel, K.E., Veronesi, E., Oura, C.A.L., Singh, K.P., Nomikou, K., Potgieter, A.C., Attoui, H., van Rooij, E., van Rijn, P., De Clercq, K., Vandenbussche, F., Zientara, S., Bréard, E., Sailleau, C., Beer, M., Hoffman, B., Mellor, P.S., Mertens, P.P.C., 2008. Sequence analysis of bluetongue virus serotype

8 from the Netherlands 2006 and comparison to other European strains. Virology 377, 308–318. https://doi.org/10.1016/j.virol.2008.04.028.
Maan, S., Maan, N.S., van Rijn, P.A., van Gennip, R.G.P., Sanders, A., Wright, I.M., Batten, C., Hoffmann, B., Eschbaumer, M., Oura, C.A.L., Potgieter, A.C., Nomikou, K., Mertens, P.P.C., 2010. Full genome characterisation of bluetongue virus serotype 6 from the Netherlands 2008 and comparison to other field and vaccine strains. PLoS One 5, e10323. https://doi.org/10.1371/journal.pone.0010323.
Maan, S., Maan, N.S., Nomikou, K., Batten, C., Antony, F., Belaganahalli, M.N., Samy, A.M., Reda, A.A., Al-Rashid, S.A., El Batel, M., Oura, C.A.L., Mertens, P.P.C., 2011a. Novel bluetongue virus serotype from Kuwait. Emerg. Infect. Dis. 17, 886–889. https://doi.org/10.3201/eid1705.101742.
Maan, S., Maan, N.S., Nomikou, K., Veronesi, E., Bachanek-Bankowska, K., Belaganahalli, M.N., Attoui, H., Mertens, P.P.C., 2011b. Complete genome characterisation of a novel 26th bluetongue virus serotype from Kuwait. PLoS One 6, e26147. https://doi.org/10.1371/journal.pone.0026147.
Maan, S., Maan, N.S., Guimera, M., Nomikou, K., Singh, K.P., Pullinger, G., Belaganahalli, M.N., Mertens, P.P.C., 2012. Genome sequence of a reassortant strain of bluetongue virus serotype 23 from Western India. J. Virol. 86, 7011–7012. https://doi.org/10.1128/JVI.00731-12.
Maan, S., Maan, N.S., Belaganahalli, M.N., Rao, P.P., Singh, K.P., Hemadri, D., Putty, K., Kumar, A., Batra, K., Krishnajyothi, Y., Chandel, B.S., Reddy, G.H., Nomikou, K., Reddy, Y.N., Attoui, H., Hegde, N.R., Mertens, P.P.C., 2015. Full-genome sequencing as a basis for molecular epidemiology studies of bluetongue virus in India. PLoS One 10. https://doi.org/10.1371/journal.pone.0131257.
MacLachlan, N.J., 2004. Bluetongue: pathogenesis and duration of viraemia. Vet. Ital. 40, 462–467.
Maclachlan, N.J., 2010. Global implications of the recent emergence of bluetongue virus in Europe. Vet. Clin. North Am. Food Anim. Pract. 26, 163–171. table of contents. https://doi.org/10.1016/j.cvfa.2009.10.012.
MacLachlan, N.J., Mayo, C.E., 2013. Potential strategies for control of bluetongue, a globally emerging, Culicoides-transmitted viral disease of ruminant livestock and wildlife. Antiviral Res. 99, 79–90. https://doi.org/10.1016/j.antiviral.2013.04.021.
MacLachlan, N.J., Thompson, J., 1985. Bluetongue virus-induced interferon in cattle. Am. J. Vet. Res. 46, 1238–1241.
MacLachlan, N.J., Schore, C.E., Osburn, B.I., 1984. Antiviral responses of bluetongue virus-inoculated bovine fetuses and their dams. Am. J. Vet. Res. 45, 1469–1473.
MacLachlan, N.J., Jagels, G., Rossitto, P.V., Moore, P.F., Heidner, H.W., 1990. The pathogenesis of experimental bluetongue virus infection of calves. Vet. Pathol. 27, 223–229.
MacLachlan, N.J., Nunamaker, R.A., Katz, J.B., Sawyer, M.M., Akita, G.Y., Osburn, B.I., Tabachnick, W.J., 1994. Detection of bluetongue virus in the blood of inoculated calves: comparison of virus isolation, PCR assay, and in vitro feeding of Culicoides variipennis. Arch. Virol. 136, 1–8.
MacLachlan, N.J., Conley, A.J., Kennedy, P.C., 2000. Bluetongue and equine viral arteritis viruses as models of virus-induced fetal injury and abortion. Anim. Reprod. Sci. 60–61, 643–651.
Magagula, N.B., Esona, M.D., Nyaga, M.M., Stucker, K.M., Halpin, R.A., Stockwell, T.B., Seheri, M.L., Steele, A.D., Wentworth, D.E., Mphahlele, M.J., 2015. Whole genome analyses of G1P[8] rotavirus strains from vaccinated and non-vaccinated South African children presenting with diarrhea. J. Med. Virol. 87, 79–101. https://doi.org/10.1002/jmv.23971.
Masuda, T., Nagai, M., Yamasato, H., Tsuchiaka, S., Okazaki, S., Katayama, Y., Oba, M., Nishiura, N., Sassa, Y., Omatsu, T., Furuya, T., Koyama, S., Shirai, J., Taniguchi, K.,

Fujii, Y., Todaka, R., Katayama, K., Mizutani, T., 2014. Identification of novel bovine group A 15P[14] strain from epizootic diarrhea of adult cows by de novo sequencing using a next-generation sequencer. Vet. Microbiol. 171, 66–73. https://doi.org/10.1016/j.vetmic.2014.03.009.

Mathijs, E., Vandenbussche, F., Van Borm, S., 2016. Using genomics for surveillance of veterinary infectious agents. Rev. Sci. Tech. Int. Off. Epizoot. 35, 143–157. https://doi.org/10.20506/rst.35.1.2424.

Matsuo, E., Roy, P., 2009. Bluetongue virus VP6 acts early in the replication cycle and can form the basis of chimeric virus formation. J. Virol. 83, 8842–8848. https://doi.org/10.1128/JVI.00465-09.

Matsuo, E., Roy, P., 2011. Bluetongue virus VP1 polymerase activity in vitro: template dependency, dinucleotide priming and cap dependency. PLoS One 6, e27702. https://doi.org/10.1371/journal.pone.0027702.

Menzies, F.D., McCullough, S.J., McKeown, I.M., Forster, J.L., Jess, S., Batten, C., Murchie, A.K., Gloster, J., Fallows, J.G., Pelgrim, W., Mellor, P.S., Oura, C.A.L., 2008. Evidence for transplacental and contact transmission of bluetongue virus in cattle. Vet. Rec. 163, 203–209.

Mertens, P.P., Sangar, D.V., 1985. Analysis of the terminal sequences of the genome segments of four orbiviruses. Virology 140, 55–67.

Mertens, P.P., Burroughs, J.N., Anderson, J., 1987. Purification and properties of virus particles, infectious subviral particles, and cores of bluetongue virus serotypes 1 and 4. Virology 157, 375–386.

Mertens, P.P.C., Maan, S., Samuel, A., Attoui, H., 2005. Orbiviruses, Reoviridae. In: Fauquet, C.M., Mayo, M.A., Maniloff, J., Desselberger, U., Ball, L.A. (Eds.), Virus Taxonomy: Eighth Report of the International Committee on Taxonomy of Viruses. Elsevier Academic Press, London, United Kingdom, pp. 466–483.

Meyer, G., Lacroux, C., Léger, S., Top, S., Goyeau, K., Deplanche, M., Lemaire, M., 2009. Lethal bluetongue virus serotype 1 infection in llamas. Emerg. Infect. Dis. 15, 608–610. https://doi.org/10.3201/eid1504.081514.

Minami-Fukuda, F., Nagai, M., Takai, H., Murakami, T., Ozawa, T., Tsuchiaka, S., Okazaki, S., Katayama, Y., Oba, M., Nishiura, N., Sassa, Y., Omatsu, T., Furuya, T., Koyama, S., Shirai, J., Tsunemitsu, H., Fujii, Y., Katayama, K., Mizutani, T., 2013. Detection of bovine group a rotavirus using rapid antigen detection kits, rt-PCR and next-generation DNA sequencing. J. Vet. Med. Sci. 75, 1651–1655.

Mo, C.L., Thompson, L.H., Homan, E.J., Oviedo, M.T., Greiner, E.C., González, J., Sáenz, M.R., 1994. Bluetongue virus isolations from vectors and ruminants in Central America and the Caribbean. Interamerican Bluetongue Team. Am. J. Vet. Res. 55, 211–215.

Mohd Jaafar, F., Belhouchet, M., Belaganahalli, M., Tesh, R.B., Mertens, P.P.C., Attoui, H., 2014. Full-genome characterisation of Orungo, Lebombo and Changuinola viruses provides evidence for co-evolution of orbiviruses with their arthropod vectors. PLoS One 9, e86392. https://doi.org/10.1371/journal.pone.0086392.

Nason, E.L., Rothagel, R., Mukherjee, S.K., Kar, A.K., Forzan, M., Prasad, B.V.V., Roy, P., 2004. Interactions between the inner and outer capsids of bluetongue virus. J. Virol. 78, 8059–8067. https://doi.org/10.1128/JVI.78.15.8059-8067.2004.

Noad, R., Roy, P., 2009. Bluetongue vaccines. Vaccine 27 (Suppl. 4), D86–89. https://doi.org/10.1016/j.vaccine.2009.08.037.

Nolan, D.V., Dallas, J.F., Piertney, S.B., Mordue Luntz, A.J., 2008. Incursion and range expansion in the bluetongue vector Culicoides imicola in the Mediterranean basin: a phylogeographic analysis. Med. Vet. Entomol. 22, 340–351. https://doi.org/10.1111/j.1365-2915.2008.00744.x.

Nomikou, K., Dovas, C.I., Maan, S., Anthony, S.J., Samuel, A.R., Papanastassopoulou, M., Maan, N.S., Mangana, O., Mertens, P.P.C., 2009. Evolution and phylogenetic analysis of full-length VP3 genes of Eastern Mediterranean bluetongue virus isolates. PLoS One 4, e6437. https://doi.org/10.1371/journal.pone.0006437.

Nomikou, K., Hughes, J., Wash, R., Kellam, P., Breard, E., Zientara, S., Palmarini, M., Biek, R., Mertens, P., 2015. Widespread reassortment shapes the evolution and epidemiology of bluetongue virus following European invasion. PLoS Pathog. 11. https://doi.org/10.1371/journal.ppat.1005056.

Oberst, R.D., Stott, J.L., Blanchard-Channell, M., Osburn, B.I., 1987. Genetic reassortment of bluetongue virus serotype 11 strains in the bovine. Vet. Microbiol. 15, 11–18.

Osburn, B.I., 1994. The impact of bluetongue virus on reproduction. Comp. Immunol. Microbiol. Infect. Dis. 17, 189–196.

Osburn, B.I., MacLachlan, N.J., Terrell, T.G., 1982. Ontogeny of the immune system. J. Am. Vet. Med. Assoc. 181, 1049–1052.

Owens, R.J., Limn, C., Roy, P., 2004. Role of an arbovirus nonstructural protein in cellular pathogenesis and virus release. J. Virol. 78, 6649–6656. https://doi.org/10.1128/JVI.78.12.6649-6656.2004.

Potgieter, A.C., Steele, A.D., van Dijk, A.A., 2002. Cloning of complete genome sets of six dsRNA viruses using an improved cloning method for large dsRNA genes. J. Gen. Virol. 83, 2215–2223. https://doi.org/10.1099/0022-1317-83-9-2215.

Potgieter, A.C., Page, N.A., Liebenberg, J., Wright, I.M., Landt, O., van Dijk, A.A., 2009. Improved strategies for sequence-independent amplification and sequencing of viral double-stranded RNA genomes. J. Gen. Virol. 90, 1423–1432. https://doi.org/10.1099/vir.0.009381-0.

Purse, B.V., Mellor, P.S., Rogers, D.J., Samuel, A.R., Mertens, P.P.C., Baylis, M., 2005. Climate change and the recent emergence of bluetongue in Europe. Nat. Rev. Microbiol. 3, 171–181. https://doi.org/10.1038/nrmicro1090.

Rao, P.P., Reddy, Y.N., Ganesh, K., Nair, S.G., Niranjan, V., Hegde, N.R., 2013. Deep sequencing as a method of typing bluetongue virus isolates. J. Virol. Methods 193, 314–319. https://doi.org/10.1016/j.jviromet.2013.06.033.

Rasmussen, L.D., Savini, G., Lorusso, A., Bellacicco, A., Palmarini, M., Caporale, M., Rasmussen, T.B., Belsham, G.J., Bøtner, A., 2013. Transplacental transmission of field and rescued strains of BTV-2 and BTV-8 in experimentally infected sheep. Vet. Res. 44, 75. https://doi.org/10.1186/1297-9716-44-75.

Ratinier, M., Shaw, A.E., Barry, G., Gu, Q., Di Gialleonardo, L., Janowicz, A., Varela, M., Randall, R.E., Caporale, M., Palmarini, M., 2016. Bluetongue virus NS4 protein is an interferon antagonist and a determinant of virus virulence. J. Virol. 90, 5427–5439. https://doi.org/10.1128/JVI.00422-16.

Reber, A., Kreienbrock, L., Casati, S., Chaignat, V., Schwermer, H., 2012. Putative risk factors for infections with Toggenburg Orbivirus in goat herds in Southern Switzerland (Canton of Ticino). Vet. Microbiol. 160, 29–34. https://doi.org/10.1016/j.vetmic.2012.05.027.

Richards, R.G., MacLachlan, N.J., Heidner, H.W., Fuller, F.J., 1988. Comparison of virologic and serologic responses of lambs and calves infected with bluetongue virus serotype 10. Vet. Microbiol. 18, 233–242.

Rossitto, P.V., MacLachlan, N.J., 1992. Neutralizing epitopes of the serotypes of bluetongue virus present in the United States. J. Gen. Virol. 73 (Pt. 8), 1947–1952.

Roy, P., 1989. Bluetongue virus genetics and genome structure. Virus Res. 13, 179–206.

Russell, H., O'Toole, D.T., Bardsley, K., Davis, W.C., Ellis, J.A., 1996. Comparative effects of bluetongue virus infection of ovine and bovine endothelial cells. Vet. Pathol. 33, 319–331.

Saegerman, C., Berkvens, D., Mellor, P.S., 2008. Bluetongue epidemiology in the European Union. Emerg. Infect. Dis. 14, 539–544. https://doi.org/10.3201/eid1404.071441.

Sailleau, C., Bréard, E., Viarouge, C., Gorlier, A., Quenault, H., Hirchaud, E., Touzain, F., Blanchard, Y., Vitour, D., Zientara, S., 2017. Complete genome sequence of bluetongue virus serotype 4 that emerged on the French island of Corsica in December 2016. Transbound. Emerg. Dis. 1–4. https://doi.org/10.1111/tbed.12660.

Sailleau, C., Bréard, E., Viarouge, C., Vitour, D., Romey, A., Garnier, A., Fablet, A., Lowenski, S., Gorna, K., Caignard, G., Pagneux, C., Zientara, S., 2015a. Re-emergence of bluetongue virus serotype 8 in France, 2015. Transbound. Emerg. Dis. https://doi.org/10.1111/tbed.12453.

Sailleau, C., Viarouge, C., Bréard, E., Perrin, J.B., Doceul, V., Vitour, D., Zientara, S., 2015b. Emergence of bluetongue virus serotype 1 in French Corsica Island in September 2013. Transbound. Emerg. Dis. 62, e89–91. https://doi.org/10.1111/tbed.12207.

Samal, S.K., el-Hussein, A., Holbrook, F.R., Beaty, B.J., Ramig, R.F., 1987a. Mixed infection of Culicoides variipennis with bluetongue virus serotypes 10 and 17: evidence for high frequency reassortment in the vector. J. Gen. Virol. 68 (Pt. 9), 2319–2329. https://doi.org/10.1099/0022-1317-68-9-2319.

Samal, S.K., Livingston, C.W., McConnell, S., Ramig, R.F., 1987b. Analysis of mixed infection of sheep with bluetongue virus serotypes 10 and 17: evidence for genetic reassortment in the vertebrate host. J. Virol. 61, 1086–1091.

Sanger, F., Coulson, A.R., 1975. A rapid method for determining sequences in DNA by primed synthesis with DNA polymerase. J. Mol. Biol. 94, 441–448.

Santman-Berends, I.M.G.A., van Wuijckhuise, L., Vellema, P., van Rijn, P.A., 2010. Vertical transmission of bluetongue virus serotype 8 virus in Dutch dairy herds in 2007. Vet. Microbiol. 141, 31–35. https://doi.org/10.1016/j.vetmic.2009.08.010.

Savini, G., MacLachlan, N.J., Sánchez-Vizcaino, J.-M., Zientara, S., 2008. Vaccines against bluetongue in Europe. Comp. Immunol. Microbiol. Infect. Dis. 31, 101–120. https://doi.org/10.1016/j.cimid.2007.07.006.

Schulz, C., Bréard, E., Sailleau, C., Jenckel, M., Viarouge, C., Vitour, D., Palmarini, M., Gallois, M., Höper, D., Hoffmann, B., Beer, M., Zientara, S., 2016. Bluetongue virus serotype 27: detection and characterization of two novel variants in Corsica, France. J. Gen. Virol. 97, 2073–2083. https://doi.org/10.1099/jgv.0.000557.

Schwartz-Cornil, I., Mertens, P.P.C., Contreras, V., Hemati, B., Pascale, F., Bréard, E., Mellor, P.S., MacLachlan, N.J., Zientara, S., 2008. Bluetongue virus: virology, pathogenesis and immunity. Vet. Res. 39, 46. https://doi.org/10.1051/vetres:2008023.

Sealfon, R.S., Lin, M.F., Jungreis, I., Wolf, M.Y., Kellis, M., Sabeti, P.C., 2015. FRESCo: finding regions of excess synonymous constraint in diverse viruses. Genome Biol. 16, 38. https://doi.org/10.1186/s13059-015-0603-7.

Sellers, R.F., 1975. Bluetongue in Cyprus. Aust. Vet. J. 51, 198–203.

Sellers, R.F., Pedgley, D.E., Tucker, M.R., 1978. Possible windborne spread of bluetongue to Portugal, June-July 1956. J. Hyg. 81, 189–196.

Sellers, R.F., Gibbs, E.P., Herniman, K.A., Pedgley, D.E., Tucker, M.R., 1979. Possible origin of the bluetongue epidemic in Cyprus, August 1977. J. Hyg. 83, 547–555.

Shaw, A.E., Ratinier, M., Nunes, S.F., Nomikou, K., Caporale, M., Golder, M., Allan, K., Hamers, C., Hudelet, P., Zientara, S., Breard, E., Mertens, P., Palmarini, M., 2013. Reassortment between two serologically unrelated bluetongue virus strains is flexible and can involve any genome segment. J. Virol. 87, 543–557. https://doi.org/10.1128/JVI.02266-12.

Spreull, J., 1905. Malarial catarrhal fever (bluetongue) of sheep in South Africa. J. Comp. Pathol. Ther. 18, 321–337. https://doi.org/10.1016/S0368-1742(05)80073-6.

Stewart, M., Hardy, A., Barry, G., Pinto, R.M., Caporale, M., Melzi, E., Hughes, J., Taggart, A., Janowicz, A., Varela, M., Ratinier, M., Palmarini, M., 2015. Characterization of a second open reading frame in genome segment 10 of bluetongue virus. J. Gen. Virol. 96, 3280–3293. https://doi.org/10.1099/jgv.0.000267.

Stuart, D.I., Gouet, P., Grimes, J., Malby, R., Diprose, J., Zientara, S., Burroughs, J.N., Mertens, P.P., 1998. Structural studies of orbivirus particles. Arch. Virol. Suppl. 14, 235–250.

Sun, E.C., Huang, L.P., Xu, Q.Y., Wang, H.X., Xue, X.M., Lu, P., Li, W.J., Liu, W., Bu, Z.G., Wu, D.L., 2016. Emergence of a novel bluetongue virus serotype, China 2014. Transbound. Emerg. Dis. 63, 585–589. https://doi.org/10.1111/tbed.12560.

Sutton, G., Grimes, J.M., Stuart, D.I., Roy, P., 2007. Bluetongue virus VP4 is an RNA-capping assembly line. Nat. Struct. Mol. Biol. 14, 449–451. https://doi.org/10.1038/nsmb1225.

Szmaragd, C., Wilson, A.J., Carpenter, S., Wood, J.L.N., Mellor, P.S., Gubbins, S., 2010. The spread of bluetongue virus serotype 8 in great Britain and its control by vaccination. PLoS One 5. https://doi.org/10.1371/journal.pone.0009353.

Takamatsu, H., Burroughs, J.N., Wade-Evans, A.M., Mertens, P.P., 1990. Identification of a bluetongue virus serotype 1-specific ovine helper T-cell determinant in outer capsid protein VP2. Virology 177, 396–400.

Theuns, S., Conceição-Neto, N., Zeller, M., Heylen, E., Roukaerts, I.D.M., Desmarets, L.M.B., Van Ranst, M., Nauwynck, H.J., Matthijnssens, J., 2016. Characterization of a genetically heterogeneous porcine rotavirus C, and other viruses present in the fecal virome of a non-diarrheic Belgian piglet. Infect. Genet. Evol. 43, 135–145. https://doi.org/10.1016/j.meegid.2016.05.018.

Van Borm, S., Wang, J., Granberg, F., Colling, A., 2016. Next-generation sequencing workflows in veterinary infection biology: towards validation and quality assurance. Rev. Sci. Tech. Int. Off. Epizoot. 35, 67–81. https://doi.org/10.20506/rst.35.1.2418.

van der Sluijs, M., Timmermans, M., Moulin, V., Noordegraaf, C.V., Vrijenhoek, M., Debyser, I., de Smit, A.J., Moormann, R., 2011. Transplacental transmission of Bluetongue virus serotype 8 in ewes in early and mid gestation. Vet. Microbiol. 149, 113–125. https://doi.org/10.1016/j.vetmic.2010.11.002.

van der Sluijs, M.T.W., Schroer-Joosten, D.P.H., Fid-Fourkour, A., Vrijenhoek, M.P., Debyser, I., Moulin, V., Moormann, R.J.M., de Smit, A.J., 2013. Transplacental transmission of Bluetongue virus serotype 1 and serotype 8 in sheep: virological and pathological findings. PLoS One 8, e81429. https://doi.org/10.1371/journal.pone.0081429.

Vandenbussche, F., Vanbinst, T., Verheyden, B., Van Dessel, W., Demeestere, L., Houdart, P., Bertels, G., Praet, N., Berkvens, D., Mintiens, K., Goris, N., De Clercq, K., 2008. Evaluation of antibody-ELISA and real-time RT-PCR for the diagnosis and profiling of bluetongue virus serotype 8 during the epidemic in Belgium in 2006. Vet. Microbiol. 129, 15–27. https://doi.org/10.1016/j.vetmic.2007.10.029.

Vandenbussche, F., Sailleau, C., Rosseel, T., Desprat, A., Viarouge, C., Richardson, J., Eschbaumer, M., Hoffmann, B., De Clerq, K., Bréard, E., Zientara, S., 2015. Full-genome sequencing of four bluetongue virus serotype 11 viruses. Transbound. Eemrg. Dis. 62, 565–571. https://doi.org/10.1111/tbed.12178.

Vitour, D., Doceul, V., Ruscanu, S., Chauveau, E., Schwartz-Cornil, I., Zientara, S., 2014. Induction and control of the type I interferon pathway by Bluetongue virus. Virus Res. 182, 59–70. https://doi.org/10.1016/j.virusres.2013.10.027.

Vögtlin, A., Hofmann, M.A., Nenniger, C., Renzullo, S., Steinrigl, A., Loitsch, A., Schwermer, H., Kaufmann, C., Thür, B., 2013. Long-term infection of goats with bluetongue virus serotype 25. Vet. Microbiol. 166, 165–173. https://doi.org/10.1016/j.vetmic.2013.06.001.

Vreede, F.T., Cloete, M., Napier, G.B., van Dijk, A.A., Viljoen, G.J., 1998. Sequence-independent amplification and cloning of large dsRNA virus genome segments by poly(dA)-oligonucleotide ligation. J. Virol. Methods 72, 243–247.

Webb, C.R., Floyd, T., Brien, S., Oura, C.A.L., Wood, J.L.N., 2011. Bluetongue serotype 8 vaccine coverage in northern and south-eastern England in 2008. Vet. Rec. 168, 428. https://doi.org/10.1136/vr.c6927.

Whetter, L.E., Maclachlan, N.J., Gebhard, D.H., Heidner, H.W., Moore, P.F., 1989. Bluetongue virus infection of bovine monocytes. J. Gen. Virol. 70 (Pt. 7), 1663–1676. https://doi.org/10.1099/0022-1317-70-7-1663.

Willgert, K.J.E., Schroedle, B., Schwermer, H., 2011. Spatial analysis of bluetongue cases and vaccination of Swiss cattle in 2008 and 2009. Geospat. Health 5, 227–237. https://doi.org/10.4081/gh.2011.175.

Wirblich, C., Bhattacharya, B., Roy, P., 2006. Nonstructural protein 3 of bluetongue virus assists virus release by recruiting ESCRT-I protein Tsg101. J. Virol. 80, 460–473. https://doi.org/10.1128/JVI.80.1.460-473.2006.

Worwa, G., Hilbe, M., Ehrensperger, F., Chaignat, V., Hofmann, M.A., Griot, C., Maclachlan, N.J., Thuer, B., 2009. Experimental transplacental infection of sheep with bluetongue virus serotype 8. Vet. Rec. 164, 499–500.

Worwa, G., Hilbe, M., Chaignat, V., Hofmann, M.A., Griot, C., Ehrensperger, F., Doherr, M.G., Thür, B., 2010. Virological and pathological findings in Bluetongue virus serotype 8 infected sheep. Vet. Microbiol. 144, 264–273. https://doi.org/10.1016/j.vetmic.2010.01.011.

Wright, I.M., 2014. Serological and Genetic Characterisation of Putative new Serotypes of Bluetongue Virus and Epizootic Haemorrhagic Disease Virus Isolated from an Alpaca. (Thesis).

Yang, H., Zhu, J., Li, H., Xiao, L., Wang, J., Li, N., Zhang, N., Kirkland, P.D., 2012. Full genome sequence of bluetongue virus serotype 4 from China. J. Virol. 86, 13122–13123. https://doi.org/10.1128/JVI.02393-12.

Yang, H., Xiao, L., Wang, J., Meng, J., Lv, M., Liao, D., Song, J., Gao, L., Xiong, H., He, Y., Niu, B., Chuang, X., Li, H., 2016. Phylogenetic characterization genome segment 2 of Bluetongue virus strains belonging to serotypes 5, 7 and 24 isolated for the first time in China during 2012 to 2014. Transbound. Emerg. Dis. 1317–1321. https://doi.org/10.1111/tbed.12479.

Zanella, G., Durand, B., Sellal, E., Breard, E., Sailleau, C., Zientara, S., Batten, C.A., Mathevet, P., Audeval, C., 2012. Bluetongue virus serotype 8: abortion and transplacental transmission in cattle in the Burgundy region, France, 2008-2009. Theriogenology 77, 65–72. https://doi.org/10.1016/j.theriogenology.2011.07.015.

Zanella, G., Martinelle, L., Guyot, H., Mauroy, A., De Clercq, K., Saegerman, C., 2013. Clinical pattern characterization of cattle naturally infected by BTV-8. Transbound. Emerg. Dis. 60, 231–237. https://doi.org/10.1111/j.1865-1682.2012.01334.x.

Zhou, E.M., Ridd, D., Riva, J., Fernando, L., Clavijo, A., 2001. Development and evaluation of an IgM-capture ELISA for detection of recent infection with bluetongue viruses in cattle. J. Virol. Methods 91, 175–182.

Zhu, J., Yang, H., Li, H., Xiao, L., Wang, J., Li, N., Zhang, N., 2013. Full-genome sequence of bluetongue virus serotype 1 (BTV-1) strain Y863, the first BTV-1 isolate of Eastern origin found in China. Genome Announc. 1https://doi.org/10.1128/genomeA.00403-13.

Zientara, S., Sailleau, C., Dauphin, G., Roquier, C., Rémond, E.M., Lebreton, F., Hammoumi, S., Dubois, E., Agier, C., Merle, G., Bréard, E., 2002. Identification of bluetongue virus serotype 2 (Corsican strain) by reverse-transcriptase PCR reaction analysis of segment 2 of the genome. Vet. Rec. 150, 598–601.

Zientara, S., Amat, J.P., Sailleau, C., Viarouge, C., Desprat, A., Vitour, D., Bréard, E., 2012. Difficulties in the interpretation of bluetongue RT-PCR results in France. Vet. Rec. 170, 599. https://doi.org/10.1136/vr.100485.

Zientara, S., Sailleau, C., Viarouge, C., Höper, D., Beer, M., Jenckel, M., Hoffmann, B., Romey, A., Bakkali-Kassimi, L., Fablet, A., Vitour, D., Bréard, E., 2014. Novel bluetongue virus in goats, Corsica, France, 2014. Emerg. Infect. Dis. 20, 2123–2125. https://doi.org/10.3201/eid2012.140924.

CHAPTER EIGHT

The Recently Discovered Bokeloh Bat Lyssavirus: Insights Into Its Genetic Heterogeneity and Spatial Distribution in Europe and the Population Genetics of Its Primary Host

Elisa Eggerbauer*^,2, Cécile Troupin†,2, Karsten Passior‡, Florian Pfaff§, Dirk Höper§, Antonie Neubauer-Juric¶, Stephanie Haberl¶, Christiane Bouchier||, Thomas C. Mettenleiter*, Hervé Bourhy†, Thomas Müller*,1, Laurent Dacheux†, Conrad M. Freuling*

*Institute of Molecular Virology and Cell Biology, WHO Collaborating Centre for Rabies Surveillance and Research, Friedrich-Loeffler-Institut, Greifswald-Insel Riems, Germany
†Institut Pasteur, Unit Lyssavirus Dynamics and Host Adaptation, WHO Collaborating Centre for Reference and Research on Rabies, Paris, France
‡Local Bat Conservation, Lower Saxony, Nordstemmen, Germany
§Institute of Diagnostic Virology, Friedrich-Loeffler-Institut, Greifswald-Insel Riems, Germany
¶Bavarian Health and Food Safety Authority, Oberschleißheim, Germany
||Institut Pasteur, Genomics Platform, Paris, France
[1]Corresponding author: e-mail address: thomas.mueller@fli.de

Contents

[2] Contributed equally.

Advances in Virus Research, Volume 99
ISSN 0065-3527
https://doi.org/10.1016/bs.aivir.2017.07.004

Abstract

In 2010, a novel lyssavirus named Bokeloh bat lyssavirus (BBLV) was isolated from a Natterer's bat (*Myotis nattereri*) in Germany. Two further viruses were isolated in the same country and in France in recent years, all from the same bat species and all found in moribund or dead bats. Here we report the description and the full-length genome sequence of five additional BBLV isolates from Germany ($n=4$) and France ($n=1$). Interestingly, all of them were isolated from the Natterer's bat, except one from Germany, which was found in a common Pipistrelle bat (*Pipistrellus pipistrellus*), a widespread and abundant bat species in Europe. The latter represents the first case of transmission of BBLV to another bat species. Phylogenetic analysis clearly demonstrated the presence of two different lineages among this lyssavirus species: lineages A and B. The spatial distribution of these two lineages remains puzzling, as both of them comprised isolates from France and Germany; although clustering of isolates was observed on a regional scale, especially in Germany. Phylogenetic analysis based on the mitochondrial cytochrome *b* (CYTB) gene from positive Natterer's bat did not suggest a circulation of the respective BBLV sublineages in specific Natterer's bat subspecies, as all of them were shown to belong to the *M. nattereri* sensu stricto clade/subspecies and were closely related (German and French positive bats). At the bat host level, we demonstrated that the distribution of BBLV at the late stage of the disease seems large and massive, as viral RNA was detected in many different organs.

1. INTRODUCTION

1.1 Rabies and Bats

Rabies is an acute, progressive incurable viral encephalitis that is transmitted following bites or scratches by infected mammals. While the name rabies is derived from the Latin name for "madness," the old Greeks derived the word lyssa, from lud or "violent"; this root is used in the name of the genus *Lyssavirus* (Jackson, 2013). Lyssaviruses, the causative agents of rabies, are negative single-strand RNA viruses of the *Rhabdoviridae* family, and alongside with the *Bornaviridae*, *Filoviridae*, *Paramyxoviridae*, *Pneumoviridae*, *Mymonaviridae*, *Nyamiviridae*, and *Sunviridae* forms the order Mononegavirales (Afonso et al., 2016). They are currently classified into 14 officially recognized (ICTV, 2015) and 2 yet unassigned species (Arechiga Ceballos et al., 2013;

Gunawardena et al., 2016). All lyssaviruses appear or are suspected to follow a rabies-like pathogenesis (Davis et al., 2015). With the advent of novel molecular techniques, the *Rhabdoviridae* family was extended (Dietzgen et al., 2017). In this chapter, we summarize the current situation in the genus *Lyssavirus* and with Bokeloh bat lyssavirus (BBLV) provide a recent example how the genus diversifies.

While rabies caused by the archetypical rabies virus (RABV, belonging to *Rabies lyssavirus* species) has been known for millennia (Jackson, 2013), causing tens of thousands of human deaths per year (Hampson et al., 2015); the association of rabies with bats (*Chiroptera*) was first demonstrated in the New World in the early 20th century (Pawan, 1936). Since the appearance of first descriptions of RABV in bats in the Americas, other divergent lyssavirus species have been detected in a wide range of chiropteran hosts, and bats are the recognized reservoirs for the majority of lyssaviruses (Banyard et al., 2013). Furthermore, there is strong ecological as well as phylogenetic support that bat-derived viruses have evolved long before those RABV of terrestrial carnivore origin (Badrane and Tordo, 2001; Troupin et al., 2016). However, the evolutionary history of lyssaviruses within bats is unresolved. For instance, the absence of non-RABV lyssaviruses in bats in the Americas, and the diversity of bat lyssaviruses elsewhere in the Old World remain enigmatic and may indicate an ancient coevolution and dispersal of lyssaviruses and their reservoir hosts, i.e., primarily bats (Rupprecht et al., 2017).

1.2 Bat Rabies in the Americas

Rabies in bats in the Americas is in some aspects different to the Old World. As introduced before, the sole causative agents are exclusively variants of classical RABV. However, RABV is found in many species of bat in the Americas (Banyard et al., 2013; Constantine, 2009; Sodre et al., 2010), and it was demonstrated that particular bat species are associated with specific RABV variants (Hughes et al., 2005; Streicker et al., 2010; Velasco-Villa et al., 2006). Among bats, cross-species transmissions (CSTs) seem to be related to the relative genetic distance of the host (Streicker et al., 2010). Also, bat RABV variants caused a historical sustained CST in raccoons (Velasco-Villa et al., 2008), and more recent CST in striped skunks (*Mephitis mephitis*), red foxes (*Vulpes vulpes*), and gray foxes (*Urocyon cinereoargenteus*) (Daoust et al., 1996; Kuzmin et al., 2012; Leslie et al., 2006). A bat rabies variant was also responsible for rabies cases in white-nosed coati (*Nasua narica*) (Arechiga-Ceballos et al., 2010).

In Latin America, besides insectivorous and frugivorous bat species, the common vampire bat (*Desmodus rotundus*), one of three blood-feeding or hematophagous bats found exclusively in this region, acts as a reservoir for rabies (Johnson et al., 2014). This behavior facilitates virus transmission not only to livestock, particularly cattle (Arellano-Sota, 1988; Johnson et al., 2014), but also causes human cases, particularly in remote areas with limited access to medical intervention (Condori-Condori et al., 2013; Stoner-Duncan et al., 2014; Streicker et al., 2012; Meynard et al., 2012).

1.3 Bat Rabies in Asia and Australia

Rabies in bats in the Asian part of Russia were described already in the 1970s. However, tracing back of these early isolates could not confirm the presence of bat-associated lyssaviruses (Kuzmin et al., 2006a). The first well documented and verified isolation of a lyssavirus from a bat occurred in 1991, when a novel lyssavirus, named Aravan virus (ARAV) was isolated from a lesser mouse-eared bat (*Myotis blythii*) in Kyrghyzstan. The discovery of Khujand virus (KHUV) in a whiskered bat (*Myotis mystacinus*) followed in 2001 (Kuzmin et al., 2003). One year later another novel lyssavirus virus, Irkut virus (IRKV) was discovered in the Irkutsk province of Russia, when a bat caught indoors died approximately 10 days later after developing clinical sings (Botvinkin et al., 2003). IRKV is insofar distinctive from the other Asian lyssaviruses as further cases were reported in China in a greater tubenosed bat (*Murina leucogaster*) (Liu et al., 2013), as well as a human rabies case in 2007 in the Russian Far East (Leonova et al., 2009). Most recently, Gannoruwa bat lyssavirus was isolated from Indian flying foxes (*Pteropus medius*) in Sri Lanka, after 62 grounded and dead found bats were tested for the presence of lyssaviruses (Gunawardena et al., 2016). Further surveillance and search for bat pathogens including lyssaviruses will likely expand the diversity and range of bat-associated lyssaviruses in Asia. In fact, serological surveys of bats indicate the presence of lyssaviruses in several countries in Southeast Asia, i.e., the Philippines (Arguin et al., 2002), Cambodia (Reynes et al., 2004), Thailand (Lumlertdacha et al., 2005), Bangladesh (Kuzmin et al., 2006b), China (Jiang et al., 2010), and Vietnam (Nguyen et al., 2014).

Australian bat lyssavirus (ABLV) was initially discovered in the brain of a black flying fox (*Pteropus alecto*) and is predominantly associated with fruit-eating bat species (Fraser et al., 1996; Weir et al., 2014). Interestingly, ABLV infection has only been confirmed by virus isolation in one insectivorous species, the yellow-bellied sheath-tailed bat (*Saccolaimus flaviventris*).

Genetically, ABLV is also delineated into two lineages associated with either fruit bats or insectivorous bats, respectively (Barrat, 2004). ABLV infection caused three known human cases in Queensland, Australia (Johnson et al., 2010), and spill-over infections were identified in two horses with neurological disease (Annand and Reid, 2014).

1.4 Bat Rabies in Africa

Of note, the first indication of the circulation of lyssaviruses other than RABV was demonstrated by early virological investigations in Africa (Boulger and Porterfield, 1958; Shope et al., 1970), and the virus isolated from a straw-colored fruit bat (*Eidolon helvum*) in Nigeria in 1956 was named Lagos bat virus (LBV). LBV infections were reported from several African countries, including detections in other bat species, dogs, cats, and one water mongoose (Markotter et al., 2006; Sabeta et al., 2007). Analysis of the virus phylogeny suggests the existence of four major LBV genetic lineages. The genetic distances between some lineages are greater than those established as a demarcation criterion for individual lyssavirus species (Markotter et al., 2008). Serosurveys revealed a high seroprevalence of antibodies against LBV, and thus indicated the circulation of LBV in various fruit bat species across parts of Africa (Dzikwi et al., 2010; Freuling et al., 2015; Hayman et al., 2008, 2012; Kalemba et al., 2017; Kuzmin et al., 2008a).

Closely related to LBV is Shimoni bat virus (SHIBV), which was isolated from a dead Commerson's leaf-nosed bat (*Hipposideros commersoni*) during a search for bat-associated pathogens in Kenya (Kuzmin et al., 2010). Comparative serological surveys in *Rousettus aegyptiacus* bats and *H. commersoni* bats in sympatric roosts in Kenya suggested that *H. commersoni* is a primary host species of SHIBV (Kuzmin et al., 2011).

Duvenhage virus (DUVV) was isolated in 1970 in South Africa following the death of a human that was bitten by an insectivorous bat (Meredith et al., 1971). Sporadically, DUVV was reported from insectivorous bats in Southern Africa (Paweska et al., 2006; Weyer et al., 2011) and caused two additional fatal human infections in South Africa (Paweska et al., 2006) and in the Netherlands (following a contamination in Kenya) (Van Thiel et al., 2008).

1.5 Bat Rabies in Europe

In Europe, the first rabid bat was diagnosed in 1954 in Germany (Mohr, 1957). Other and only sporadic reports indicated the presence of bat-associated rabies in Europe until the 1980s (Müller et al., 2007), when two human rabies cases due to bat contact sparked surveillance efforts in many European countries (Johnson et al., 2010). From 1977 until 2016, bat rabies was confirmed in 1180 cases during bat rabies surveillance activities across Europe (Source: Rabies Bulletin Europe database, found at: http://www.who-rabies-bulletin.org/, accessed May 2017). A recent study evaluating all available data on bat rabies surveillance across Europe concluded that bat rabies surveillance is not evenly conducted (Schatz et al., 2013). Therefore, while cases detected in bats indicate the circulation among bats in that region, the absence of evidence is not the evidence of absence.

Molecular characterization initially demonstrated that viruses isolated from European bats and human cases belong to two genetically diverse lyssaviruses, European bat lyssavirus type 1 (EBLV-1) and European bat lyssavirus type 2 (EBLV-2), belonging to *European bat 1 lyssavirus* and *European bat 2 lyssavirus* species, respectively (Amengual et al., 1997; Bourhy et al., 1993; Davis et al., 2005). While serotine bats (*Eptesicus serotinus* and *Eptesicus isabellinus*) are considered the primary reservoir hosts for EBLV-1, rabies cases in Daubenton's bats (*Myotis daubentonii*) and pond bats (*Myotis dasycneme*) were genetically characterized as EBLV-2 (Banyard et al., 2013). Besides the European bat lyssaviruses, two other unrelated bat lyssaviruses were also isolated from common bent-winged bats (*Miniopterus schreibersii*) in the Caucasus region (West Caucasian bat lyssavirus—WCBV—belonging to *West Caucasian bat lyssavirus species*, Kuzmin et al., 2008b) and in Spain (Lleida bat lyssavirus—LLEBV—(Arechiga Ceballos et al., 2013) belonging to the *Lleida bat lyssavirus* species), respectively.

In contrast to the Americas, CSTs are relatively rare in Europe. However, EBLV-1 CSTs have been documented in a number of occasions, including in a Dutch colony of captive Egyptian fruit bats (*R. aegyptiacus*) (Wellenberg et al., 2002). Also, EBLV-1 has been detected in sheep from Denmark (Ronsholt, 2002), in a stone marten (*Martes foina*) in Germany (Müller et al., 2004) and two domestic cats in France (Dacheux et al., 2009).

Sporadic human rabies cases following a bat bite have been described in Europe. In 1977, the first human rabies case associated with a bat bite in Europe was reported in the Ukraine (Kuzmin et al., 2006a). Another confirmed case in Russia that has been characterized was transmitted from a bat and occurred in 1985 (Selimov et al., 1989). The virus responsible for this infection was isolated and shown to be EBLV-1 (Amengual et al., 1997).

A Swiss biologist who had multiple bat bites died in Finland in 1985. Rabies as a diagnosis was confirmed by laboratory tests and EBLV-2 was isolated for the first time (Lumio et al., 1986). The second confirmed human rabies case with this lyssavirus was a 56-year-old bat conservationist from Angus, Scotland who died in November 2002 (Fooks et al., 2003). In all cases, the individual infected had a history of close contact with bats and none had received vaccination against rabies.

2. BBLV: A NOVEL BAT LYSSAVIRUS ISOLATED FIRST IN GERMANY

In 2010, another novel lyssavirus named BBLV was isolated from a Natterer's bat (*Myotis nattereri*) in Germany. The presence of lyssavirus antigens in the diseased bat was confirmed using fluorescence antibody test and immunohistochemistry with standard rabies biologics (Freuling et al., 2011). Two further viruses were isolated in Germany and in France, respectively, 2 years later, all from the same bat species and all found in moribund or dead bats (Freuling et al., 2013; Picard-Meyer et al., 2013).

Animal experimental studies demonstrated that BBLV was pathogenic for mice via intracranial and intramuscular inoculation routes, causing fatal encephalitis irrespective of the isolate used (Nolden et al., 2014). As typical to lyssaviruses, during infection BBLV formed intracytoplasmic inclusions, detected by staining with FITC-conjugated antinucleocapsid monoclonal antibodies. Kaplan–Meyer survival plots were significantly different to EBLV-2 and similar to EBLV-1, although BBLV is on a genetic level more closely related to the former (Nolden et al., 2014). The distinction between BBLV and other lyssavirus species was also evident in the antigenic patterns in reactions with antinucleocapsid monoclonal antibodies. This difference was supported by antigenic cartography where BBLV was separated from all characterized lyssaviruses, but related to phylogroup I viruses (Nolden et al., 2014). While the genome organization and sequence relationships are consistent with the classification as a lyssavirus, BBLV nucleotide sequences demonstrate 80% and 79% identity to the sequence of the most similar viruses KHUV and EBLV-2, respectively. Altogether, genetic distance, phylogenetic reconstructions, antigenic patterns, and ecologic features were sufficiently different to existing lyssavirus species, and the creation of the *Bokeloh bat lyssavirus* species was approved (ICTV, 2015).

2.1 The Natterer's Bat: The Presumed Primary Host of BBLV

For better understanding of virus–host interactions, i.e., between the BBLV and the Natterer's bat, knowledge on their ecological and phylogenetic characteristics are essential. *M. nattereri* (Kuhl, 1818), synonym *Myotis escalerae* (Cabrera, 1904), is a middle-sized, crevice-dwelling, nocturnal, insectivorous European bat species of the Western Palearctic belonging to the genus *Myotis*, subfamily *Myotinae*, chiropteran family *Vespertilionidae* of the suborder Vespertilioniformes. This bat was first described by the German naturalist and zoologist Heinrich Kuhl in 1818, who named it in honor of his Austrian colleague Johann Natterer (Beolens et al., 2009). Ancestors of *M. nattereri* can be traced back millions of years ago till the Pliocene (Sigé and Legendre, 1983). Paleontological evidence suggests an evolutionary segregation and subsequent development and adaptation of several subspecies at the beginning of the Holocene–Pleistocene (Horacek and Hanak, 1983; Hutterer et al., 2012). Its geographic range extends throughout most of Europe, from the Atlantic Ocean and the Mediterranean basin up to latitude of 60 degree north (Sweden, Finland) into western Russia, and Ukraine (Fig. 1). Populations of *M. nattereri* also occur in western and southwestern Asia Minor, the Caucasus region, the coastal regions of the Near East, the Kopetdag Mountains (Turkmenia, Iran), and northern Kazakhstan. Its historic range included Norway, in which for reasons unknown it is no longer found anymore. Presumably, Natterer's bats do not occur on the islands of Sardinia, Malta, and Kreta (Hutson et al., 2008; Topal, 2011). It can also be found in the western Atlas and coastal zones northwest Africa, although few records from north Africa indicate that the population is likely to be relatively small (Hutson et al., 2008).

Although *M. nattereri* occurs in various habitats from sea level up to an altitude of about 2000 m, in the middle European inland it is predominantly found in open countryside with scattered broad-leaved woodland and close to water bodies such as tree-lined river corridors (Smith, 2001) (http://www.eurobats.org). Natterer's bats are generally considered a sedentary species (Schober and Grimmberger, 1998). In general, roost switching seems to be very common among *M. nattereri*, especially in summer times (Smith, 2001) with transboundary migration flows in border areas of European countries. Bat banding data from 12 European countries reveal that migration distances between maternity, swarming, and hibernation roosts usually vary according to region and climatic conditions (Hutterer et al., 2005). While in southern Europe due to the moderate climate during winter times populations of Natterer's bats are believed not to be forced to move from one roost to another (DePaz et al., 1986), migration distances between those roost sites of their northern conspecifics are reported to range between 14 and 90 km, on average (Topal, 2011). However, there is increasing evidence that this species is at least facultative migrant suggesting that this bat

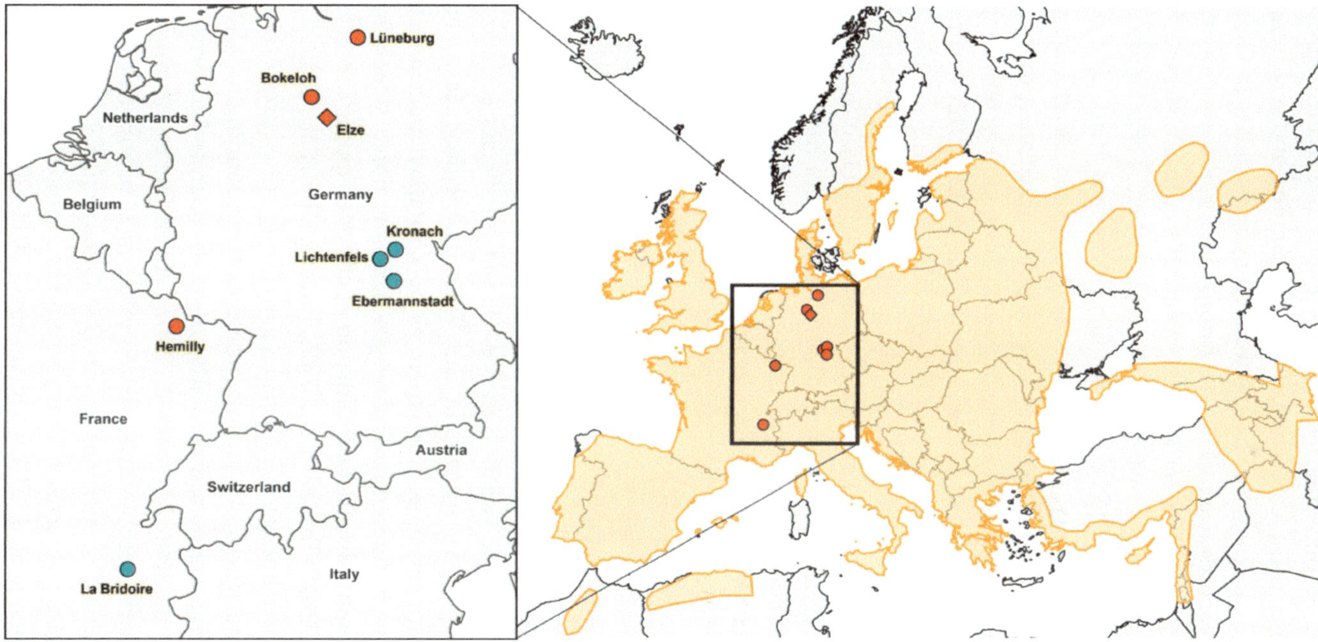

Fig. 1 Geographic origin of the BBLV isolates (*left*) in combination with the range (*right* in *yellow*) of the presumed reservoir species, the Natterer's bat (*Myotis nattereri*). The case associated with a pipistrelle bat (*Pipistrellus pipistrellus*) is indicated (*diamond*). Also, the delineation into the two genetic lineages A (*red*) and B (*blue*) is shown in the left box. *Source for the range data:* Myotis nattereri *In: IUCN, 2012.* IUCN Red List of Threatened Species. *Version 2012.1. http:/www.iucnredlist.org. Downloaded on 30 April, 2017.*

species is more a regional migrant than a true sedentary species (Hutterer et al., 2005). Based on registered one-way flights and return flights from winter to summer roost and between winter, summer, and swarming sites some individuals are known to have covered longer distances. In Germany, long distance records (return flight from wintering roost) of 157 km (Haensel, 2004) and up to 266 km and even 327 km in a male and female for one-way flights, respectively, have been described. These unusual long movements are believed to be dispersal flights (Steffens et al., 2005).

In the course of history, for the Natterer's bat species, several subspecies and other potentially related species have been identified and proposed (for review, see Topal, 2011). However, differentiating closely related species from another can represent a challenge (Padial et al., 2010; Shaffer and Thomson, 2007), because traditional taxonomy based on morphological characteristics can be misleading by the presence of phenotypic plesiomorphism or adaptative convergence (Salicini et al., 2011). With the advent of sophisticated genetic tools, molecular phylogenetics have been widely used to survey biological diversity of mammals, in particular chiroptera, with results shedding new light on molecular systematics and taxonomy of species and their biogeography. In fact, the genetic biodiversity of the Natterer's bat seems to be much more complex and is still not fully understood. Recent phylogenetic analysis based on nuclear and mitochondrial microsatellite loci revealed new phylogenetic relationships within the *M. nattereri* species complex (also referred to as *M. nattereri* sensu *lato*) and among closely related species (Puechmaille et al., 2012). Evidence is mounting that there is much higher genetic variation in the *M. nattereri* species complex than previously thought. *M. nattereri* in the Western Mediterranean, France, and Germany, for example, was shown to be a paraphyletic group that is composed of well-differentiated lineages across regions, which have a narrower geographic distribution with potentially overlapping contact zones (Halczok et al., 2017; Ibanez et al., 2006; Puechmaille et al., 2012; Salicini et al., 2011). These lineages match clades partially described previously, e.g., *M. nattereri* sensu stricto, *M. escalerae*, clade A corresponding to *Myotis* sp. A and clade B corresponding to *Myotis* sp. B (Garcia-Mudarra et al., 2009; Ibanez et al., 2006; Mayer et al., 2007). According to the unified species concept which is proposing separately evolving metapopulation lineages as the only necessary property of species (De Queiroz, 2007) some of these novel phylogenetic *M. nattereri* lineages could be considered as new cryptic species (Puechmaille et al., 2012; Salicini et al., 2011, 2013). More research is needed, however, to further substantiate these observations.

3. FURTHER INVESTIGATION IN THE PRESENCE OF BBLV IN BATS IN GERMANY AND IN FRANCE

3.1 Bat Lyssavirus Surveillance

In both countries, France and Germany rabies surveillance in bats are based on a passive surveillance system. In Germany, due to the federal structure, animal disease detection including routine bat rabies surveillance is performed at regional veterinary laboratories focusing on clinically suspect animals and especially those with human contact. Furthermore, in order to increase submission numbers an enhanced passive surveillance was initiated in 1998 (Schatz et al., 2014), whereby dead bats from different sources and collections, e.g., conservation agencies, universities, private collections, or museums were acquired. The submitting bat handlers morphologically delineated all submitted bats into species using phenotypical features (Dietz and von Helversen, 2004). In France, all bats associated with human contact are submitted for analysis to the National Reference Center for Rabies (NRC-R), at Institut Pasteur, Paris, whereas dead found bats without human contact are investigated by the National Reference Laboratory for Rabies, Anses-Malzeville.

3.2 Postmortem Diagnosis of Rabies in Bats

Similar to other rabies suspected animals, the *postmortem* diagnosis of rabies in bats is based on standard virological techniques. With the advent of molecular diagnostics, the rapid detection, confirmation, and differentiation of bat rabies isolates became possible and is additionally applied (Box 1).

3.3 Characterization of Positive Specimens

Historically, lyssaviruses were delineated based on reaction patterns of antinucleocapsid monoclonal antibodies (Schneider et al., 1985). The initial BBLV isolate was also subject to typing, confirming the antigenic distinction between BBLV and other lyssavirus species (Freuling et al., 2011). Furthermore, sequence comparisons allow for the inference of phylogenetic relationships and have been applied for lyssaviruses, including BBLV (Freuling et al., 2011; Picard-Meyer et al., 2013). Besides classical Sanger sequencing for bat species verification, next-generation sequencing (NGS) was applied to obtain full-length genome sequences of all available BBLVs (Box 2).

BOX 1 Rabies diagnostics

Postmortem rabies diagnosis of bats is usually performed on brain samples using the fluorescent antibody test (FAT) as described (Dean et al., 1996). In brief, brain smears are fixed on slides and stained with fluorescein isothiocyanate (FITC) conjugates (Bio-Rad, Marnes-la-Coquette, France and SIFIN, Berlin, Germany). All FAT results in France and positive or questionable results in Germany are confirmed by the rabies tissue culture infection test (RTCIT) using mouse neuroblastoma cell lines (Neuro-2a, ATCC CCL-131, or MNA 42/13, FLI Cat. No. 0229, respectively), essentially as described (Bourhy et al., 1989; Webster and Casey, 1996). RTCIT was performed using clarified brain suspension and homogenates from various other organs where available (Table 2).

Viral RNA detection: RNA isolation in France was performed following manufacturer's recommendations, from the brain suspensions or directly from organs following a digestion step (Dacheux et al., 2008), after immersion into 1 mL of TriReagent or 0.8 mL of TriReagent LS (Molecular Research Center), respectively. Diluted RNA (1:10, in nuclease-free water) was subjected to real-time RT-PCR (RT-qPCR) as previously described (Dacheux et al., 2016). In short, two RT-qPCRs were performed. The first was a probe-based (TaqMan) RT-qPCR for detecting the RABV species (pan-RABV RT-qPCR) and in the second RT-qPCR an intercalating dye (SYBR Green) was used for the detection of other lyssavirus species (pan-lyssa RT-qPCR). Furthermore, two conventional hemi-nested RT-PCRs targeting the nucleoprotein and the polymerase genes were performed as described (Dacheux et al., 2008; Davis et al., 2005), followed by Sanger sequencing of the respective amplicons for genotyping. In Germany, 250 μL of the brain and organ homogenates were immersed into 750 μL of peqGOLD TriFast (peqlab) and RNA extraction was performed according to manufacturer's instructions. The resulting RNA was eluted in 20 μL of nuclease-free water and subjected to multiplex R14 RT-qPCR as described elsewhere (Fischer et al., 2014) using the AgPath-IDTM One-Step RT-PCR Kit (Applied-Biosystems).

3.4 Confirmation of Further Cases of BBLV

In Germany, from 1998 until 2016 a total of 5591 bats were tested in the frame of enhanced passive surveillance, with the most frequently submitted bat species being the common pipistrelle bat (*Pipistrellus pipistrellus*) and the noctule bat (*Nyctalus noctula*). Of all samples, 67 tested positive for lyssavirus antigen representing primarily EBLV-1 isolated from its reservoir host, the serotine bat. Also, a few infections in other bat species, i.e., the common pipistrelle bat, the Nathusius' pipistrelle bat (*Pipistrellus nathusii*), and the brown long-eared bat (*Plecotus auritus*) were found as reported before

BOX 2 Sequencing

For high quality full-length genome sequencing of the BBLV isolates the protocol described by Nolden et al. (2016) can be applied. Briefly, extracted RNA is transcribed into cDNA using the cDNA synthesis system kit (Roche) together with random hexamer primers (Roche). Subsequently, either Illumina or Ion Torrent compatible libraries are generated using the respective adapters. After size selection, libraries are quantified and sequenced either on an Illumina MiSeq (Illumina) or Ion Torrent PGM (Life Technologies) instrument using MiSeq reagent kit v3 (Illumina) and HiQ-View chemistry (Life Technologies), respectively. Application of an iterative mapping and de novo assembly approach using 454 Sequencing Systems Software [version 3.0; Roche] generates reliable BBLV consensus sequences from each isolate. Subsequently, the consensus sequences are annotated for instance using the Geneious software package (Kearse et al., 2012).

An alternative sequencing protocol was described (Dacheux et al., 2010; Joffret et al., 2013). Here, total RNA is reverse transcribed using Superscript III reverse transcriptase with random hexamers (Invitrogen) and amplified using QuantiTect Whole Transcriptome kit (Qiagen). Amplified cDNA is fragmented and libraries are prepared using KAPA Library Preparation Kit (Kapa Biosystems). The resulting libraries are subsequently sequenced on Illumina MiSeq running in 325 nt single-end modus. The obtained sequencing reads are preprocessed to remove low quality or artificial bases using AlienTrimmer (Criscuolo and Brisse, 2013) as implemented in Galaxy (Goecks et al., 2010). To obtain viral genome sequences, trimmed reads are mapped along appropriate reference sequences using CLC Genomics Assembly Cell (CLC bio) as implemented in Galaxy. Annotation of the derived consensus sequence is performed using Sequencher 5.2.4 software (Gene Codes Corporation).

For bat species determination, partial or full cytochrome *b* (CYTB) sequences are used. These are usually obtained by Sanger sequencing essentially as described previously (Bickham et al., 2004; Schatz et al., 2014). To this end, wing membrane samples or brain material is used which are lysed overnight. Alternatively, CYTB-DNA is extracted from a patagium biopsy using the using DNeasy Blood and Tissue kit (Qiagen). CYTB-DNA is then amplified using specific primer pairs (CytB Uni fw and CytB Uni rev or LGL-765-F and LGL-766-R). The obtained PCR fragments are separated in a 1% agarose gel stained with ethidium bromide, purified using the QIAquick Gel Extraction Kit (Qiagen) and sequenced.

Regardless of the protocol used, sequences are made publicly available by submission to one of the INSDC (International Nucleotide Sequence Database Collaboration) databases. From these databases, the sequences can be retrieved using their individual accessions (see for instance Table 1). Alternatively, at the European Nucleotide Archive (part of the INSDC) sequences can combined into studies and be retrieved using a single study accession (see for instance PRJEB20392 for a number of BBLV and related CYTB sequences).

(Schatz et al., 2014). During the same period, five cases of BBLV infection were observed in Natterer's bats (*M. nattereri*, Table 1, Fig. 1) and an additional case of BBLV infection was observed in the common pipistrelle bat, representing the first case of BBLV in a species other than a Natterer's bat. The bat was initially found injured during daytime in December 2015 and was taken to a regional bat conservationist for rehabilitation and eventual return to the wild, but died 3 days later without clear signs of a disease. The bat was subject of sampling in the frame of retrospective bat lyssavirus surveillance, and initial fluorescent antibody test results were corroborated by real-time RT-PCR specific for BBLV.

In France during a 28-year period (from 1989 to 2016), a total of 546 dead bat specimens suspected of human exposure were received at the NRC-R for rabies diagnosis. Of those, 17 were confirmed to be rabid after postmortem diagnosis and found to be infected by EBLV-1, with 4 EBLV subtype a and 12 EBLV subtype b. All of them were serotine bats (*E. serotinus*), except for one in 2005 which was identified as a common pipistrelle bat (*P. pipistrellus*) and for which viral RNA (EBLV-1b) and viral antigens by ELISA were detected (data from the NRC-R). In 2013, a case of BBLV infection was diagnosed in a Natterer's bat (*M. nattereri*) in La Bridoire (Table 1, Fig. 1). It represented the second case of BBLV infection in France in a Natterer's bat, the first one being detected in 2012 (Picard-Meyer et al., 2013). Similar to the German BBLV isolates, all these isolates were detected using classical techniques for the postmortem diagnosis of rabies.

3.5 Virus Distribution in BBLV-Infected Bats

Different tissue samples were tested for virus isolation by RTCIT and for viral RNA detection by RT-qPCR from BBLV-infected bat specimens. Viable virus was isolated from all tested organs from the bat from Kronach (39608) and in 5/8 organs of the common pipistrelle bat from Elze (39068) (Table 2). In contrast, virus isolation was successful in four out of 10 organs from the bat from Ebermannstadt (42052). Similarly to virus isolation, lyssavirus RNA was detected by RT-qPCR in all tested organs in the bat from Kronach (39607) and Elze (39068), and in 8/10 tested organs for bats from Ebermannstadt (42052) (Table 2). For the BBLV positive bat from La Bridoire (13001FRA), virus could only be isolated from brain and lyssavirus RNA were detected in brain and kidney (Table 2).

Table 1 Details on BBLV Isolates Including, e.g., Host, Clinical Signs, and Accession Numbers

Isolate No.	Species	Date	Location	Clinical Signs	Duration of Clinical Stage	Virus Accession Numbers	References
21961	*Myotis nattereri*	Feb. 10	Bokeloh, Germany	Aggressive, directly approaching any moving object, vigorously trying to bite, and screaming ferociously, general weakness, lethargy, and paralysis. After the first 3 days of the clinical course, the bat stopped drinking and eating	10 days	JF311903	Freuling et al. (2011)
KC169985	*Myotis nattereri*	Jun. 12	Hemilly, France	Paralysis, weight loss	>2 days	KC169985	Picard-Meyer et al. (2013)
29008/29028	*Myotis nattereri*	Oct. 12	Lichtenfels, Germany	Prominent biting behavior, aggressiveness, tremor of the wings, and was unwilling to drink	6 days	KF245925	Freuling et al. (2013)
13001FRA	*Myotis nattereri*	Aug. 13	La Bridoire, France	Found dead by two children, who received postexposure prophylaxis	N/A	MF043188	This chapter
39608	*Myotis nattereri*	May 15	Kronach, Germany	Animal was found grounded and was transferred to a rehabilitation center where it showed overt aggressive behavior. Eventually, the animal died and was submitted for rabies testing	N/A	LT839617	

Continued

Table 1 Details on BBLV Isolates Including, e.g., Host, Clinical Signs, and Accession Numbers—cont'd

Isolate No.	Species	Date	Location	Clinical Signs	Duration of Clinical Stage	Virus Accession Numbers	References
39068	*Pipistrellus pipistrellus*	Dec. 15	Elze, Germany	Animal was found grounded during daytime and was cared for in a rehabilitation center, animal refused to eat and drink and died without other clinical signs of rabies	>3 days	LT839642	
41021	*Myotis nattereri*	Aug. 16	Lüneburg, Germany	Difficulty in swallowing, refused to eat and drink, motion stimulated vocalization, defensive wing beating	9 days	LT839644	
42052/42053	*Myotis nattereri*	Aug. 16	Ebermannstadt, Germany	Animal was found grounded and died the same day, following refusal to drink and eat	>1 day	LT839643	

Table 2 Virological Investigations on Tissue Samples for Different Bats Infected With BBLV (RT-PCR/RTCIT)

	Isolate			
Organ	**39068 (Elze)**	**39608 (Kronach)**	**13001FRA (La Bridoire)**	**42052 (Ebermannstadt)**
Brain	+/+	+/+	+/+	+/+
Tongue	+/+	+/+	−/−	+/+
Kidney	+/−	+/+	+/−	+/−
Pectoral muscle	+/+	+/+	−/−	+/+
Lung	+/+	+/+	−/−	+/−
Heart	+/+	+/+	−/−	+/−
Liver	+/−	+/+	n.a.	+/−
Spleen	+/−	+/+	n.a.	−/−
Salivary gland	n.a.	+/+	n.a.	−/+
Bladder	n.a.	+/+	n.a.	n.a.
Intestine	n.a.	n.a.	−/−	+/−

n.a., not analyzed.

3.6 Sequence and Phylogenetic Analyses

Full-length genome sequences were obtained using NGS for the five new BBLV isolates (Box 2) and were analyzed together with the three sequences already available (Table 1). Phylogenetic analysis of these eight genomes showed a differentiation of BBLV sequences into two lineages. Lineage A included BBLV isolates 21961, 39068, and 41021 from lower Saxony, Germany, and the isolate KC169985 from Hemilly, France, whereas lineage B included BBLV isolates 29008, 39608, and 42052 from Bavaria, Germany, and the isolate 13001FRA from La Bridoire, France (Figs. 1–3). Nucleotide identity within lineage A was above 98.8% and within lineage B above 96.8%. The heterogeneity of 3.2% in lineage B is primarily caused by the genetically divergent isolate 13001FRA from La Bridoire, France, whereas the BBLV isolates from Bavaria (29008, 39608, and 42052) have a nucleotide identity of above 99.2%. The same applies to lineage A with the BBLV isolates from lower Saxony (21961, 39068, and 41021) having a nucleotide

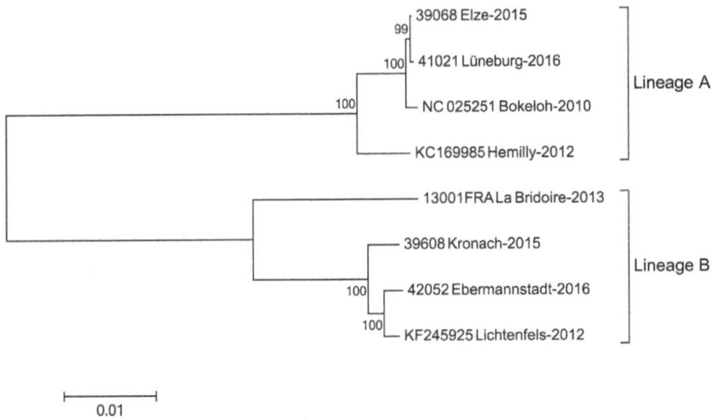

Fig. 2 Phylogenetic relationship of BBLV isolates from available full-length genome sequences using maximum likelihood method. Following an alignment with MAFFT (Katoh and Standley, 2013), the genetic distances were calculated using GTR+G as the optimal substitution model with 1000 bootstrap replications as implemented in MEGA6 (Tamura et al., 2013). The tree is drawn to scale, with branch lengths measured in the number of substitutions per site. BBLV can be differentiated into two lineages: lineage A included BBLV isolates 21961, 39068, and 41021 from lower Saxony, Germany and the isolate from Hemilly, France, whereas lineage B included BBLV isolates 29008, 39608, and 42052 from Bavaria, Germany and the isolate 13001FRA from La Bridoire, France.

identity above 99.7%. Nucleotide identities between the two lineages varied between 92.7% and 93.2%, depending on the isolates.

4. DISCUSSION

Since its first and recent discovery in 2010 in Bokeloh, Germany, five additional cases of BBLV were found in Germany and two in France. In contrast, most other recently discovered lyssaviruses from insectivorous bats, i.e., ARAV, KHUV, West Caucasian bat virus (WCBV), SHIBV, and LLEBV, have all been detected only once (Banyard et al., 2013). Diagnostically, as with other lyssaviruses, all BBLV isolates have been detected by techniques used for the *postmortem* diagnosis of rabies, based on antinucleocapsid antibodies conjugated with fluorescent dyes. Evidently, modern molecular tools offer advantages, e.g., increased speed, sensitivity, and possibilities for genetic characterization (Fischer et al., 2012; Fooks et al., 2012). Specifically, RT-qPCRs dedicated to large spectrum detection of lyssaviruses are now available and suitable for the diagnosis of emerging lyssaviruses such as BBLV (Dacheux et al., 2016). In addition, multiplex

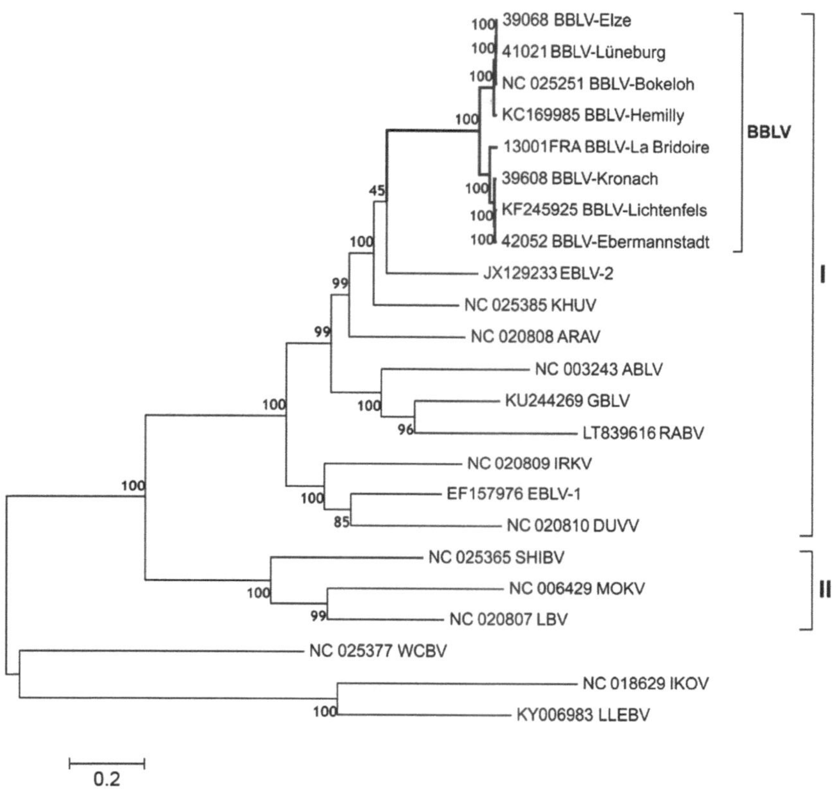

Fig. 3 Phylogenetic relationship of all lyssavirus species representatives and all BBLV isolates inferred using the maximum likelihood method based on the GTR model. A discrete Gamma distribution was used to model evolutionary rate differences among sites. The tree is drawn to scale, with branch lengths measured in the number of substitutions per site, and bootstrap values (1000 replication) shown next to the branches. Evolutionary analyses were conducted in MEGA6 (Tamura et al., 2013). Virus species' associations to phylogroups I and II are also indicated.

RT-qPCRs have improved the differentiation of bat rabies into the individual bat lyssavirus species (Fischer et al., 2014). Also, it was demonstrated before that undirected molecular detection methods were useful in identifying novel BBLV cases. The bat brain from the BBLV case from Lichtenfels, Germany (29008) was subjected to microarray analysis and in parallel to NGS (Freuling et al., 2013). Similarly, the brain sample of the bat infected with the French isolate 13001FRA was tested using a novel high-density resequencing microarray (called VirID), based on the two previous versions of the PathogenID microarray (Dacheux et al., 2010) and was successfully detected (data not shown). As for the previous versions, this detection was possible with the presence of conserved nucleotide region in the polymerase genes of prototype viruses comprising seven species of lyssavirus (namely RABV, LBV, MOKV, DUVV, EBLV-1, EBLV-2, and ABLV) and an artificial consensus sequence based on all of these species.

Phylogenetic analysis performed on the full-length genome sequences of all BBLV isolates known to date clearly demonstrated the presence of two different lineages among this lyssavirus species: lineages A and B, with nucleotide identities between them above 92%. The spatial distribution of these two lineages is puzzling, as both lineages comprise isolates from France and Germany. However, on a regional scale all isolates from Bavaria cluster closely together, and the same can be observed for the isolates from Lower Saxony, while the French viruses appear more distinct (Figs. 1 and 2).

At the bat host level, the distribution of BBLV at the late stage of the disease seems large and massive, because viral RNA and infectious virus were detected in several tissues and organs, including tongue, kidney, pectoral muscle, lung, heart, liver, spleen, salivary glands, bladder, and intestine, confirming earlier studies in one bat from Lichtenfels, Germany (Freuling et al., 2013). This large distribution was found for Natterer's bats as well as for the common pipistrelle bat. An exception was observed with the French bat, for which the presence of viral RNA was observed only in kidney and brain. This is probably the consequence of the poor general condition of this bat carcass.

The sudden emergence of BBLV is somewhat puzzling as surveillance activities across Europe and particularly in Germany and in France have not much changed in the last few years. In fact, in Europe, more than 1000 bat rabies cases have been reported, and no significant changes in reported numbers have been observed in recent years (Fig. 4).

Although surveillance efforts were further encouraged by the Eurobats agreement in 2006 (Anon, 2006), bat rabies surveillance in Europe remained heterogeneous (Schatz et al., 2013). This may have contributed to the fact that all BBLV isolates were discovered in France and Germany, two countries in which continuous bat rabies surveillance have been performed for decades (Picard-Meyer et al., 2014; Schatz et al., 2014). However, heterogeneous surveillance on a spatial level does not explain the number of novel cases of BBLV observed. Also, there is no indication that bat rabies cases in Natterer's bats were found in the past that had not been characterized (Brass, 1994; Kappeler, 1989; Schatz et al., 2013).

In Germany, besides routine surveillance targeting suspect animals and public health-related incidents, in 1998 enhanced passive surveillance was

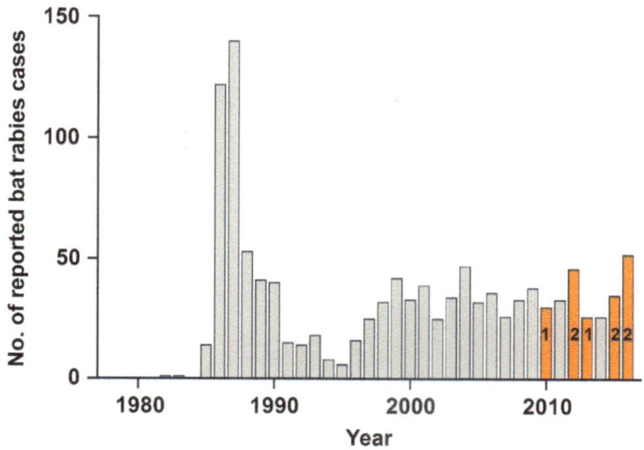

Fig. 4 Graph showing the annual number of bat rabies cases in Europe as reported to the database of the WHO Rabies Bulletin Europe. The years in which BBLV was detected are highlighted and the number of BBLV cases per year are indicated.

started to include bats, which would otherwise be missed for lyssavirus surveillance. In total, the combined surveillance efforts resulted in sampling of 8365 bats, with the most submitted bat species being the common pipistrelle. Although the pipistrelle bat is not considered as a primary host for any known bat lyssavirus species, three lyssavirus infections were discovered in this bat species. One was a case of EBLV-1 infection as described before (Schatz et al., 2014). A similar infection of EBLV-1 in common pipistrelle bat was observed in France (data from the NRC-R). The other was the case of BBLV in a common pipistrelle from Elze described here. The latter represents the first evidence of BBLV in another bat species apart from the Natterer's bat. All other cases of BBLV were discovered in Natterer's bats, leading to the assumption that this bat species is the primary host for BBLV. When material was available, bats species were determined by partial CYTB gene sequencing, since bat species determination due to morphological features can be challenging, especially when dealing with cryptic species. Also, when carcasses arrive in the laboratory, they may be incomplete or in a state of degradation, thus making it difficult to verify the bat species. Indeed, the Natterer's bat from Ebermannstadt, Germany was initially thought to be a Daubenton's bat and only phylogenetic analysis of partial CYTB sequence revealed that it was in fact a Natterer's bat (Fig. 5). Besides bat speciation, the combined molecular analysis of both host and virus is also essential for drawing further epidemiological conclusions, as exemplified for RABV in arctic foxes from Greenland (Hanke et al., 2016).

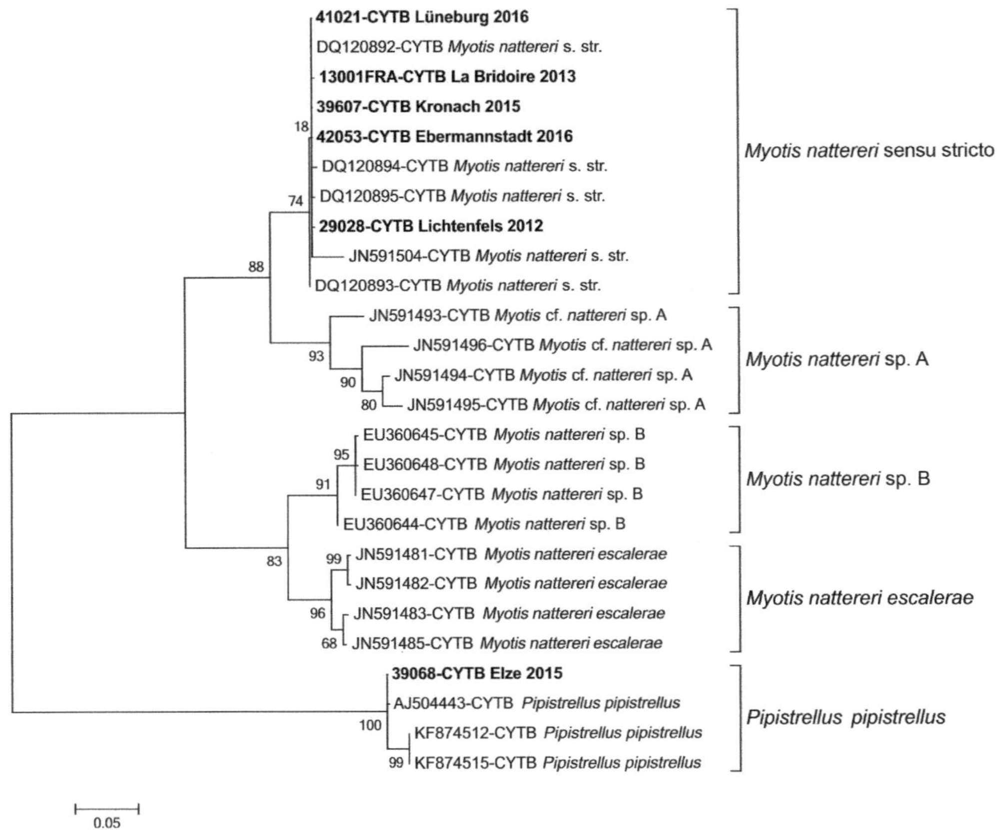

Fig. 5 See legend on opposite page.

As indicated previously, full-length analyses demonstrated that two distinct lineages of BBLV circulate among European Natterer's bats. On the host side, recent phylogenetic analyses revealed a much higher genetic variation in the *M. nattereri* species complex than previously thought. In Central Europe, *M. nattereri* was shown to be a paraphyletic group that is composed of four well-differentiated lineages, i.e., *M. nattereri* sensu stricto, *M. nattereri escalerae*, *M. nattereri* sp. A, and *M. nattereri* sp. B (Halczok et al., 2017; Ibanez et al., 2006; Puechmaille et al., 2012; Salicini et al., 2011, 2013). All Natterer's bats found positive for BBLV belong to the subspecies *M. nattereri* sensu stricto (Fig. 5), which is not surprising, as this subspecies is known to occur in Germany and in France (Salicini et al., 2011, 2013). However, the distinct two lineages of BBLV do not seem to be related to a circulation in specific Natterer's bat subspecies.

In a recent study correlating genetic distances of Natterer's bats and their associated astroviruses, identical astrovirus haplotypes were not shared between different sampling regions suggesting that astroviruses are mostly transmitted among host colonies at regional scale. The fact that some virus haplotypes found in several different regions had high genetic similarity implied that occasional transmission across regions occurred (Halczok et al., 2017). However, against the background that this bat species is a regional migrant species (Hutterer et al., 2005), the same can be assumed for BBLV, where the respective outliers could be a result of yet not documented long-distance migration. Evidently, for BBLV this analysis is restricted to the available dataset and further isolations may support this delineation. Interestingly, the same observation was made with EBLV-1 in serotine bats where almost identical nucleotide sequences from within certain geographical regions of the Netherlands and Germany indicate genomic stability during the transmission cycle of these virus variants but with occasional geographic spread or intermixing (Freuling et al., 2012;

Fig. 5 Phylogenetic relationship of partial CYTB sequences (484 nt) of bats found positive for BBLV and representatives of *Myotis nattereri* clades (Salicini et al., 2011, 2013). While the bat from Elze (Lower Saxony) was identified as a common pipistrelle bat (*Pipistrellus pipistrellus*), all other bats found positive for BBLV were closely related and classified as Natterer's bats (*Myotis nattereri*), and more precisely belong to the *Myotis nattereri* sensu stricto clade/subspecies. Phylogenetic analysis was inferred by using the maximum likelihood method based on the Hasegawa–Kishino–Yano model (HKY) with invariant sites (G) conducted in MEGA6 (Tamura et al., 2013). The tree is drawn to scale, with branch lengths measured in the number of substitutions per site, and bootstrap values (1000 replication) shown next to the branches.

Van der Poel et al., 2005). However, on a European scale there seems to be no clear correlation between migration behavior of lyssavirus reservoir species and the phylogeographical clustering of virus isolates. For EBLV-1, the primary host serotine bat is a rather sedentary species, which is not reflected in the clustering of isolates (Davis et al., 2005). On the other hand, the Daubenton's bat as the reservoir for EBLV-2 can migrate over larger distances. However, sequence analysis for EBLV-2 revealed a strong geographical clustering (Jakava-Viljanen et al., 2010). Further research, including and combining host biology, migration data, as well as molecular characterization of host and virus is essential to understand transmission dynamics and gain further insights into the epidemiology of bat–virus interaction. This may contribute to predict virus spread and establish risk assessments for virus transmission to humans.

BBLV was shown to be as pathogenic in mice as EBLV-1 (Nolden et al., 2014) and it can be assumed that BBLV infections in humans potentially lead to an encephalitis eventually causing death. Therefore, BBLV alongside with the other known bat lyssaviruses in Europe represents a low but undeniable public health threat. Unlike other remote regions in the world where people do not get into close contacts with bats, in Europe research and species conservation programs require the handling of bats by bat workers. While in the eastern parts of Germany, the cumulative number of bat handlings from 2000 to 2010 for serotine bats was 1260, those for *M. daubentonii* and *M. dasycneme* were 22,245 and 1086, respectively, with 37,140 by far the most handlings were recorded for the Natterer's bat (Bat Marking Centre, Dresden). This underlines the importance of adequate prophylaxis for bat handlers. In addition, accidental exposure to bats could occur for individuals other than bat handlers. For example in 2015, a total of 128 patients consulted an antirabic clinic (ARC) in France after being exposed to a bat in France metropolitan, representing 2.6% of the total number of patients consulting such clinic ($n = 8336$) (data from NRC-R, available at Pasteur, 2017). Some of these patients could be exposed to laboratory confirmed rabid bats, as it was the case for the two children who were in contact with the BBLV-infected bat from La Bridoire, France. Considering the relatedness of BBLV to other members of the phylogroup I lyssaviruses (Fig. 3), it appears likely that persons who received rabies prophylaxis will be protected. In fact, a recent study showed that available human inactivated vaccines elicit an antibody response that cross-neutralizes BBLV and therefore, are considered to confer protection against BBLV and other European bat lyssaviruses (Malerczyk et al., 2014; Nolden et al., 2014). Similarly, serum samples from French patients preventively vaccinated against rabies were analyzed for the

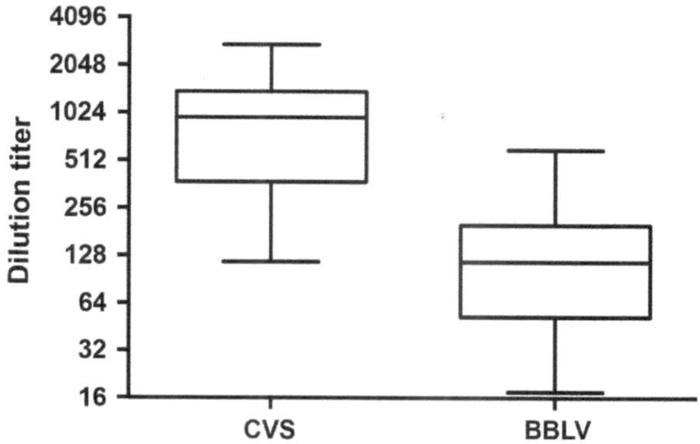

Fig. 6 Comparison of seroneutralization against RABV (CVS) and BBLV (13001FRA) with 37 sera of patients vaccinated against rabies using a modified RFFIT (Feyssaguet et al., 2007). The dilution value to obtain a 50% viral inhibition is visualized as box plots, with the first, third quartiles are indicated at the bottom, and top of the box, respectively, and the median is indicated with the band. The neutralizing activity was found to be significantly lower (paired Student *t*-test, *P*-value < 0.05) against BBLV (mean value $= 157$, standard deviation $= 137$) compared to CVS (mean value $= 907$, standard deviation $= 623$), with a 5.8-fold decrease in the titer dilution.

potency of neutralization against the CVS rabies strain (used as the reference virus) and 13001FRA isolate adapted to cell culture. Seroneutralization of BBLV was observed, but with a 5.8-fold decrease compared to CVS (Fig. 6). In any case, as there are currently no other vaccines available (Evans et al., 2012), consistent preventive vaccination of all persons working with bats and PEP of all persons exposed to bats should be given priority as recommended in France (Haut Conseil de la santé publique, 2013; Lafeuille et al., 2005) and in Germany (Delere et al., 2011).

Despite the recent discovery of a novel bat lyssavirus, lyssaviruses have been characterized in only a small proportion of recognized species (Banyard et al., 2013). Future surveillance of bat-associated pathogens using highly sensitive and novel high-throughput technologies, e.g., NGS (Dacheux et al., 2014) or pan viral microarrays, will likely increase the number of detected pathogens including novel lyssavirus species. This may also lead to further discoveries of BBLV in countries along the distribution of the Natterer's bat.

ACKNOWLEDGMENTS

We thank Patrick Zitzow, Jeannette Kliemt, and Patrick Wysocki at the Friedrich-Loeffler-Institut and Marion Tanguy, Pascal Cozette, Lauriane Kergoat, Birgit von Kölln-Braun, and Rachel Lavenir for excellent technical assistance. Furthermore, we thank Roland Heuser for submitting bats from Lower Saxony to retrospective bat rabies surveillance. We are also grateful to the staff of the ARCs of Chambery and Paris, France, who were in-charge of the two French exposed patients to BBLV, and especially to Philippe Poujol for providing the serum samples from those patients. This work was supported by an intramural collaborative research grant on Lyssaviruses at the Friedrich-Loeffler-Institut and by European Union Seventh Framework Programme PREDEMICS (grant number 278433). There is no conflict of interest in relation to this chapter.

REFERENCES

Afonso, C.L., Amarasinghe, G.K., Banyai, K., Bao, Y., Basler, C.F., Bavari, S., Bejerman, N., Blasdell, K.R., Briand, F.X., Briese, T., Bukreyev, A., Calisher, C.H., Chandran, K., Cheng, J., Clawson, A.N., Collins, P.L., Dietzgen, R.G., Dolnik, O., Domier, L.L., Durrwald, R., Dye, J.M., Easton, A.J., Ebihara, H., Farkas, S.L., Freitas-Astua, J., Formenty, P., Fouchier, R.A., Fu, Y., Ghedin, E., Goodin, M.M., Hewson, R., Horie, M., Hyndman, T.H., Jiang, D., Kitajima, E.W., Kobinger, G.P., Kondo, H., Kurath, G., Lamb, R.A., Lenardon, S., Leroy, E.M., Li, C.X., Lin, X.D., Liu, L., Longdon, B., Marton, S., Maisner, A., Muhlberger, E., Netesov, S.V., Nowotny, N., Patterson, J.L., Payne, S.L., Paweska, J.T., Randall, R.E., Rima, B.K., Rota, P., Rubbenstroth, D., Schwemmle, M., Shi, M., Smither, S.J., Stenglein, M.D., Stone, D.M., Takada, A., Terregino, C., Tesh, R.B., Tian, J.H., Tomonaga, K., Tordo, N., Towner, J.S., Vasilakis, N., Verbeek, M., Volchkov, V.E., Wahl-Jensen, V., Walsh, J.A., Walker, P.J., Wang, D., Wang, L.F., Wetzel, T., Whitfield, A.E., Xie, J.T., Yuen, K.Y., Zhang, Y.Z., Kuhn, J.H., 2016. Taxonomy of the order Mononegavirales: update 2016. Arch. Virol. 161, 2351–2360.

Amengual, B., Whitby, J.E., King, A., Cobo, J.S., Bourhy, H., 1997. Evolution of European bat lyssaviruses. J. Gen. Virol. 78, 2319–2328.

Annand, E.J., Reid, P.A., 2014. Clinical review of two fatal equine cases of infection with the insectivorous bat strain of Australian bat lyssavirus. Aust. Vet. J. 92, 324–332.

Anon, 2006. Agreement on the Conservation of Populations of Bats in Europe (EUROBATS). In: 5th Session of the Meeting of Parties, Resolution 5.2, Bat Rabies. Available at: http://www.eurobats.org/sites/default/files/documents/pdf/Meeting_of_Parties/MoP5_Record_Annex5_Res5_2_bat_rabies.pdf.

Arechiga Ceballos, N., Moron, S.V., Berciano, J.M., Nicolas, O., Lopez, C.A., Juste, J., Nevado, C.R., Setien, A.A., Echevarria, J.E., 2013. Novel lyssavirus in bat, Spain. Emerg. Infect. Dis. 19, 793–795.

Arechiga-Ceballos, N., Velasco-Villa, A., Shi, M., Flores-Chavez, S., BarroN, B., Cuevas-Dominguez, E., Gonzalez-Origel, A., Aguilar-Setien, A., 2010. New rabies virus variant found during an epizootic in white-nosed coatis from the Yucatan Peninsula. Epidemiol. Infect. 138, 1586–1589.

Arellano-Sota, C., 1988. Biology, ecology, and control of the vampire bat. Rev. Infect. Dis. 10, 615–619.

Arguin, P.M., Murray-Lillibridge, K., Miranda, M.E.G., Smith, J.S., Calaor, A.B., Rupprecht, C.E., 2002. Serologic evidence of Lyssavirus infections among bats, the Philippines. Emerg. Infect. Dis. 8, 258–262.

Badrane, H., Tordo, N., 2001. Host switching in lyssavirus history from the Chiroptera to the Carnivora orders. J. Virol. 75, 8096–8104.

Banyard, A.C., Hayman, D., Freuling, C.M., Müller, T., Fooks, A.R., Johnson, N., 2013. Bat rabies. In: Jackson, A.C. (Ed.), Rabies: Scientific Basis of the Disease and Its Management. Academic Press, New York, pp. 215–267.

Barrat, J., 2004. Australian Bat Lyssavirus. The University of Queensland, St. Lucia.

Beolens, B., Watkins, M., Grayson, M., 2009. The Eponym Dictionary of Mammals. Johns Hopkins University Press, Baltimore, MD (574 p).

Bickham, J.W., Patton, J.C., Schlitter, D.A., Rautenbach, I.L., Honeycutt, R.L., 2004. Molecular phylogenetics, karyotypic diversity, and partition of the genus Myotis (Chiroptera: Vespertilionidae). Mol. Phylogenet. Evol. 33, 333–338.
Botvinkin, A.D., Poleschuk, E.M., Kuzmin, I., Borisova, T.I., Gazaryan, S.V., Yager, P., Rupprecht, C.E., 2003. Novel lyssaviruses isolated from bats in Russia. Emerg. Infect. Dis. 9, 1623–1625.
Boulger, L.R., Porterfield, J.S., 1958. Isolation of a virus from Nigerian fruit bats. Trans. R. Soc. Trop. Med. Hyg. 52, 421–424.
Bourhy, H., Rollin, P.E., Vincent, J., Sureau, P., 1989. Comparative field evaluation of the fluorescent-antibody test, virus isolation from tissue culture, and enzyme immunodiagnosis for rapid laboratory diagnosis of rabies. J. Clin. Microbiol. 27, 519–523.
Bourhy, H., Kissi, B., Tordo, N., 1993. Molecular diversity of the lyssavirus genus. Virology 194, 70–81.
Brass, D.A., 1994. Rabies-related viruses from bats in Europe and Asia. In: Rabies in Bats: Natural History and Public Health Implications. Livia Press, Ridgefield.
Cabrera, A., 1904. Ensayo monográfico sobre los quirópteros de España. Mem. Soc. Esp. Hist. Nat. 2, 249–292.
Condori-Condori, R.E., Streicker, D.G., Cabezas-Sanchez, C., Velasco-Villa, A., 2013. Enzootic and epizootic rabies associated with vampire bats, Peru. Emerg. Infect. Dis. 19, 1463–1469.
Constantine, D.G., 2009. Bat Rabies and Other Lyssavirus Infections. US Geological Survey Circular 1329, Reston.
Criscuolo, A., Brisse, S., 2013. AlienTrimmer: a tool to quickly and accurately trim off multiple short contaminant sequences from high-throughput sequencing reads. Genomics 102, 500–506.
Dacheux, L., Reynes, J.-M., Buchy, P., Sivuth, O., Diop, B.M., Rousset, D., Rathat, C., Jolly, N., Dufourcq, J.-B., Nareth, C., Diop, S., Iehlé, C., Rajerison, R., Sadorge, C., Bourhy, H., 2008. A reliable diagnosis of human rabies based on analysis of skin biopsy specimens. Clin. Infect. Dis. 47, 1410–1417.
Dacheux, L., Larrous, F., Mailles, A., Boisseleau, D., Delmas, O., Biron, C., Bourchier, C., Ilari, F., Lefranc, T., Raffi, F., Goudal, M., Bourhy, H., 2009. European bat lyssavirus transmission among cats, Europe. Emerg. Infect. Dis. 15, 280–284.
Dacheux, L., Wacharapluesadee, S., Hemachudha, T., Meslin, F.X., Buchy, P., Reynes, J.M., Bourhy, H., 2010. More accurate insight into the incidence of human rabies in developing countries through validated laboratory techniques. PLoS Negl. Trop. Dis. 4, e765.
Dacheux, L., Cervantes-Gonzalez, M., Guigon, G., Thiberge, J.M., Vandenbogaert, M., Maufrais, C., Caro, V., Bourhy, H., 2014. A preliminary study of viral metagenomics of French bat species in contact with humans: identification of new mammalian viruses. PLoS One 9, e87194.
Dacheux, L., Larrous, F., Lavenir, R., Lepelletier, A., Faouzi, A., Troupin, C., Nourlil, J., Buchy, P., Bourhy, H., 2016. Dual combined real-time reverse transcription polymerase chain reaction assay for the diagnosis of lyssavirus infection. PLoS Negl. Trop. Dis. 10, e0004812.
Daoust, P.Y., Wandeler, A.I., Casey, G.A., 1996. Cluster of rabies cases of probable bat origin among red foxes in Prince Edward Island, Canada. J. Wildl. Dis. 32, 403–406.
Davis, P.L., Holmes, E.C., Larrous, F., van der Poel, W.H., Tjornehoj, K., Alonso, W., Bourhy, H., 2005. Phylogeography, population dynamics, and molecular evolution of European bat lyssaviruses. J. Virol. 79, 10487–10497.
Davis, B.M., Rall, G.F., Schnell, M.J., 2015. Everything you always wanted to know about rabies virus (but were afraid to ask). Annu. Rev. Virol. 2, 451–471.
De Queiroz, K., 2007. Species concepts and species delimitation. Syst. Biol. 56, 879–886.

Dean, D.J., Abelseth, M.K., Athanasiu, P., 1996. The fluorescence antibody test. In: Meslin, F.X., Kaplan, M.M., Koprowski, H. (Eds.), Laboratory Techniques in Rabies. World Health Organization, Geneva, pp. 88–93.
Delere, Y., Wichmann, O., Müller, T., Freuling, C.M., Roggendorf, M., Ross, R.S., 2011. Tollwut in Deutschland: Gelöstes Problem oder versteckte Gefahr? Epidemiol. Bull. 8, 57–61.
DePaz, O., Fernández, R., Benzai, J., 1986. El allimaniento chiropteros en el centro de la peninsula Iberica durante el periodo 1977–1986. Bol. Estc. Cent. Ecol. 15, 113–138.
Dietz, C., von Helversen, O., 2004. Identification Key to the Bats of Europe—Electronical Publication, Version 1.0. URL, www.uni-tuebingen.de/tierphys/Kontakt/mitarbeiter_seiten/dietz.htm.
Dietzgen, R.G., Kondo, H., Goodin, M.M., Kurath, G., Vasilakis, N., 2017. The family Rhabdoviridae: mono- and bipartite negative-sense RNA viruses with diverse genome organization and common evolutionary origins. Virus Res. 227, 158–170.
Dzikwi, A.A., Kuzmin, I.I., Umoh, J.U., Kwaga, J.K., Ahmad, A.A., Rupprecht, C.E., 2010. Evidence of Lagos bat virus circulation among Nigerian fruit bats. J. Wildl. Dis. 46, 267–271.
Evans, J.S., Horton, D.L., Easton, A.J., Fooks, A.R., Banyard, A.C., 2012. Rabies virus vaccines: is there a need for a pan-lyssavirus vaccine? Vaccine 30, 7447–7454.
Feyssaguet, M., Dacheux, L., Audry, L., Compoint, A., Morize, J.L., Blanchard, I., Bourhy, H., 2007. Multicenter comparative study of a new ELISA, PLATELIA RABIES II, for the detection and titration of anti-rabies glycoprotein antibodies and comparison with the rapid fluorescent focus inhibition test (RFFIT) on human samples from vaccinated and non-vaccinated people. Vaccine 25, 2244–2251.
Fischer, M., Hoffmann, B., Freuling, C., Müller, T., Beer, M., 2012. Perspectives on molecular detection methods of lyssaviruses. Berl. Munch. Tierarztl. Wochenschr. 125, 264–271.
Fischer, M., Freuling, C.M., Muller, T., Wegelt, A., Kooi, E.A., Rasmussen, T.B., Voller, K., Marston, D.A., Fooks, A.R., Beer, M., Hoffmann, B., 2014. Molecular double-check strategy for the identification and characterization of European lyssaviruses. J. Virol. Methods 203, 23–32.
Fooks, A.R., McElhinney, L.M., Pounder, D.J., Finnegan, C.J., Mansfield, K., Johnson, N., Brookes, S.M., Parsons, G., White, K., McIntyre, P.G., Nathwani, D., 2003. Case report: isolation of a European bat lyssavirus type 2a from a fatal human case of rabies encephalitis. J. Med. Virol. 71, 281–289.
Fooks, A.R., McElhinney, L., Horton, D., Banyard, A.C., Johnson, N., Marston, D.A., Freuling, C.M., Hoffmann, B., Tu, C., Fehlner-Gardiner, C., Sabeta, C., Cliquet, F., Müller, T., Rupprecht, C., 2012. Molecular tools for rabies diagnosis in animals. In: Fooks, A.R., Müller, T. (Eds.), Rabies Control—Towards Sustainable Prevention at the Source. Compendium of the OIE Global Conference on Rabies Control, Incheon-Seoul, 7–9 September 2011, Republic of Korea. OIE, Paris, pp. 75–86.
Fraser, G.C., Hooper, P.T., Lunt, R.A., Gould, A.R., Gleeson, L.J., Hyatt, A.D., Russell, G.M., Kattenbelt, J.A., 1996. Encephalitis caused by a lyssavirus in fruit bats in Australia. Emerg. Infect. Dis. 2, 327–331.
Freuling, C.M., Beer, M., Conraths, F.J., Finke, S., Hoffmann, B., Keller, B., Kliemt, J., Mettenleiter, T.C., Muhlbach, E., Teifke, J.P., Wohlsein, P., Muller, T., 2011. Novel lyssavirus in Natterer's bat, Germany. Emerg. Infect. Dis. 17, 1519–1522.
Freuling, C., Hoffmann, B., Selhorst, T., Conraths, F.J., Kliemt, A., Schatz, J., Müller, T., 2012. New insights into the genetics of EBLV-1 from Germany. Berl. Munch. Tierarztl. Wochenschr. 125, 259–263.
Freuling, C.M., Abendroth, B., Beer, M., Fischer, M., Hanke, D., Hoffmann, B., Höper, D., Just, F., Mettenleiter, T.C., Schatz, J., Müller, T., 2013. Molecular diagnostics for the

detection of Bokeloh bat lyssavirus in a bat from Bavaria, Germany. Virus Res. 177, 201–204.
Freuling, C.M., Binger, T., Beer, M., Adu-Sarkodie, Y., Schatz, J., Fischer, M., Hanke, D., Hoffmann, B., Höper, D., Mettenleiter, T.C., Oppong, S.K., Drosten, C., Müller, T., 2015. Lagos bat virus transmission in an Eidolon helvum bat colony, Ghana. Virus Res. 210, 42–45.
Garcia-Mudarra, J.L., Ibanez, C., Juste, J., 2009. The Straits of Gibraltar: barrier or bridge to Ibero-Moroccan bat diversity? Biol. J. Linn. Soc. 96, 434–450.
Goecks, J., Nekrutenko, A., Taylor, J., 2010. Galaxy Team, Galaxy: a comprehensive approach for supporting accessible, reproducible, and transparent computational research in the life sciences. Genome Biol. 11, R86.
Gunawardena, P.S., Marston, D.A., Ellis, R.J., Wise, E.L., Karawita, A.C., Breed, A.C., McElhinney, L.M., Johnson, N., Banyard, A.C., Fooks, A.R., 2016. Lyssavirus in Indian flying foxes, Sri Lanka. Emerg. Infect. Dis. 22, 1456–1459.
Haensel, J., 2004. Zum saisonbedingten Ortswechsel der Fransenfledermaus (Myotis nattereri) im Raum Berlin/Brandenburg unter besonderer Berücksichtigung des Schwärmverhaltens. Nyctalus 9, 305–327.
Halczok, T.K., Fischer, K., Gierke, R., Zeus, V., Meier, F., Tress, C., Balkema-Buschmann, A., Puechmaille, S.J., Kerth, G., 2017. Evidence for genetic variation in Natterer's bats (Myotis nattereri) across three regions in Germany but no evidence for co-variation with their associated astroviruses. BMC Evol. Biol. 17, 5.
Hampson, K., Coudeville, L., Lembo, T., Sambo, M., Kieffer, A., Attlan, M., Barrat, J., Blanton, J.D., Briggs, D.J., Cleaveland, S., Costa, P., Freuling, C.M., Hiby, E., Knopf, L., Leanes, F., Meslin, F.X., Metlin, A., Miranda, M.E., Muller, T., Nel, L.H., Recuenco, S., Rupprecht, C.E., Schumacher, C., Taylor, L., Vigilato, M.A., Zinsstag, J., Dushoff, J., Global Alliance for Rabies Control Partners for Rabies, Prevention, 2015. Estimating the global burden of endemic canine rabies. PLoS Negl. Trop. Dis. 9, e0003709.
Hanke, D., Freuling, C.M., Fischer, S., Hueffer, K., Hundertmark, K., Nadin-Davis, S., Marston, D., Fooks, A.R., Bötner, A., Mettenleiter, T.C., Beer, M., Rasmussen, T.B., Müller, T.F., Höper, D., 2016. Spatio-temporal analysis of the genetic diversity of Arctic rabies viruses and their reservoir hosts in Greenland. PLoS Negl. Trop. Dis. 10, e0004779.
Haut Conseil de la santé publique, 2013. Vaccination antirabique préventive, traitement post-exposition et suivi sérologique des personnes régulièrement exposées au virus de la rage (voyageurs, professionnels, chiroptérologues). In: Vaccinations contre la rage et prophylaxie post-exposition Recommandations. Haut Conseil de la santé publique. Collection Avis et Rapports.
Hayman, D.T., Fooks, A.R., Horton, D., Suu-Ire, R., Breed, A.C., Cunningham, A.A., Wood, J.L., 2008. Antibodies against Lagos bat virus in megachiroptera from West Africa. Emerg. Infect. Dis. 14, 926–928.
Hayman, D.T., Fooks, A.R., Rowcliffe, J.M., McCrea, R., Restif, O., Baker, K.S., Horton, D.L., Suu-Ire, R., Cunningham, A.A., Wood, J.L., 2012. Endemic Lagos bat virus infection in Eidolon helvum. Epidemiol. Infect. 140, 2163–2171.
Horacek, I., Hanak, V., 1983. Comments on the systematic and phylogeny of Myotis nattereri (Kuhl, 1818). Myotis 21–22, 20–29.
Hughes, G.J., Orciari, L., Rupprecht, C.E., 2005. Evolutionary timescale of rabies virus adaptation to North-American bats inferred from the substitution rate of the nucleoprotein gene. J. Gen. Virol. 86, 1467–1474.
Hutson, A.M., Aulagnier, S., Spitzenberger, F., 2008. Myotis nattereri. IUCN Red List of Threatened Species. Version 2013.1. http://www.iucnredlist.org/details/14135/0.
Hutterer, R., Ivanova, T., Meyer-Cords, C., Rodrigues, L., 2005. Natterer's bat (Myotis nattereri). In: Bat Migrations in Europe: A Review of Banding Data and Literature.

Schriftenreihe Naturschutz und Biologische Vielfalt, German Agency for Nature Conservation, Bonn, pp. 83–84.
Hutterer, R., Montermann, C., Weigt, M., 2012. A Holocene bat fauna from the Eifel Mountains, Germany. Vespertilio 16, 159–164.
Ibanez, C., García-Mudarra, J.L., Ruedi, M., Stadelmann, B., Juste, J., 2006. The Iberian contribution to cryptic diversity in European bats. Acta Chiropt. 8, 277–297.
ICTV, 2015. Virus Taxonomy: 2015 Release. International Committee on Taxonomy of Viruses (ICTV). https://talk.ictvonline.org/taxonomy/.
Jackson, A.C., 2013. History of rabies research. In: Jackson, A.C., Wunner, W. (Eds.), Rabies. Academic Press, New York, pp. 1–13.
Jakava-Viljanen, M., Lilley, T., Kyheroinen, E.M., Huovilainen, A., 2010. First encounter of European bat lyssavirus type 2 (EBLV-2) in a bat in Finland. Epidemiol. Infect. 138, 1581–1585.
Jiang, Y., Wang, L., Lu, Z., Xuan, H., Han, X., Xia, X., Zhao, F., Tu, C., 2010. Seroprevalence of rabies virus antibodies in bats from Southern China. Vector Borne Zoonotic Dis. 10, 177–181.
Joffret, M.L., Bouchier, C., Grandadam, M., Zeller, H., Maufrais, C., Bourhy, H., Despres, P., Delpeyroux, F., Dacheux, L., 2013. Genomic characterization of Sebokele virus 1 (SEBV1) reveals a new candidate species among the genus Parechovirus. J. Gen. Virol. 94, 1547–1553.
Johnson, N., Vos, A., Freuling, C., Tordo, N., Fooks, A.R., Müller, T., 2010. Human rabies due to lyssavirus infection of bat origin. Vet. Microbiol. 142, 151–159.
Johnson, N., Arechiga-Ceballos, N., Aguilar-Setien, A., 2014. Vampire bat rabies: ecology, epidemiology and control. Viruses 6, 1911–1928.
Kalemba, L.N., Niezgoda, M., Gilbert, A.T., Doty, J.B., Wallace, R.M., Malekani, J.M., Carroll, D.S., 2017. Exposure to lyssaviruses in bats of the Democratic Republic of the Congo. J. Wildl. Dis. 53, 408–410.
Kappeler, A., 1989. Bat rabies surveillance in Europe. Rabies Bull. Eur. 13, 12–13.
Katoh, K., Standley, D.M., 2013. MAFFT multiple sequence alignment software version 7: improvements in performance and usability. Mol. Biol. Evol. 30, 772–780.
Kearse, M., Moir, R., Wilson, A., Stones-Havas, S., Cheung, M., Sturrock, S., Buxton, S., Cooper, A., Markowitz, S., Duran, C., Thierer, T., Ashton, B., Meintjes, P., Drummond, A., 2012. Geneious Basic: an integrated and extendable desktop software platform for the organization and analysis of sequence data. Bioinformatics 28, 1647–1649.
Kuzmin, I.V., Orciari, L.A., Arai, Y.T., Smith, J.S., Hanlon, C.A., Kameoka, Y., Rupprecht, C.E., 2003. Bat lyssaviruses (Aravan and Khujand) from Central Asia: phylogenetic relationships according to N, P and G gene sequences. Virus Res. 97, 65–79.
Kuzmin, I.V., Botvinkin, A.D., Poleschuk, E.M., Orciari, L.A., Rupprecht, C.E., 2006a. Bat rabies surveillance in the former Soviet Union. Dev. Biol. (Basel) 125, 273–282.
Kuzmin, I.V., Niezgoda, M., Carroll, D.S., Keeler, N., Hossain, M.J., Breiman, R.F., Ksiazek, T.G., Rupprecht, C.E., 2006b. Lyssavirus surveillance in bats, Bangladesh. Emerg. Infect. Dis. 12, 486–488.
Kuzmin, I.V., Niezgoda, M., Franka, R., Agwanda, B., Markotter, W., Beagley, J.C., Urazova, O.Y., Breiman, R.F., Rupprecht, C.E., 2008a. Lagos bat virus in Kenya. J. Clin. Microbiol. 46, 1451–1461.
Kuzmin, I.V., Wu, X., Tordo, N., Rupprecht, C.E., 2008b. Complete genomes of Aravan, Khujand, Irkut and West Caucasian bat viruses, with special attention to the polymerase gene and non-coding regions. Virus Res. 136, 81–90.
Kuzmin, I.V., Mayer, A.E., Niezgoda, M., Markotter, W., Agwanda, B., Breiman, R.F., Rupprecht, C.E., 2010. Shimoni bat virus, a new representative of the lyssavirus genus. Virus Res. 149, 197–210.

Kuzmin, I.V., Turmelle, A.S., Agwanda, B., Markotter, W., Niezgoda, M., Breiman, R.F., Rupprecht, C.E., 2011. Commerson's leaf-nosed bat (Hipposideros commersoni) is the likely reservoir of Shimoni bat virus. Vector Borne Zoonotic Dis. 11, 1465–1470.

Kuzmin, I.V., Shi, M., Orciari, L.A., Yager, P.A., Velasco-Villa, A., Kuzmina, N.A., Streicker, D.G., Bergman, D.L., Rupprecht, C.E., 2012. Molecular inferences suggest multiple host shifts of rabies viruses from bats to mesocarnivores in Arizona during 2001–2009. PLoS Pathog. 8, e1002786.

Lafeuille, H., Lerasle, S., Astoul, J., Bourhy, H., Cliquet, F., Dufour, B., Goudal, M., Mailles, A., Morer, I., Moutou, F., Rotivel, Y., 2005. Rapport du groupe de travail du conseil superieur d'hygiene publique de France. Avis du Haut Conseil de Santé Publique, Paris.

Leonova, G.N., Belikov, S.I., Kondratov, I.G., Krylova, N.V., Pavlenko, E.V., Romanova, E.V., Chentsova, I.V., Petukhova, S.A., 2009. A fatal case of bat lyssavirus infection in Primorye Territory of the Russian Far East. Rabies Bull. Eur. 33, 5–8.

Leslie, M.J., Messenger, S., Rohde, R.E., Smith, J., Cheshier, R., Hanlon, C., Rupprecht, C.E., 2006. Bat-associated rabies virus in skunks. Emerg. Infect. Dis. 12, 1274–1277.

Liu, Y., Zhang, S.F., Zhao, J.H., Zhang, F., Hu, R.L., 2013. Isolation of Irkut virus from a Murina leucogaster bat in China. PLoS Negl. Trop. Dis. 7.

Lumio, J., Hillbom, M., Roine, R., Ketonen, L., Haltia, M., Valle, M., Neuvonen, E., Lähdewirta, J., 1986. Human rabies of bat origin in Europe. Lancet 327, 378.

Lumlertdacha, B., Boongird, K., Wanghongsa, S., Wacharapluesadee, S., Chanhome, L., Khawplod, P., Hemachudha, T., Kuzmin, I., Rupprecht, C.E., 2005. Survey for bat lyssaviruses, Thailand. Emerg. Infect. Dis. 11, 232–236.

Malerczyk, C., Freuling, C., Gniel, D., Giesen, A., Selhorst, T., Müller, T., 2014. Cross-neutralization of purified Chick embryo cell vaccine (PCECV) against different lyssavirus species. Hum. Vaccin. Immunother. (in press).

Markotter, W., Kuzmin, I., Rupprecht, C.E., Randles, J., Sabeta, C.T., Wandeler, A.I., Nel, L.H., 2006. Isolation of Lagos bat virus from water mongoose. Emerg. Infect. Dis. 12, 1913–1918.

Markotter, W., Kuzmin, I., Rupprecht, C.E., Nel, L.H., 2008. Phylogeny of Lagos bat virus: challenges for lyssavirus taxonomy. Virus Res. 135, 10–21.

Mayer, F., Dietz, C., Kiefer, A., 2007. Molecular species identification boosts bat diversity. Front. Zool. 4, 4.

Meredith, C.D., Prossouw, A.P., Koch, H.P., 1971. An unusual case of human rabies thought to be of chiropteran origin. S. Afr. Med. J. 45, 767–769.

Meynard, J.B., Flamand, C., Dupuy, C., Mahamat, A., Eltges, F., Queuche, F., Renner, J., Fontanella, J.M., Hommel, D., Dussart, P., Grangier, C., Djossou, F., Dacheux, L., Goudal, M., Berger, F., Ardillon, V., Krieger, N., Bourhy, H., Spiegel, A., 2012. First human rabies case in French Guiana, 2008: epidemiological investigation and control. PLoS Negl. Trop. Dis. 6 (2), e1537.

Mohr, W., 1957. Die Tollwut. Med. Klin. 52, 1057–1060.

Müller, T., Cox, J., Peter, W., Schäfer, R., Johnson, N., McElhinney, L.M., Geue, J.L., Tjornehoj, K., Fooks, A.R., 2004. Spill-over of European bat lyssavirus type 1 into a stone marten (Martes foina) in Germany. J. Veterinary Med. Ser. B 51, 49–54.

Müller, T., Johnson, N., Freuling, C.M., Fooks, A.R., Selhorst, T., Vos, A., 2007. Epidemiology of bat rabies in Germany. Arch. Virol. 152, 273–288.

Nguyen, A.T., Nguyen, T.T., Noguchi, A., Nguyen, D.V., Ngo, G.C., Thong, V.D., Olowokure, B., Inoue, S., 2014. Bat lyssaviruses, northern Vietnam. Emerg. Infect. Dis. 20, 161–163.

Nolden, T., Banyard, A.C., Finke, S., Fooks, A.R., Hanke, D., Höper, D., Horton, D., Mettenleiter, T.C., Müller, T., Teifke, J.P., Freuling, C.M., 2014. Comparative studies on genetic, antigenic and pathogenic characteristics of Bokeloh bat lyssavirus. J. Gen. Virol. 95, 1647–1653.

Nolden, T., Pfaff, F., Nemitz, S., Freuling, C.M., Höper, D., Müller, T., Finke, S., 2016. Reverse genetics in high throughput: rapid generation of complete negative strand RNA virus cDNA clones and recombinant viruses thereof. Sci. Rep. 6, 23887.

Padial, J.M., Miralles, A., De la Riva, I., Vences, M., 2010. The integrative future of taxonomy. Front. Zool. 7, 16.

Pasteur, I. 2017. https://www.pasteur.fr/fr/sante-publique/cnr/les-cnr/rage/rapports-d-activite (Institut Pasteur).

Pawan, J.L., 1936. The transmission of paralytic rabies in Trinidad by the vampire bat (Desmodus rotunduns murinus Wagner, 1840). Ann. Trop. Med. Parasitol. 30, 101–130.

Paweska, J.T., Blumberg, L.H., Liebenberg, C., Hewlett, R.H., Grobbelaar, A.A., Leman, P.A., Croft, J.E., Nel, L.H., Nutt, L., Swanepoel, R., 2006. Fatal human infection with rabies-related Duvenhage virus, South Africa. Emerg. Infect. Dis. 12, 1965–1967.

Picard-Meyer, E., Servat, A., Robardet, E., Moinet, M., Borel, C., Cliquet, F., 2013. Isolation of Bokeloh bat lyssavirus in Myotis nattereri in France. Arch. Virol. 1–8.

Picard-Meyer, E., Robardet, E., Arthur, L., Larcher, G., Harbusch, C., Servat, A., Cliquet, F., 2014. Bat rabies in France: A 24-year retrospective epidemiological study. PLoS One 9, e98622.

Puechmaille, S.J., Allegrini, B., Boston, E.S.M., Dubourg-Savage, M.J., Evin, A., Knochel, A., Le Bris, Y., Lecoq, V., Lemaire, M., Rist, D., Teeling, E.C., 2012. Genetic analyses reveal further cryptic lineages within the Myotis nattereri species complex. Mamm. Biol. 77, 224–228.

Reynes, J.M., Molia, S., Audry, L., Hout, S., Ngin, S., Walston, J., Bourhy, H., 2004. Serologic evidence of lyssavirus infection in bats, Cambodia. Emerg. Infect. Dis. 10, 2231–2234.

Ronsholt, L., 2002. A new case of European bat lyssavirus (EBL) infection in Danish sheep. Rabies Bull. Eur. 26, 15.

Rupprecht, C., Kuzmin, I., Meslin, F., 2017. Lyssaviruses and rabies: current conundrums, concerns, contradictions and controversies. F1000Res 6, 184.

Sabeta, C., Markotter, W., Mohale, D., Shumba, W., Wandeler, A., Nel, L., 2007. Mokola virus in domestic mammals, South Africa. Emerg. Infect. Dis. 13, 1371–1373.

Salicini, I., Ibanez, C., Juste, J., 2011. Multilocus phylogeny and species delimitation within the Natterer's bat species complex in the Western Palearctic. Mol. Phylogenet. Evol. 61, 888–898.

Salicini, I., Ibáñez, C., Juste, J., 2013. Deep differentiation between and within Mediterranean glacial refugia in a flying mammal, the Myotis nattereri bat complex. J. Biogeogr. 40, 1182–1193.

Schatz, J., Fooks, A.R., McElhinney, L., Horton, D., Echevarria, J., Vazquez-Moron, S., Kooi, E.A., Rasmussen, T.B., Müller, T., Freuling, C.M., 2013. Bat rabies surveillance in Europe. Zoonoses Public Health 60, 22–34.

Schatz, J., Freuling, C.M., Auer, E., Goharriz, H., Harbusch, C., Johnson, N., Kaipf, I., Mettenleiter, T.C., Mühldorfer, K., Mühle, R.-U., Ohlendorf, B., Pott-Dörfer, B., Prüger, J., Ali, H.S., Stiefel, D., Teubner, J., Ulrich, R.G., Wibbelt, G., Müller, T., 2014. Enhanced passive bat rabies surveillance in indigenous bat species from Germany—a retrospective study. PLoS Negl. Trop. Dis. 8, e2835.

Schneider, L.G., Barnard, B.J.H., Schneider, H.P., Odegaard, O.A., Mueller, J., Selimov, M., Cox, J.H., Wandeler, A.I., Blancou, J., Meyer, S., 1985. Application of monoclonal antibodies for epidemiological investigations and oral vaccination studies. In: Kuwert, E., Merieux, C. (Eds.), Rabies in the Tropics. Springer Verlag, Berlin, pp. 47–59.

Schober, W., Grimmberger, E., 1998. Die Fledermäuse Europas—Kennen, bestimmen, schützen, second ed. Franckh-Kosmos Verlags-GmbH, Stuttgart.
Selimov, M.A., Tatarov, A.G., Botvinkin, A.D., Klueva, E.V., Kulikova, L.G., Khismatullina, N.A., 1989. Rabies-related Yuli virus; identification with a panel of monoclonal antibodies. Acta Virol. 33, 542–546.
Shaffer, H.B., Thomson, R.C., 2007. Delimiting species in recent radiations. Syst. Biol. 56, 896–906.
Shope, R.E., Murphy, F.A., Harrison, A.K., Causey, O.R., Kemp, G.E., Simpson, D.I.H., Moore, D.L., 1970. Two African viruses serologically and morphologically related to rabies virus. J. Virol. 6, 690–692.
Sigé, B., Legendre, S., 1983. L'histoire des peuplements de chiroptères du Bassin Méditerranéen: l'apport comparé des reemplissages karstiques et des dépots fluviolacustres. Mém. Biospéol. 10, 209–225.
Smith, P.G., 2001. Habitat Preference, Range Use and Roosting Ecology in Natterer's Bats, Myotis nattereri, in a Grasland-Woodland Landscape. University of Aberdeen, Aberdeen.
Sodre, M.M., Gama, A.R.D., Almeida, M.F.D., 2010. Updated list of bat species positive for rabies in Brazil. Rev. Inst. Med. Trop. Sao Paulo 52, 75–81.
Steffens, R., Zöphel, U., Brockmann, D., 2005. 40 Jahre Fledermausmarkierungszentrale Dresden—methodische Hinweise und Ergebnisübersicht. Sächsisches Landesamt für Umwelt und Geologie, Dresden, Germany.
Stoner-Duncan, B., Streicker, D.G., Tedeschi, C.M., 2014. Vampire bats and rabies: toward an ecological solution to a public health problem. PLoS Negl. Trop. Dis. 8, e2867.
Streicker, D., Turmelle, A.S., Vonhof, M.J., Kuzmin, I., McCracken, G.F., Rupprecht, C.E., 2010. Host phylogeny constrains cross-species emergence and establishment of rabies virus in bats. Science 329, 676–679.
Streicker, D.G., Recuenco, S., Valderrama, W., Gomez Benavides, J., Vargas, I., Pacheco, V., Condori Condori, R.E., Montgomery, J., Rupprecht, C.E., Rohani, P., Altizer, S., 2012. Ecological and anthropogenic drivers of rabies exposure in vampire bats: implications for transmission and control. Proc. Biol. Sci. 279, 3384–3392.
Tamura, K., Stecher, G., Peterson, D., Filipski, A., Kumar, S., 2013. MEGA6: molecular evolutionary genetics analysis version 6.0. Mol. Biol. Evol. 30, 2725–2729.
Topal, G., 2011. Myotis nattereri (Kuhl, 1818)—Fransenfledermaus. In: Krapp, F. (Ed.), Die Fledermäuse Europas. AULA-Verlag, Wiebelsheim, pp. 405–442.
Troupin, C., Dacheux, L., Tanguy, M., Sabeta, C., Blanc, H., Bouchier, C., Vignuzzi, M., Duchene, S., Holmes, E.C., Bourhy, H., 2016. Large-scale phylogenomic analysis reveals the complex evolutionary history of rabies virus in multiple carnivore hosts. PLoS Pathog. 12, e1006041.
Van der Poel, W.H., Van der Heide, R., Verstraten, E.R.A.M., Takumi, K., Lina, P.H.C., Kramps, J.A., 2005. European bat lyssaviruses, the Netherlands. Emerg. Infect. Dis. 11, 1854–1859.
Van Thiel, P.P., van den Hoek, J.A., Eftiov, F., Tepaske, R., Zaarijer, H.J., Spanjaard, L., de Boer, H.E., van Doornum, G.J., Schuten, M., Osterhaus, A.D., Kager, P.A., 2008. Fatal case of human rabies (Duvenhage virus) from a bat in Kenya: The Netherlands, December 2007. Euro. Surveill. 13 (2), pii: 8007.
Velasco-Villa, A., Orciari, L.A., Juarez-Islas, V., Gomez-Sierra, M., Padilla-Medina, I., Flisser, A., Souza, V., Castillo, A., Franka, R., Escalante-Mane, M., Sauri-Gonzalez, I., Rupprecht, C.R., 2006. Molecular diversity of rabies viruses associated with bats in Mexico and other countries of the Americas. J. Clin. Microbiol. 44, 1697–1710.
Velasco-Villa, A., Reeder, S.A., Orciari, L.A., Yager, P.A., Franka, R., Blanton, J.D., Zuckero, L., Hunt, P., Oertli, E.H., Robinson, L.E., Rupprecht, C.E., 2008. Enzootic

rabies elimination from dogs and reemergence in wild terrestrial carnivores, United States. Emerg. Infect. Dis. 14, 1849–1854.

Webster, W.A., Casey, G.A., 1996. Virus isolation in neuroblastoma cell culture. In: Meslin, F.X., Kaplan, M.M., Koprowski, H. (Eds.), Laboratory Techniques in Rabies. World Health Organization, Geneva, pp. 93–104.

Weir, D.L., Annand, E.J., Reid, P.A., Broder, C.C., 2014. Recent observations on Australian bat lyssavirus tropism and viral entry. Virus 6, 909–926.

Wellenberg, G.J., Audry, L., Ronsholt, L., van der Poel, W.H.M., Bruschke, C.J.M., Bourhy, H., 2002. Presence of European bat lyssavirus RNAs in apparently healthy Rousettus aegyptiacus bats. Arch. Virol. 147, 349–361.

Weyer, J., Szmyd-Potapczuk, A.V., Blumberg, L.H., Leman, P.A., Markotter, W., Swanepoel, R., Paweska, J.T., Nel, L.H., 2011. Epidemiology of human rabies in South Africa, 1983–2007. Virus Res. 155, 283–290.

CHAPTER NINE

Software Dedicated to Virus Sequence Analysis "Bioinformatics Goes Viral"

Martin Hölzer[*,†], **Manja Marz**[*,†,‡,1]
[*]RNA Bioinformatics and High Throughput Analysis, Faculty of Mathematics and Computer Science, Friedrich Schiller University Jena, Jena, Germany
[†]European Virus Bioinformatics Center (EVBC), Jena, Germany
[‡]FLI Leibniz Institute for Age Research, Jena, Germany
[1]Corresponding author: e-mail address: manja@uni-jena.de

Contents

Abstract

Computer-assisted technologies of the genomic structure, biological function, and evolution of viruses remain a largely neglected area of research. The attention of bioinformaticians to this challenging field is currently unsatisfying in respect to its medical and biological importance. The power of new genome sequencing technologies, associated with new tools to handle "big data", provides unprecedented opportunities to address fundamental questions in virology. Here, we present an overview of the current technologies, challenges, and advantages of Next-Generation Sequencing (NGS) in relation to the field of virology. We present how viral sequences can be detected de novo out of current short-read NGS data. Furthermore, we discuss the challenges and applications of viral quasispecies and how secondary structures, commonly shaped by RNA viruses, can be computationally predicted. The phylogenetic analysis of viruses, as another ubiquitous field in virology, forms an essential element of describing viral epidemics and challenges current algorithms. Recently, the first specialized virus-bioinformatic organizations have been established. We need to bring together virologists and bioinformaticians and provide a platform for the implementation of interdisciplinary collaborative projects at local and international scales. Above all, there is an urgent need for dedicated software tools to tackle various challenges in virology.

Advances in Virus Research, Volume 99
ISSN 0065-3527
https://doi.org/10.1016/bs.aivir.2017.08.004

"Big data" has been awarded to be the second-best Anglicism in 2014.[a] Although microorganisms and particularly viruses are tiny, the standard properties of big data apply: volume, variety, velocity, and veracity. The biodiversity of viruses with its coverage of multiple scales and its high complexity is a big challenge for algorithm and software development in the big data field (Beckstein et al., 2014).

Recently, we have started to explore the virus' and host's genomes, transcriptomes, metabolome, proteome, and metagenome but also their phenotype, occurrence, and environment. Linking such raw heterogeneous data with current data, e.g., collected from social networks on cumulative occurrences of disease-carrying mosquitoes, is a challenging task. For example, such a task might be solved by combining georeference photos from mobile phones with an automatic determination software, allowing better decisions on overarching questions (Graham et al., 2011).

The storage of such data is essential and currently a computationally unsolved problem. Additionally, calculations on computational cluster machines have annual electricity costs of one third of its acquisition costs. Medical data are usually only semianonymous and therefore cannot be stored and computed in clouds.[b] In the future, we will need novel, qualitatively different computational methods and paradigms. We will witness the rapid extension of computational pan-genomics, a new subarea of research in bioinformatics. A prominent example for a computational paradigm shift is the transition from the representation of single reference genomes as strings to cloud-like representations as graphs (Marschall et al., 2016). Especially, viruses are notorious mutation machines. Therefore, a viral quasispecies is a cloud of viral haplotypes that surround a given master virus (Qin et al., 2012).

Interestingly, already the storage of simple linear viral genomes is complicated. For instance, although most viral genomes are stored in the NCBI, many virologists refuse to integrate their data due to the generality of the database: One of the first questions during the upload process is "What chromosome is this?" Therefore, virus-specific databases are necessary, however, only a few exist so far (Table 1), and a general database for all viruses needs to be urgently developed.

[a] http://www.anglizismusdesjahres.de/anglizismen-des-jahres/adj-2014/.

[b] Cloud storage or cloud computing refers to shared computer processing resources and data on demand.

Table 1 Virus-Specific Databases Besides the General NCBI Database

Tool	Description	Ref.
ViPR	ViPR database integrates genomes and various other types of data for multiple virus families belonging to the Arenaviridae, Bunyaviridae, Caliciviridae, Coronaviridae, Flaviviridae, Filoviridae, Hepeviridae, Herpesviridae, Paramyxoviridae, Picornaviridae, Poxviridae, Reoviridae, Rhabdoviridae, and Togaviridae families.	Pickett et al. (2012)
EpiFlu™	GISAID EpiFlu™ is the world's most complete collection of genetic sequence data of influenza viruses and related clinical and epidemiological data. EpiFlu™ is tailored to the needs of influenza researchers from both the human and the veterinary fields. The data is publicly accessible but not public domain (GISAID does not remove nor waive any preexisting rights).	Shu and McCauley (2017)
HIV	The HIV database contains data on HIV genetic sequences and immunological epitopes. The website also provides an access to several tools that can be used for analysis and visualization.	Druce et al. (2016)
HCV	HCV is a comprehensive database of the hepatitis C virus (HCV).	Kuiken et al. (2005)
ViralZone	ViralZone is a web-resource from the Swiss Institute of Bioinformatics for all viral genus and families, providing general molecular and epidemiological information, along with virion and genome figures. Each virus or family page gives an easy access to UniProtKB/Swiss-Prot viral protein entries.	Hulo et al. (2011)
VVR	The virus variation resource (VVR) is a selection of web retrieval interfaces, analysis, and visualization tools for virus sequence datasets.	Hatcher et al. (2017)

1. NEXT-GENERATION SEQUENCING

Next-Generation Sequencing (NGS) has dramatically increased the accessibility of genetic information, generating in only a few hours massive amounts of genome and transcriptome data that is rapidly changing the landscape of many life science disciplines (Goodwin et al., 2016). In April 2003, the complete human genome was announced and the project succeeded after spending $3-billion with a high-quality human reference genome (Schmutz et al., 2004). Although the assembly of such a huge genome is still a very challenging task, nowadays the sequencing can be done in just a few days and for only some thousands of dollars (Goodwin et al., 2016) by utilizing the still emerging NGS technologies.

In recent years, DNA sequencing (DNA-Seq) based on novel NGS technologies (Table 2) became the most sophisticated method for the sequencing of full genomes. A general DNA-Seq workflow starts with the library

Table 2 Commonly Used Next-Generation Sequencing (NGS) Technologies and Their Major Specifications

Platform	Length (bp)	Throughput	Number of Reads	Error	Cost Per Gb
Short-read NGS					
Sequencing by synthesis: SNA					
454 Pyrosequencing	400–1000	35–700 Mb	0.1–1 M	1%, indel	$10–40,000
Ion Torrent	200–400	100 Mb–15 Gb	2–80 M	1%, indel	$500–2000
Sequencing by synthesis: CRT					
Illumina Solexa	25–300	2–900 Gb	10 M–4 B	0.1%, subst.	$7–1000
Qiagen GeneReader	100	NA	10 M–4 B	0.1%, subst.	NA
Sequencing by ligation					
SOLiD	60–100	10–320 Gb	700 M–1.4 B	0.1%, AT bias	$100
Long-read SMRT NGS					
Pacific BioSciences	up to 40 Kb	0.5–7 Gb	~55 k	13% (single)	$1000
				1% (circular)	
Oxford Nanopore (MinION)	up to 200 Kb	up to 1.5 Gb	>100 k	12%, indel	$750

Generally, NGS technologies can be divided in *short-read* and *long-read* approaches, depending on the length of the produced reads. SNA, single-nucleotide addition; CRT, cyclic reversible termination; SMRT, single-molecule real-time sequencing; indel, nucleotide insertion–deletion; subst., nucleotide substitution. This table is mainly based on recent reviews (Goodwin et al., 2016; Mardis, 2017).

preparation including the fragmentation (chemically, physically) of the DNA molecules. After amplification and sequencing millions of short subsequences, so-called *reads*, are produced. In general, methods like Illumina and Ion Torrent produce reads with a length between 50 and 500 bp, depending on the setup and machine used (Goodwin et al., 2016). Next to that short read producing NGS technologies more and more long read NGS approaches are emerging. Very popular is the single-molecule real-time sequencing (SMRT) introduced by Pacific Biosciences (Rhoads and Au, 2015) (PacBio) producing reads with an average length of 15,000 bp and a maximum of >40,000 bases. However, PacBio produces only ~50,000 reads per SMRT cell, whereas Illumina yields ~180 million reads on one HiSeq2500 lane (Goodwin et al., 2016). It is clearly important to produce longer reads to improve the results of various analyzes like the de novo assembly of highly repetitive, large or fast mutating genomes.

Nanopore sequencing is another recent incumbent in the SMRT area: the way nanopore-based sequencing works is by pulling a nucleotide strand (DNA or RNA) through a kind of molecular channel isolated from a bacterium. While passing through the pore, the nucleotide sequence produces a small change in the applied voltage, which can be reinterpreted as the familiar sequence of the bases A, C, T/U, and G, including also modifications such as methylation (Jain et al., 2016). Because each pore produces its own signal, this technology can be highly parallelized. For example, with the current USB-sized MinION sequencer, 2048 pores are situated on a membrane of the size of a finger nail. The sequencer itself costs a fraction of the aforementioned ones. Furthermore, each pore's signal can be detected in real time (Gardy et al., 2015), allowing unprecedented speed and mobility in sequence-based diagnostics, as exemplary demonstrated in field trials during the 2014 Ebola outbreak (Quick et al., 2016). Furthermore, nanopore sequencing is currently the only technique that does (in theory) not technologically limit the potential read length, which means an entire viral genome can be sequenced in one part at an intact pore. No additional assembly step would be required. The current read length maximum is >900 Kbp (personal communication with N. Loman). The MinION's throughput has been shown to provide up to 15 Gb in 48 h with a protocol-dependent error rate of 5%–15%.

Besides the sequencing of genomic DNA, RNA sequencing (RNA-Seq) emerged as a powerful method for discovering, profiling, and quantifying RNA transcripts or viral RNA genomes (Mortazavi et al., 2008). However, with currently available short-read NGS techniques such as Illumina it is not possible to directly sequence RNA molecules—first the RNA must be reversely transcribed to complementary DNA (cDNA) for sequencing. Strikingly, nanopore just recently announced a sequencing kit that should allow for the direct sequencing of RNA molecules (and therefore also RNA viruses).

Importantly, within each NGS project one should consider the need and amount of replication, different protocols for molecule selection and library preparation, the achieved throughput and length of the reads and further specific parameters like strand-specificity and the insertion size between paired-end reads.

2. DETECTION OF DE NOVO VIRUSES

Within the last decade numerous genomes of previously unknown viruses have been identified. However, it is still a challenging task to discriminate an outnumbered amount of viral sequences from the majority of host reads. Genome assemblers specifically designed for viral genomes are rare (Table 3) and cannot overcome an uneven or incomplete coverage of viral genomes.

Many assembly tools and software suites have been developed for the complete genome assembly in general, such as `Velvet` (Zerbino and Birney, 2008), `ABySS` (Simpson et al., 2009), or `Geneious` (Kearse et al., 2012) (Fig. 1). These common tools often fail to assemble full viral genomes, due to a low and uneven read coverage (Peng et al., 2012), as well as repetitive elements in the viral UTR regions. However, algorithms developed for single-cell sequencing like `SPAdes` (Bankevich et al., 2012) or `IDBA-UD` (Peng et al., 2012) perform very well for tested samples and outperform assembly tools like `VICUNA` (Yang et al., 2012), especially designed for viral data (Fig. 1).

For an efficient viral de novo assembly we suggest enriching of the viruses by, e.g., ultracentrifugation or FACS prior to the library preparation step. After the sequencing, a standard read quality control should be conducted followed by a host genome filter step, if possible. Finally, the assembly step can be performed based on *de Bruijn* graphs or overlapping layout consensus (OLC) approaches. If possible, the usage of multiple *k*-mer values is recommended. The final assembly can be used for annotation and identification of contigs from viral origin. Fig. 2 shows the viral assembly workflow as used in the `VrAP` assembly pipeline (Fricke et al., 2017).

Table 3 De Novo Assembly Tools Suitable for the Assembly of Viral Genomes

Tool	Description	Ref.
AV454	AV454 is a de novo consensus assembler designed for small and nonrepetitive genomes sequenced at high depth.	Henn et al. (2012)
RIEMS	RIEMS is a software for the sensitive and reliable analysis of metagenomic datasets.	Scheuch et al. (2015)
V-FAT	V-FAT is a tool to perform automated computational finishing and annotation of de novo viral assemblies.	Charlebois et al. (n.d.)
VICUNA	VICUNA is a de novo assembly tool targeting populations with high mutation rates.	Yang et al. (2012)
VrAP	The VrAP (Viral Assembly Pipeline) is based on the genome assembler SPAdes (Bankevich et al., 2012) combined with an additional read correction and several filter steps. The pipeline classifies the contigs (contiguous sequences constructed from short reads) to distinguish host from viral sequences. VrAP can identify viruses without any sequence homology to known references.	Fricke et al. (2017)

3. VIRAL QUASISPECIES

The above described de novo assembly methods can reconstruct viral genomes. However, to yield a small number of contigs, the algorithms usually include a step that calls a consensus on a given sequence position. This consensus is implemented to reduce the noise in the raw assembly. However, in the context of viral haplotype variants, this step is misleading, because it effectively ignores low-frequency variants and technical errors (Marz et al., 2014).

To gain insights into viral haplotypes, the reads should be mapped either to a known reference genome or to the contigs that were generated during assembly. This "classification" can be used to infer the viral population structure of each individual species in the sample, thereby increasing the resolution of the diversity estimate. (Intrahost) viral populations consist of many

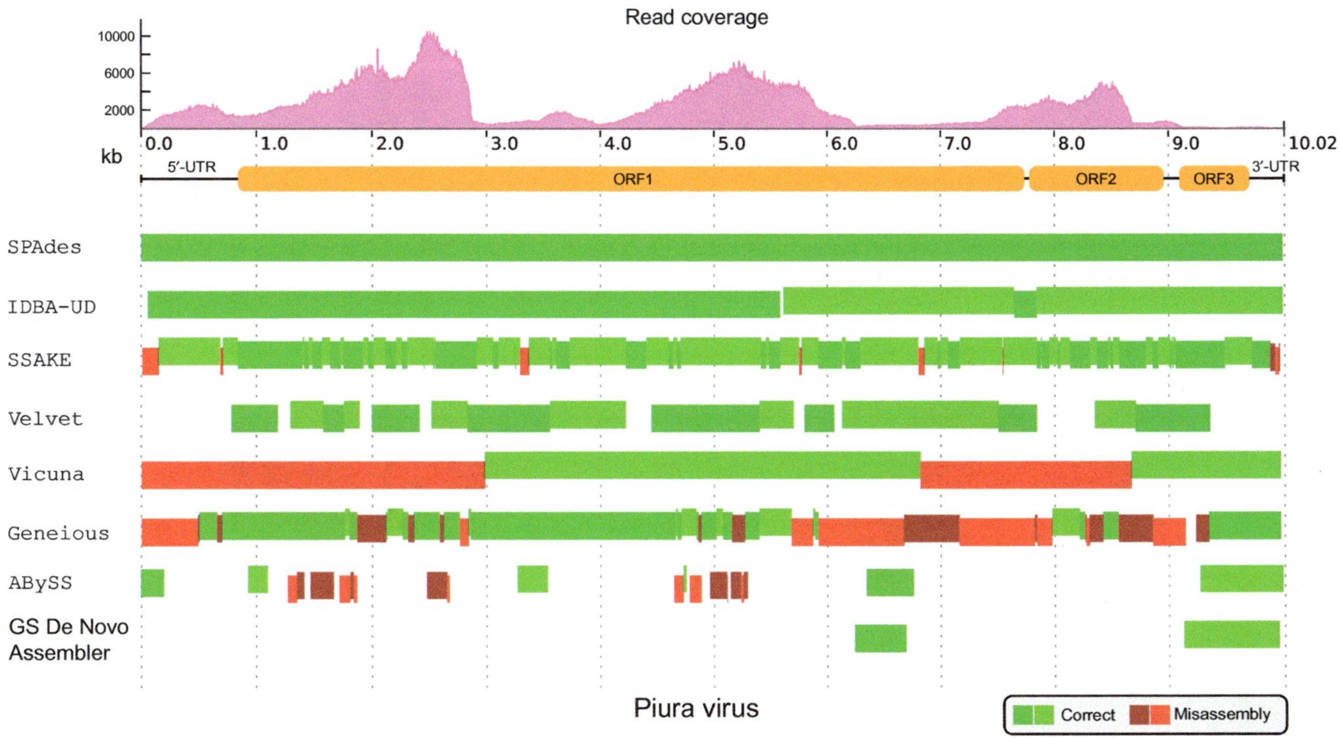

Fig. 1 Comparison of eight assembly tools based on a sequenced C6/36 cell, infected with a Piura virus strain from Mexico. The figure depicts an alignment of de novo assembled contigs (rectangles) to the reference genome of Piura virus (KM249340.1). `SPAdes` assembles the full viral genome without any difficulties. All other assemblers fail to build a continuous single contig. *Green*—contigs that align correctly. *Red*—misassemblies. The different *color shades* are only for a better visualization of adjacent contigs. The alignment plot was created with `Quast` (Gurevich et al., 2013). *Figure adapted from Fricke, M., Zirkel, F., Drosten, C., Junglen, S., Marz, M., 2017. VrAP: full length de novo genome assembly of unknown RNA viruses (submitted for publication).*

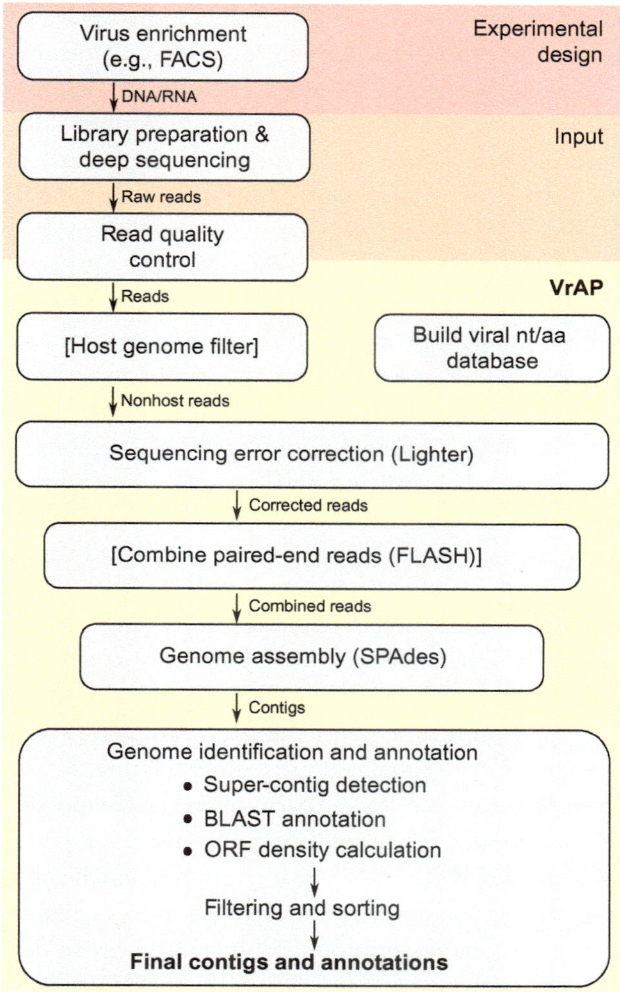

Fig. 2 Workflow of the viral de novo assembly pipeline `VrAP`. The pipeline requires (preprocessed) reads as input. The output consists of final contigs and an annotation list. The pipeline combines multiple read corrections with `SPAdes`, a super-contig construction and a contig classification. `VrAP` comes as an easy to use command-line tool (http://www.rna.uni-jena.de/en/vrap/). All steps in *square bracket* are optional. *FACS*, fluorescent activated cell sorting. *Figure adapted from Fricke, M., Zirkel, F., Drosten, C., Junglen, S., Marz, M. 2017. VrAP: full length de novo genome assembly of unknown RNA viruses (submitted for publication).*

related virions, generated by mutation, recombination, and selection. The resulting diversity is especially large for RNA viruses (Holmes, 2009). Even low-frequency variants can be of great interest, for example, because they may harbor drug resistance mutations (Barzon et al., 2011), facilitate immune escape (Luciani et al., 2012), or affect virulence (Töpfer et al., 2013). Estimating intrahost viral genetic diversity and reconstructing the individual haplotype sequences relies on both error correction and read assembly (Pulido-Tamayo et al., 2015). It can be performed on different spatial scales, including single sites of the genome (single-nucleotide-variant calling), small sliding windows (local reconstruction), or complete genomes (global reconstruction). Viral haplotype reconstruction tools can quantify viral diversity from NGS data (e.g., Beerenwinkel et al., 2012). It was shown that haplotypes differ enough, current NGS reads are not too short and the coverage is high enough to assemble accurate viral haplotype genomes (Zagordi et al., 2012). A common prerequisite for these tools is a high-quality alignment of the reads (e.g., Töpfer et al., 2014). However, tools exist that allow haplotype calling without a reference genome as presented in Gregor et al. (2016). Nevertheless, the short-read-based discovery of viral sequences in mixed samples remains challenging (Marschall et al., 2016) because most analysis steps are not easily automated and various technical or biological limitations exist (Fricke et al., 2017). There is a need for an integrated workflow combining the different processing steps in viral diversity studies to discover the underlying virus populations that can be used on a daily basis by clinicians and virologists. The advent of SMRT sequencing provides new opportunities. One of the main limitations of the past was the limited length of the sequenced nucleotide fragments. Currently, it is not possible to write cDNA longer than a few thousand of nucleotides (e.g., ~2000 nucleotides for the wheat stripe rust pathogen (Ling et al., 2007)). However, even if the cDNA transcription would be no limiting factor, current short-read sequencing technologies such as Illumina are only able to sequence small fragments of several hundred nucleotides. Nanopore sequencing lifts these two constraints: it is now possible to sequence much longer fragments (as described above) and to sequence the RNA directly, without the need of a cDNA intermediate, advancing the detection of viral quasispecies.

4. SECONDARY STRUCTURES OF RNA VIRUSES

RNA viruses are flanked by highly structured 5′- and 3′-untranslated regions (UTRs), which are indispensable for translation and replication of the viral genome (Liu et al., 2009; Lohmann, 2013).

Table 4 A Selection of Tools for the Detection of Secondary Structures in RNA Viruses

Tool	Description	Alignment	Ref.
RNAfold	RNAfold is a tool to predict secondary structures of single stranded RNA or DNA sequences.	No	Gruber et al. (2008)
mfold	mfold is a web server that provides easy access to RNA and DNA folding and hybridization software.	No	Zuker (2003)
RNAalifold	RNAalifold is a tool for calculating secondary structures for a set of aligned RNAs. It is part of the Vienna RNA Package.	Yes	Hofacker (2007)
LocARNA	LocARNA is a multiple alignment tool based on the calculation of sequence and structure simultaneously.	Yes	Will et al. (2007)
LRIscan	LRIscan is a tool for the prediction of long-range interactions in full viral genomes based on a multiple genome alignment. LRIscan is able to find interactions spanning thousands of nucleotides.	Yes	Fricke and Marz (2016)

Standard RNA secondary structure prediction tools such as mfold and RNAfold (Table 4) are based on the calculation of the minimum free energy (MFE) and can fold reliably on small local windows of up to 300 nt. Secondary structures of larger genomic segments or interactions spanning larger regions, including pseudogenes, are still bioinformatically challenging. Foldings based on not only one but also multiple sequences are generally more reliable due to following the footsteps of evolution by compensatory mutations. Viruses usually come along with a high mutation rate and therefore with a bunch of similar sequences perfect for a large alignment and predicting secondary structures.

For example, LocARNA creates a multiple alignment based on sequence and structure simultaneously. Based on this tool larger genomic regions up to 800 nt can be reliable predicted as shown for coronaviruses (Fig. 3) (Madhugiri et al., 2014) and HCV (Fig. 4) (Fricke et al., 2015). Nowadays, long-range interactions (LRIs) are computationally predictable by tools such as LRIscan (Fricke and Marz, 2016), suggesting circularizations of viruses during replication.

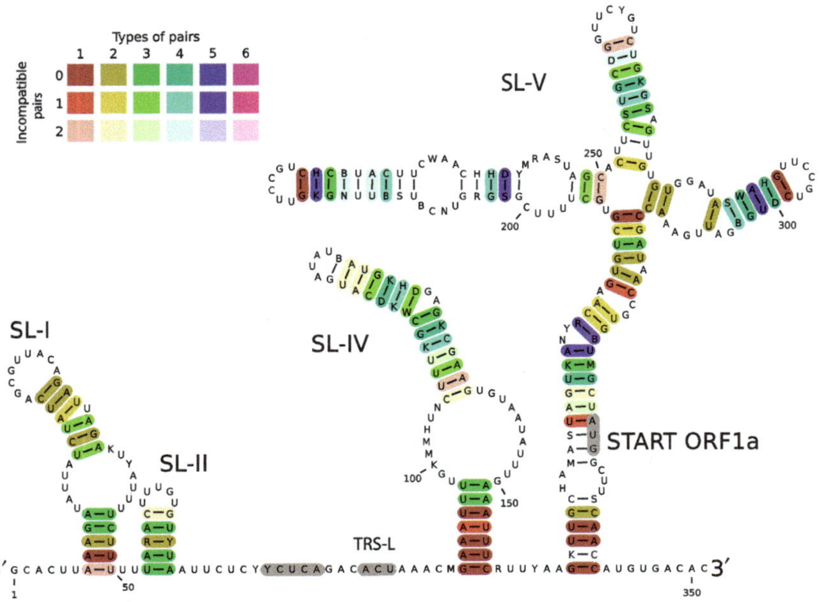

Fig. 3 Alignment-based secondary structure prediction of 5′ genome regions of alphacoronaviruses. The viruses included in this analysis represent all currently recognized species in the genus *Alphacoronavirus*. The alignment (not shown) was calculated by `LocARNA` (Will et al., 2007) and the structure by `RNAalifold` (Hofacker, 2007). The consensus sequence is represented using the IUPAC code. *Colors* are used to indicate conserved base pairs: from *red* (conservation of only one base pair type) to *purple* (all six base pair types are found); from *dark* (all sequences contain this base pair) to *light colors* (one or two sequences are unable to form this base pair). To refine the alignment, an anchor at the highly conserved core TRS-L was used. *Figure adapted from Madhugiri, R., Fricke, M., Marz, M., Ziebuhr, J., 2014. RNA structure analysis of alphacoronavirus terminal genome regions. Virus Res. 194, 76–89. doi:10. 1016/j.virusres.2014.10.001.*

5. ANALYSIS OF TRANSCRIPTOMIC HOST REACTIONS TO VIRAL INFECTIONS

The general workflow of a short-read RNA-Seq experiment involves: (1) the extraction of total RNA from a biological sample of interest, (2) the purification of the sample to enrich a certain type of RNA such as mRNAs or microRNAs, and (3) the preparation of a library ready for short-read NGS. The generation of the library may involve steps like the fragmentation of longer RNA molecules, followed by the reverse transcription of the RNA to cDNA, ligation of adapters to the 5′- and/or 3′-ends of the cDNA fragments and PCR amplification to enrich the library for correctly ligated

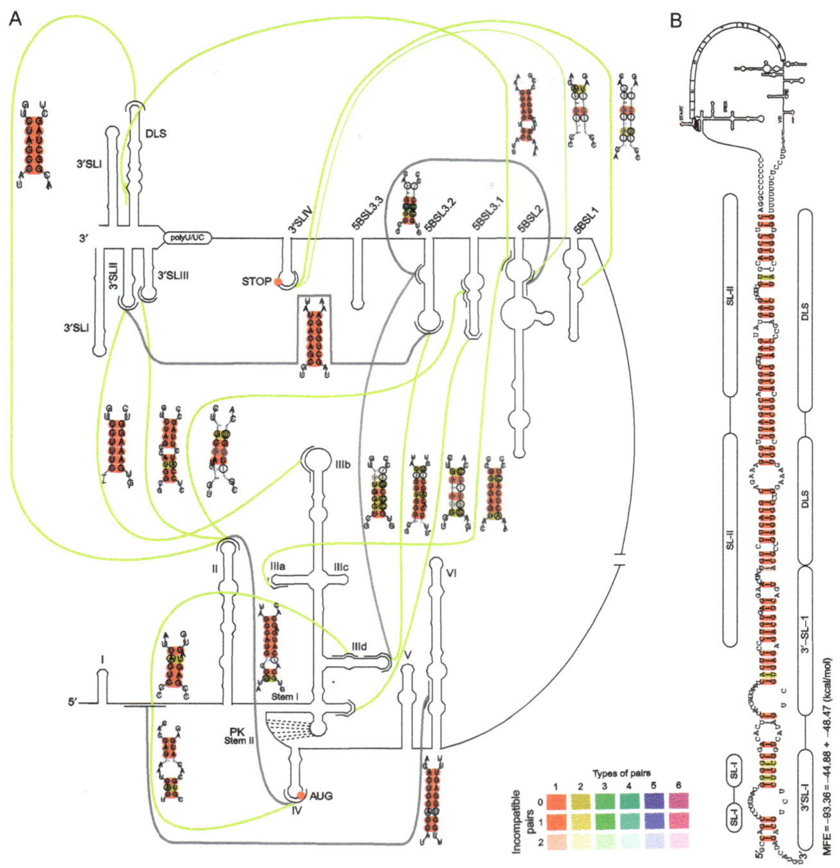

Fig. 4 Long-range interactions in 5′-UTR, CRE, VR, and X-tail of HCV (Fricke et al., 2015). (A) Overview and possible interactions for all tested HCV sequences. *Gray lines*—known interactions derived from literature, validated by this analysis for all examined isolates; *Green lines*—novel interactions (based on new calculations). The detailed interactions are shown on the right side next to each corresponding interaction line. The leftmost interaction can be extended for a possible circularization of HCV. (B) Possible circularization of HCV. Interacting loops of SLII and DLS of the HCV plus-strand can be extended to at least 62 bp in all available 19 isolates.

cDNA fragments (Corney, 2013). The resulting reads from an RNA-Seq experiment can be used to estimate the abundances of certain transcripts within each sequenced sample. If different conditions are sequenced, the obtained transcript abundances can be further used to identify differential expressed genes. Before RNA-Seq came up, gene expression studies were performed with hybridization-based microarrays. Contrasting the microarray technology, RNA-Seq allows for the identification of novel transcripts and does not necessarily need a sequenced reference genome. Furthermore, RNA-Seq allows for the genome-wide analysis of transcripts at a single-nucleotide resolution and therefore includes the identification of single-nucleotide variants, gene fusions, allele-specific expression, and alternative splicing events (Corney, 2013).

However, besides all its advantages, RNA-Seq is still an expensive technology. Therefore, in most RNA-Seq studies the number of biological replicates is limited (only 3–5 replicates per condition are quite common) contrasting the comparative high number of genes that are simultaneously tested.

A typical RNA-Seq experiment, involving an eukaryotic cell line and involving two different conditions (untreated, infected), three time points and four biological replicates already results in the sequencing of 24 samples. The current Ensembl annotation of the human genome (v85) consists of 58,051 genes comprising 19,961 genes coding for proteins. In a differential gene expression study, all expressed genes can be compared between different conditions and time points, resulting in an overwhelming amount of data. Genes can be further analyzed for differential expressed isoforms and clustered according to their function. With a de novo gene prediction, one of the huge advantages of RNA-Seq in comparison to microarrays, an incomplete annotation can be further extended and even more genes are possibly involved. The use of different library preparation protocols can extend the complexity of such an RNA-Seq study even further.

Therefore, the statistical analysis of RNA-Seq data with the final goal to define significantly differential expressed genes is a challenging task. Especially, if a high number of reads originating from viral transcripts is involved, outshining the expression of host genes. Furthermore, the generation of a sensible number of biological replicates can be difficult when working with such deadly viruses like Ebola. The analysis can become even more complicated when no reference genome for mapping and quantification of the RNA-Seq reads is available. In this case, a de novo transcriptome assembly can be constructed and annotated from scratch.

To tackle these difficulties, profoundly occurring when working with virus infected RNA-Seq data, different tools and parameter settings should be conducted and combined to achieve a comprehensive overview picture of the host's transcriptional reaction to a viral infection. An exemplary pipeline combining different tools for mapping and assembly and working on a genomic and transcriptomic context as well is given in Fig. 5.

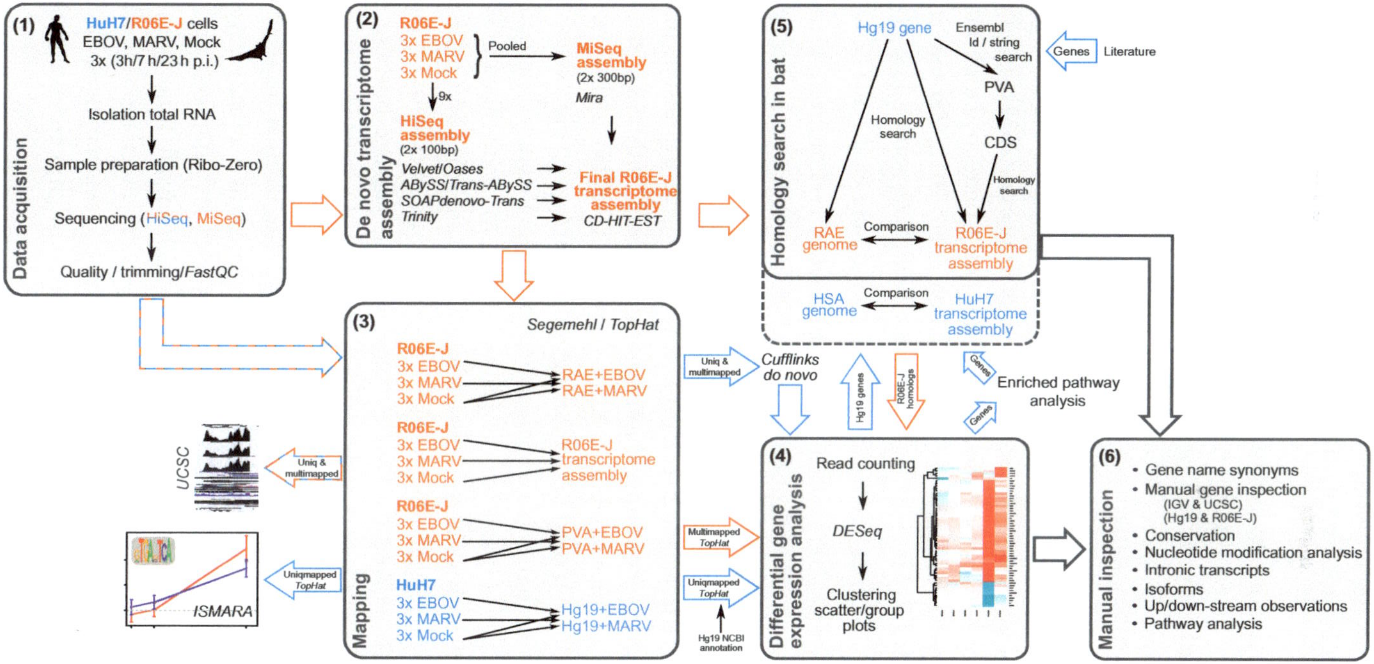

Fig. 5 Host–virus RNA-Seq methods pipeline for the detection of differential expressed genes (see text for details). *Modified from Hölzer , M., Krähling, V., Amman, F., Barth, E., Bernhart, S.H., Carmelo, V.A.O., Collatz, M., Doose, G., Eggenhofer, F., Ewald, J., Fallmann, J., Feldhahn, L.M., Fricke, M., Gebauer, J., Gruber, A.J., Hufsky, F., Indrischek, H., Kanton, S., Linde, J., Mostajo, N., Ochsenreiter, R., Riege, K., Rivarola- Duarte, L., Sahyoun, A.H., Saunders, S.J., Seemann, S.E., Tanzer, A., Vogel, B., Wehner, S., Wolfinger, M.T., Backofen, R., Gorodkin, J., Grosse, I., Hofacker, I., Hoffmann, S., Kaleta, C., Stadler, P.F., Becker, S., Marz, M., 2016. Differential transcriptional responses to Ebola and Marburg virus infection in bat and human cells. Sci. Rep. 6, 34589. doi:10.1038/srep34589.*

The overall goal of the underlying study was to understand why bats can live with the Ebola virus, while humans suffer so much from this deadly infection.

In this study, performed by Hölzer et al. (2016), (1) total RNA from a human HuH7 cell line and a fruit bat cell line (R06E-J; *Rousettus aegyptiacs*) infected with either the Ebola or Marburg virus (EBOV, MARV) was harvested 3, 7, and 23 h postinfection, depleted of ribosomal RNA and sequenced on an Illumina HiSeq2500. The bat RNA was further pooled and additionally sequenced on an Illumina MiSeq system. Initial quality control and trimming of the raw data were conducted with `FastQC` (Andrews, 2010) and `PRINSEQ` (Schmieder and Edwards, 2011). (2) For bat RNA, a de novo transcriptome assembly was constructed by combining MiSeq and HiSeq data using `Velvet/Oases` (Schulz et al., 2012; Zerbino and Birney, 2008), `ABySS/Trans-ABySS` (Birol et al., 2009; Simpson et al., 2009), `SOAPdenovo-Trans` (Luo et al., 2012), `Trinity` (Grabherr et al., 2011), and `Mira` (Chevreux et al., 2004) with default parameters and multiple *k*-mer values, if possible. (3) The mapping of the RNA-Seq short-reads was performed for Mock-, EBOV-, and MARV-treated cells onto human/bat genomes and the bat transcriptome with `Segemehl` (Hoffmann et al., 2014) and `TopHat` (Kim et al., 2013). (4) A differential gene expression analysis was performed by counting uniquely mapped reads with `HTSeq-count` (Anders et al., 2015) and applying a `DESeq` (Love et al., 2014) analysis in `R`. The results were further used for clustering and scatter/group plot analyzes. (5) A homology search in bats was performed for all significantly differential expressed genes from (4) and for the genes assumed to be involved in the response to infection based on an enriched pathway analysis and the literature. The *Rousettus aegyptiacus* genome and coding sequences from *Pteropus vampyrus*, a closely related bat species, were used to validate but also to detect homologous sequences in the bat transcriptome. Detected homologs were employed for the differential gene expression analysis. (6) One huge advantage of this comprehensive study was the manual inspection of ~7.5 % of the human genes. Each candidate gene was manually investigated in the `IGV` (Thorvaldsdóttir et al., 2013) and `UCSC` (Dreszer et al., 2012) browsers for the human and bat samples from all time points. Single-nucleotide modifications (differential SNPs, posttranscriptional modifications), intronic transcripts and regulators, alternative splicing and isoforms, as well as upstream and downstream transcript characteristics were described.

6. VIRAL PHYLOGENY/COPHYLOGENY

Phylogenetic analysis is a common method in virology, forming a crucial element of investigations describing viruses or viral epidemiology. Nevertheless, many characteristics of viruses pose distinct challenges for phylogenetics: (1) strong differences in evolution rates, (2) great potential for recombination and gene transfer, (3) evolutionary relationships between viruses and their hosts, (4) lack of physical "fossil records" of viruses, and (5) the abundance of genomic viral fossils as parts of ancient viral genomes that occur within the genomes of extant species.

Today, various phylogenetic tree-building methods such as `MrBayes` (Ronquist and Huelsenbeck, 2003), `BEAST` (Drummond et al., 2012), `PhyloBayes` (Lartillot et al., 2009), and `RAxML` (Stamatakis et al., 2008) exist. However, trees cannot represent complex evolutionary relations relevant for viruses such as horizontal gene transfer, interspecific recombination, or virus–host coevolution. Different types of phylogenetic networks were developed to represent such relations (e.g., Huson et al., 2011). However, there is still a high need for research on how to reconstruct such aspects of virus phylogeny.

Genomic evolution can be already observed over the course of years or even days due the fact that the short-term evolution rates of many viruses are so high. It is important that the phylogenetic methods can include the sampling dates of the sequences for analyzing short-term evolution as implemented in `TipDate` (Rambaut, 2000). Furthermore, spatial dispersal processes play an essential role, for example, the spatial distribution of a virus within the host's body (Bloomquist et al., 2010). Moreover, the evolutionary substitution rates of viruses can differ even for short-term evolutionary scenarios. One reason is that substitution rates reflect a complex product of mutation rate, generation time, effective population size, and fitness (Jenkins et al., 2002; Sanjuan et al., 2010). Particularly in viruses, substitutions might be an artifact generated by polymerase errors and nucleotide modifications (Domingo and Holland, 1997). Thus, the classical assumption of a time-homogeneous substitution process used by different phylogeographic statistical inference methods does not hold and new approaches that can include varying evolutionary rates have been already introduced (e.g., Bielejec et al., 2014).

Another problem for viral "deep phylogeny" reconstruction is the genetic distance between viruses. The distance can be so large that reasonable alignments become impossible to calculate. To achieve biologically correct alignments, the development of advanced approaches would help, however, can only marginally alleviate the problem of saturated substitution processes. Including aspects such as genome organization or protein structure as phylogenetic characters could further improve viral alignments and phylogenies (Holmes, 2011).

Several ancient viruses have left parts of their genome (or other traces) in the genome of germ line cells of their hosts. Such parts, called endogenous viral elements (EVEs), have survived as nonfunctional, neutrally evolving pseudogenes, or even became fixed as functional. Most EVEs stem from retroviruses because they integrate into host genomes as part of their life cycle. For example, ~8% of the human genome is derived from >100,000 retroviral fossils (Lander et al., 2001). However, in recent years, EVEs from many other viruses have been found (Horie and Tomonaga, 2011). Different programs have been developed to detect EVEs in complete genome sequences such as `RepeatMasker` (Smit et al., n.d.), `LTR_STRUC` (McCarthy and McDonald, 2003), and `RetroTector` (Sperber et al., 2009). Moreover, a combination of several of these programs seems very promising for the calculation of viral phylogenies (Lerat, 2010).

Withal, associations between viruses and their hosts can influence the phylogeny of both partners. A divergence of the host can also lead to a divergence of the virus (codivergence) and thus to a (local) congruence of both phylogenies. A match of the virus phylogeny with host evolutionary events at known dates can be used to adjust the virus phylogeny or corresponding molecular clocks (Sharp and Simmonds, 2011). The ability of viruses to switch their hosts can enable viruses to replicate and spread more efficiently. This process is commonly known as an epidemic and is observed in pathogenic viruses (Weiss, 2003). Owing to the advantages conferred by the conquest of new host territory, several researchers presume host switching as an elementary component of virus evolution that might initiate viral speciation (Kitchen et al., 2011). Attributed to the fact that virologists are highly interested in the reconstruction of the common history of viruses and their hosts, several bioinformatic tools have been developed for this purpose (de Vienne et al., 2013).

However, there is still a huge amount of research questions that need to be answered based on new computational methods. For example, the inclusion of biogeographic information, ecological traits, or preferential host switching are crucial tasks (Cuthill and Charleston, 2013). A better knowledge of the timing and underlying conditions of those processes might enable projections into the future and thereby contribute to tackle one of the major issues in today's infectious diseases research: the prediction and prevention of future pandemics and outbreaks.

7. CONCLUSIONS AND FUTURE PERSPECTIVES

It is essential to bundle the expertise's of virus bioinformatics to follow with larger steps the small footsteps that were already taken. There is an urgent need for novel and specialized tools that allow the efficient detection, assembly, and classification of already known and completely new viruses in a fast and reliable way.

One big step in this direction involves the establishment of research networks between experienced scientists to facilitate the exchange of knowledge and to speed-up the development of powerful tools. DiaMETA-net is a German network which focuses on metagenomics in infection medicine. The research groups within the network devote themselves to the very broad detection and characterization of pathogens (viruses, bacteria, parasites) by means of NGS. However, the first specialized virology-bioinformatics organization, the EVBC (European Virus Bioinformatics Center), has been established rather recently on March 2017, comprising up to now 100 members from over 50 research institutions distributed across 13 European countries.

The future of virus bioinformatics clearly depends on how fast we develop specific bioinformatical tools, take first steps to establish a useful virus-specific database, and help to establish joint research projects. We must initiate and coordinate ring trials, undergraduate courses, graduate summer schools, and courses for principal investigators.

Whereas the list of bioinformatical tools presented in this section is supposed to be incomplete, they should provide a good overview and starting point to dive even deeper into the computational analysis of viral sequences.

REFERENCES

Anders, S., Pyl, P.T., Huber, W., 2015. HTSeq-a Python framework to work with high-throughput sequencing data. Bioinformatics (Oxford, England) 31, 166–169. https://doi.org/10.1093/bioinformatics/btu638.

Andrews, S., 2010. FastQC: a quality control tool for high throughput sequence data. Available online at: http://www.bioinformatics.babraham.ac.uk/projects/fastqc.

Bankevich, A., Nurk, S., Antipov, D., Gurevich, A.A., Dvorkin, M., Kulikov, A.S., Lesin, V.M., Nikolenko, S.I., Pham, S., Prjibelski, A.D., Pyshkin, A.V., Sirotkin, A.V., Vyahhi, N., Tesler, G., Alekseyev, M.A., Pevzner, P.A., 2012. SPAdes: a new genome assembly algorithm and its applications to single-cell sequencing. J. Comput. Biol. 19 (5), 455–477. https://doi.org/10.1089/cmb.2012.0021.

Barzon, L., Lavezzo, E., Militello, V., Toppo, S., Palu, G., 2011. Applications of next-generation sequencing technologies to diagnostic virology. Int. J. Mol. Sci. ISSN: 1422-0067. 12, 7861–7884. https://doi.org/10.3390/ijms12117861.

Beckstein, C., Böcker, S., Bogdan, M., H. Bruelheide, H.M.B., Denzler, J., Dittrich, P., Grosse, I., Hinneburg, A., König-Ries, B., Löffler, F., Marz, M., Müller-Hannemann, M., Winter, M., Zimmermann, W., 2014. Explorative analysis of heterogeneous, unstructured, and uncertain data: a computer science perspective on biodiversity research. In: Helfert, M., Holzinger, A., Belo, O., Francalanci, C. (Eds.), In: Proceedings of the 3rd International Conference on Data Management Technologies and Applications, DATA 2014, Vienna, Austria, August 29–31, 2014, SciTePress, pp. 251–257.

Beerenwinkel, N., Günthard, H.F., Roth, V., Metzner, K.J., 2012. Challenges and opportunities in estimating viral genetic diversity from next-generation sequencing data. Front. Microbiol. 3, 329. ISSN 1664-302X. https://doi.org/10.3389/fmicb.2012.00329.

Bielejec, F., Lemey, P., Baele, G., Rambaut, A., Suchard, M.A., 2014. Inferring heterogeneous evolutionary processes through time: from sequence substitution to phylogeography. Syst. Biol. 63 (4), 493–504.

Birol, I., Jackman, S.D., Nielsen, C.B., Qian, J.Q., Varhol, R., Stazyk, G., Morin, R.D., Zhao, Y., Hirst, M., Schein, J.E., Horsman, D.E., Connors, J.M., Gascoyne, R.D., Marra, M.A., Jones, S.J.M., 2009. *De novo* transcriptome assembly with ABySS. Bioinformatics (Oxford, England). 25, 2872–2877. ISSN 1367-4811. https://doi.org/10.1093/bioinformatics/btp367.

Bloomquist, E.W., Lemey, P., Suchard, M.A., 2010. Three roads diverged? Routes to phylogeographic inference. Trends Ecol. Evol. 25 (11), 626–632.

Charlebois, P., Yang, X., Newman, R., Henn, M., Zody, M., n.d. Vfat: A post-assembly pipeline for the finishing and annotation of viral genomesPlease provide year of publication for this reference, if available. https://www.broadinstitute.org/viral-genomics/v-fat (accessed 07.03.17).

Chevreux, B., Pfisterer, T., Drescher, B., Driesel, A.J., Müller, W.E., Wetter, T., Suhai, S., 2004. Using the miraEST assembler for reliable and automated mRNA transcript assembly and SNP detection in sequenced ESTs. Genome Res. 14 (6), 1147–1159.

Corney, D.C., 2013. RNA-seq using next generation sequencing. Mater Methods 3, 203.

Cuthill, J.H., Charleston, M.A., 2013. A simple model explains the dynamics of preferential host switching among mammal RNA viruses. Evolution. 67, 980–990. ISSN 1558-5646. https://doi.org/10.1111/evo.12064.

de Vienne, D.M., Refrégier, G., López-Villavicencio, M., Tellier, A., Hood, M.E., Giraud, T., 2013. Cospeciation vs host-shift speciation: methods for testing, evidence from natural associations and relation to coevolution. New Phytol. 198, 347–385. https://doi.org/10.1111/nph.12150.

Domingo, E., Holland, J.J., 1997. RNA virus mutations and fitness for survival. Annu. Rev. Microbiol. 51, 151–178. ISSN 0066-4227. https://doi.org/10.1146/annurev.micro.51.1.151.

Dreszer, T.R., Karolchik, D., Zweig, A.S., Hinrichs, A.S., Raney, B.J., Kuhn, R.M., Meyer, L.R., Wong, M., Sloan, C.A., Rosenbloom, K.R., Roe, G., Rhead, B., Pohl, A., Malladi, V.S., Li, C.H., Learned, K., Kirkup, V., Hsu, F., Harte, R.A., Guruvadoo, L., Goldman, M., Giardine, B.M., Fujita, P.A., Diekhans, M., Cline, M.S., Clawson, H., Barber, G.P., Haussler, D., James Kent, W., 2012. The UCSC genome browser database: extensions and updates 2011. Nucleic Acids Res. 40, D918–D923. https://doi.org/10.1093/nar/gkr1055.

Druce, M., Hulo, C., Masson, P., Sommer, P., Xenarios, I., Le Mercier, P., De Oliveira, T., 2016. Improving HIV proteome annotation: new features of BioAfrica HIV Proteomics Resource. Database (Oxford), baw045. ISSN: 1758-0463. https://doi.org/10.1093/database/baw045.

Drummond, A.J., Suchard, M.A., Xie, D., Rambaut, A., 2012. Bayesian phylogenetics with BEAUti and the BEAST 1.7. Mol. Biol. Evol. 29, 1969–1973. ISSN 1537-1719. https://doi.org/10.1093/molbev/mss075.

Fricke, M., Marz, M., 2016. Prediction of conserved long-range RNA-RNA interactions in full viral genomes. Bioinformatics (Oxford, England). 32, 2928–2935. ISSN 1367-4811. https://doi.org/10.1093/bioinformatics/btw323.

Fricke, M., Dünnes, N., Zayas, M., Bartenschlager, R., Niepmann, M., Marz, M., 2015. Conserved RNA secondary structures and long-range interactions in hepatitis C viruses. RNA 21 (7), 1219–1232. https://doi.org/10.1261/rna.049338.114.

Fricke, M., Zirkel, F., Drosten, C., Junglen, S., Marz, M., 2017. VrAP: full length *de novo* genome assembly of unknown RNA viruses, submitted for publication.

Gardy, J., Loman, N.J., Rambaut, A., 2015. Real-time digital pathogen surveillance—the time is now. Genome Biol. 16, 155. ISSN 1474-760X.

Goodwin, S., McPherson, J.D., McCombie, W.R., 2016. Coming of age: ten years of next-generation sequencing technologies. Nat. Rev. Genet. 17, 333–351. ISSN 1471-0064. https://doi.org/10.1038/nrg.2016.49.

Grabherr, M.G., Haas, B.J., Yassour, M., Levin, J.Z., Thompson, D.A., Amit, I., Adiconis, X., Fan, L., Raychowdhury, R., Zeng, Q., Chen, Z., Mauceli, E., Hacohen, N., Gnirke, A., Rhind, N., di Palma, F., Birren, B.W., Nusbaum, C., Lindblad-Toh, K., Friedman, N., Regev, A., 2011. Full-length transcriptome assembly from RNA-Seq data without a reference genome. Nat. Biotechnol. 29 (7), 644–652. https://doi.org/10.1038/nbt.1883.

Graham, E.A., Henderson, S., Schloss, A., 2011. Using mobile phones to engage citizen scientists in research. Eos Trans. AGU 92 (38), 313–315.

Gregor, I., Schönhuth, A., McHardy, A.C., 2016. Snowball: strain aware gene assembly of metagenomes. Bioinformatics (Oxford, England). 32, i649–i657. ISSN 1367-4811. https://doi.org/10.1093/bioinformatics/btw426.

Gruber, A.R., Lorenz, R., Bernhart, S.H., Neuböck, R., Hofacker, I.L., 2008. The Vienna RNA websuite. Nucleic Acids Res. 36, W70–W74. ISSN 1362-4962. https://doi.org/10.1093/nar/gkn188.

Gurevich, A., Saveliev, V., Vyahhi, N., Tesler, G., 2013. QUAST: quality assessment tool for genome assemblies. Bioinformatics 29 (8), 1072–1075. https://doi.org/10.1093/bioinformatics/btt086.

Hatcher, E.L., Zhdanov, S.A., Bao, Y., Blinkova, O., Nawrocki, E.P., Ostapchuck, Y., Schäffer, A.A., Brister, J.R., 2017. Virus Variation Resource - improved response to emergent viral outbreaks. Nucleic Acids Res. 45, D482–D490. https://doi.org/10.1093/nar/gkw1065.

Henn, M.R., Boutwell, C.L., Charlebois, P., Lennon, N.J., Power, K.A., Macalalad, A.R., Berlin, A.M., Malboeuf, C.M., Ryan, E.M., Gnerre, S., Zody, M.C., Erlich, R.L., Green, L.M., Berical, A., Wang, Y., Casali, M., Streeck, H., Bloom, A.K., Dudek, T., Tully, D., Newman, R., Axten, K.L., Gladden, A.D., Battis, L., Kemper, M., Zeng, Q., Shea, T.P., Gujja, S., Zedlack, C., Gasser, O., Brander, C., Hess, C., Günthard, H.F., Brumme, Z.L., Brumme, C.J., Bazner, S., Rychert, J., Tinsley, J.P., Mayer, K.H., Rosenberg, E., Pereyra, F., Levin, J.Z., Young, S.K., Jessen, H., Altfeld, M., Birren, B.W., Walker, B.D., Allen, T.M., 2012. Whole genome deep sequencing of HIV-1 reveals the impact of early minor variants upon immune recognition during acute infection. PLoS Pathog. 8, e1002529. ISSN 1553-7374. https://doi.org/10.1371/journal.ppat.1002529.

Hofacker, I.L., 2007. RNA consensus structure prediction with RNAalifold. Methods Mol. Biol. 395, 527–544.

Hoffmann, S., Otto, C., Doose, G., Tanzer, A., Langenberger, D., Christ, S., Kunz, M., Holdt, L.M., Teupser, D., Hackermüller, J., Stadler, P.F., 2014. A multi-split mapping algorithm for circular RNA, splicing, trans-splicing and fusion detection. Genome Biol. 15 (2), R34. https://doi.org/10.1186/gb-2014-15-2-r34.

Holmes, E.C., 2009. The Evolution and Emergence of RNA Viruses. Oxford University Press, New York.

Holmes, E.C., 2011. What does virus evolution tell us about virus origins? J. Virol. 85 (11), 5247–5251.

Hölzer, M., Krähling, V., Amman, F., Barth, E., Bernhart, S.H., Carmelo, V.A.O., Collatz, M., Doose, G., Eggenhofer, F., Ewald, J., Fallmann, J., Feldhahn, L.M., Fricke, M., Gebauer, J., Gruber, A.J., Hufsky, F., Indrischek, H., Kanton, S., Linde, J., Mostajo, N., Ochsenreiter, R., Riege, K., Rivarola-Duarte, L., Sahyoun, A.H., Saunders, S.J., Seemann, S.E., Tanzer, A., Vogel, B., Wehner, S., Wolfinger, M.T., Backofen, R., Gorodkin, J., Grosse, I., Hofacker, I., Hoffmann, S., Kaleta, C., Stadler, P.F., Becker, S., Marz, M., 2016. Differential transcriptional responses to Ebola and Marburg virus infection in bat and human cells. Sci. Rep. 6, 34589. https://doi.org/10.1038/srep34589.

Horie, M., Tomonaga, K., 2011. Non-retroviral fossils in vertebrate genomes. Viruses 3 (10), 1836–1848.

Hulo, C., de Castro, E., Masson, P., Bougueleret, L., Bairoch, A., Xenarios, I., Le Mercier, P., 2011. ViralZone: a knowledge resource to understand virus diversity. Nucleic Acids Res. 39 (Database issue), D576–D582. https://doi.org/10.1093/nar/gkq901.

Huson, D.H., Rupp, R., Scornavacca, C., 2011. Phylogenetic Networks: Concepts, Algorithms and Applications. Cambridge University Press, New York, NY. ISBN 0521755964, 9780521755962.

Jain, M., Olsen, H.E., Paten, B., Akeson, M., 2016. The Oxford Nanopore MinION: delivery of nanopore sequencing to the genomics community. Genome Biol. 17, 239. ISSN 1474-760X. https://doi.org/10.1186/s13059-016-1103-0.

Jenkins, G.M., Rambaut, A., Pybus, O.G., Holmes, E.C., 2002. Rates of molecular evolution in RNA viruses: a quantitative phylogenetic analysis. J. Mol. Evol. 54 (2), 156–165.

Kearse, M., Moir, R., Wilson, A., Stones-Havas, S., Cheung, M., Sturrock, S., Buxton, S., Cooper, A., Markowitz, S., Duran, C., Thierer, T., Ashton, B., Meintjes, P., Drummond, A., 2012. Geneious basic: an integrated and extendable desktop software platform for the organization and analysis of sequence data. Bioinformatics (Oxford, England). 28, 1647–1649. ISSN 1367-4811. https://doi.org/10.1093/bioinformatics/bts199.

Kim, D., Pertea, G., Trapnell, C., Pimentel, H., Kelley, R., Salzberg, S.L., 2013. TopHat2: accurate alignment of transcriptomes in the presence of insertions, deletions and gene fusions. Genome Biol. 14 (4), R36. https://doi.org/10.1186/gb-2013-14-4-r36.

Kitchen, A., Shackelton, L.A., Holmes, E.C., 2011. Family level phylogenies reveal modes of macroevolution in RNA viruses. Proc. Natl. Acad. Sci. 108 (1), 238–243.

Kuiken, C., Yusim, K., Boykin, L., Richardson, R., 2005. The Los Alamos hepatitis C sequence database. Bioinformatics (Oxford, England). 21, 379–384. ISSN 1367-4803. https://doi.org/10.1093/bioinformatics/bth485.

Lander, E.S., Linton, L.M., Birren, B., Nusbaum, C., Zody, M.C., Baldwin, J., Devon, K., Dewar, K., Doyle, M., FitzHugh, W., et al., 2001. Initial sequencing and analysis of the human genome. Nature 409 (6822), 860–921.

Lartillot, N., Lepage, T., Blanquart, S., 2009. PhyloBayes 3: a Bayesian software package for phylogenetic reconstruction and molecular dating. Bioinformatics (Oxford, England). 25, 2286–2288. ISSN 1367-4811. https://doi.org/10.1093/bioinformatics/btp368.

Lerat, E., 2010. Identifying repeats and transposable elements in sequenced genomes: how to find your way through the dense forest of programs. Heredity 104 (6), 520–533.
Ling, P., Wang, M., Chen, X., Campbell, K.G., 2007. Construction and characterization of a full-length cDNA library for the wheat stripe rust pathogen (Puccinia striiformis f. sp. tritici). BMC Genomics. 8, 145. ISSN 1471-2164. https://doi.org/10.1186/1471-2164-8-145.
Liu, Y., Wimmer, E., Paul, A.V., 2009. Cis-acting RNA elements in human and animal plus-strand RNA viruses. Biochim Biophys. Acta 1789 (9), 495–517.
Lohmann, V., 2013. Hepatitis C virus RNA replication. Curr. Top. Microbiol. Immunol. 369, 167–198. ISSN 0070-217X. https://doi.org/10.1007/978-3-642-27340-7_7.
Love, M.I., Huber, W., Anders, S., 2014. Moderated estimation of fold change and dispersion for RNA-seq data with DESeq2. Genome Biol. 15 (12), 550. https://doi.org/10.1186/s13059-014-0550-8.
Luciani, F., Bull, R.A., Lloyd, A.R., 2012. Next generation deep sequencing and vaccine design: today and tomorrow. Trends Biotechnol. 30, 443–452. https://doi.org/10.1016/j.tibtech.2012.05.005.
Luo, R., Liu, B., Xie, Y., Li, Z., Huang, W., Yuan, J., He, G., Chen, Y., Pan, Q., Liu, Y., Tang, J., Wu, G., Zhang, H., Shi, Y., Liu, Y., Yu, C., Wang, B., Lu, Y., Han, C., Cheung, D.W., Yiu, S.M., Peng, S., Xiaoqian, Z., Liu, G., Liao, X., Li, Y., Yang, H., Wang, J., Lam, T.W., Wang, J., 2012. SOAPdenovo2: an empirically improved memory-efficient short-read *de novo* assembler. GigaScience. 1, 18. ISSN 2047-217X. https://doi.org/10.1186/2047-217X-1-18.
Madhugiri, R., Fricke, M., Marz, M., Ziebuhr, J., 2014. RNA structure analysis of alphacoronavirus terminal genome regions. Virus Res. 194, 76–89. https://doi.org/10.1016/j.virusres.2014.10.001.
Mardis, E.R., 2017. DNA sequencing technologies: 2006–2016. Nat. Protoc. 12, 213–218. ISSN 1750-2799. https://doi.org/10.1038/nprot.2016.182.
Marschall, T., Marz, M., Abeel, T., Dijkstra, L., Dutilh, B.E., Ghaffaari, A., Kersey, P., Kloosterman, W., Makinen, V., Novak, A., et al., 2016. Computational pan-genomics: status, promises and challenges. BioRxiv 043430.
Marz, M., Beerenwinkel, N., Drosten, C., Fricke, M., Frishman, D., Hofacker, I.L., Hoffmann, D., Middendorf, M., Rattei, T., Stadler, P.F., Töpfer, A., 2014. Challenges in RNA virus bioinformatics. Bioinformatics 30 (13), 1793–1799. https://doi.org/10.1093/bioinformatics/btu105.
McCarthy, E.M., McDonald, J.F., 2003. LTR_STRUC: a novel search and identification program for LTR retrotransposons. Bioinformatics 19 (3), 362–367.
Mortazavi, A., Williams, B.A., McCue, K., Schaeffer, L., Wold, B., 2008. Mapping and quantifying mammalian transcriptomes by RNA-Seq. Nat. Methods 5 (7), 621–628.
Peng, Y., Leung, H.C.M., Yiu, S., Chin, F.Y.L., 2012. IDBA-UD: a *de novo* assembler for single-cell and metagenomic sequencing data with highly uneven depth. Bioinformatics 28 (11), 1420–1428. https://doi.org/10.1093/bioinformatics/bts174.
Pickett, B.E., Sadat, E.L., Zhang, Y., Noronha, J.M., Squires, R.B., Hunt, V., Liu, M., Kumar, S., Zaremba, S., Gu, Z., Zhou, L., Larson, C.N., Dietrich, J., Klem, E.B., Scheuermann, R.H., 2012. ViPR: an open bioinformatics database and analysis resource for virology research. Nucleic Acids Res. 40 (Database issue), D593–D598. https://doi.org/10.1093/nar/gkr859.
Pulido-Tamayo, S., Sánchez-Rodríguez, A., Swings, T., Van den Bergh, B., Dubey, A., Steenackers, H., Michiels, J., Fostier, J., Marchal, K., 2015. Frequency-based haplotype reconstruction from deep sequencing data of bacterial populations. Nucleic Acids Res. 43, e105. https://doi.org/10.1093/nar/gkv478.
Qin, J., Li, Y., Cai, Z., Li, S., Zhu, J., Zhang, F., Liang, S., Zhang, W., Guan, Y., Shen, D., et al., 2012. A metagenome-wide association study of gut microbiota in type 2 diabetes. Nature 490 (7418), 55–60.

Quick, J., Loman, N.J., Duraffour, S., Simpson, J.T., Severi, E., Cowley, L., Bore, J.A., Koundouno, R., Dudas, G., Mikhail, A., Ouédraogo, N., Afrough, B., Bah, A., Baum, J.H.J., Becker-Ziaja, B., Boettcher, J.P., Cabeza-Cabrerizo, M., Camino-Sánchez, A., Carter, L.L., Doerrbecker, J., Enkirch, T., García-Dorival, I., Hetzelt, N., Hinzmann, J., Holm, T., Kafetzopoulou, L.E., Koropogui, M., Kosgey, A., Kuisma, E., Logue, C.H., Mazzarelli, A., Meisel, S., Mertens, M., Michel, J., Ngabo, D., Nitzsche, K., Pallasch, E., Patrono, L.V., Portmann, J., Repits, J.G., Rickett, N.Y., Sachse, A., Singethan, K., Vitoriano, I., Yemanaberhan, R.L., Zekeng, E.G., Racine, T., Bello, A., Sall, A.A., Faye, O., Faye, O., Magassouba, N., Williams, C.V., Amburgey, V., Winona, L., Davis, E., Gerlach, J., Washington, F., Monteil, V., Jourdain, M., Bererd, M., Camara, A., Somlare, H., Camara, A., Gerard, M., Bado, G., Baillet, B., Delaune, D., Nebie, K.Y., Diarra, A., Savane, Y., Pallawo, R.B., Gutierrez, G.J., Milhano, N., Roger, I., Williams, C.J., Yattara, F., Lewandowski, K., Taylor, J., Rachwal, P., Turner, D.J., Pollakis, G., Hiscox, J.A., Matthews, D.A., O'Shea, M.K., Johnston, A.M., Wilson, D., Hutley, E., Smit, E., Di Caro, A., Wölfel, R., Stoecker, K., Fleischmann, E., Gabriel, M., Weller, S.A., Koivogui, L., Diallo, B., Keïta, S., Rambaut, A., Formenty, P., Günther, S., Carroll, M.W., 2016. Real-time, portable genome sequencing for Ebola surveillance. Nature. 530, 228–232. ISSN 1476-4687. https://doi.org/10.1038/nature16996.

Rambaut, A., 2000. Estimating the rate of molecular evolution: incorporating non-contemporaneous sequences into maximum likelihood phylogenies. Bioinformatics 16 (4), 395–399.

Rhoads, A., Au, K.F., 2015. PacBio sequencing and its applications. Genomics Proteomics Bioinformatics 13 (5), 278–289.

Ronquist, F., Huelsenbeck, J.P., 2003. MrBayes 3: Bayesian phylogenetic inference under mixed models. Bioinformatics (Oxford, England). 19, 1572–1574. ISSN 1367-4803.

Sanjuan, R., Nebot, M.R., Chirico, N., Mansky, L.M., Belshaw, R., 2010. Viral mutation rates. J. Virol. 84 (19), 9733–9748.

Scheuch, M., Höper, D., Beer, M., 2015. RIEMS: a software pipeline for sensitive and comprehensive taxonomic classification of reads from metagenomics datasets. BMC Bioinf. 16, 69. ISSN 1471-2105. https://doi.org/10.1186/s12859-015-0503-6.

Schmieder, R., Edwards, R., 2011. Quality control and preprocessing of metagenomic datasets. Bioinformatics. 27, 863–864. ISSN 1367-4811. https://doi.org/10.1093/bioinformatics/btr026.

Schmutz, J., Wheeler, J., Grimwood, J., Dickson, M., Yang, J., Caoile, C., Bajorek, E., Black, S., Chan, Y.M., Denys, M., et al., 2004. Quality assessment of the human genome sequence. Nature 429 (6990), 365–368.

Schulz, M.H., Zerbino, D.R., Vingron, M., Birney, E., 2012. Oases: robust *de novo* RNA-Seq assembly across the dynamic range of expression levels. Bioinformatics (Oxford, England). 28, 1086–1092. ISSN 1367-4811. https://doi.org/10.1093/bioinformatics/bts094.

Sharp, P.M., Simmonds, P., 2011. Evaluating the evidence for virus/host co-evolution. Curr. Opin. Virol. 1 (5), 436–441.

Shu, Y., McCauley, J., 2017. GISAID: global initiative on sharing all influenza data - from vision to reality. Euro surveill. 22, 30494. ISSN 1560-7917. https://doi.org/10.2807/1560-7917.ES.2017.22.13.30494.

Simpson, J.T., Wong, K., Jackman, S.D., Schein, J.E., Jones, S.J.M., Birol, I., 2009. ABySS: a parallel assembler for short read sequence data. Genome Res. 19, 1117–1123. ISSN 1088-9051. https://doi.org/10.1101/gr.089532.108.

Smit, A.F.A., Hubley, R., Green, P., n.d. Repeat masker open-3.0 http://www.repeatmasker.org. Accessed 2017.

Sperber, G., Lövgren, A., Eriksson, N.E., Benachenhou, F., Blomberg, J., 2009. Retrotector online, a rational tool for analysis of retroviral elements in small and medium size vertebrate genomic sequences. BMC Bioinf. 10 (6), S4.

Stamatakis, A., Hoover, P., Rougemont, J., 2008. A rapid bootstrap algorithm for the RAxML web servers. Syst. Biol. 57, 758–771. ISSN 1076-836X. https://doi.org/10.1080/10635150802429642.

Thorvaldsdóttir, H., Robinson, J.T., Mesirov, J.P., 2013. Integrative Genomics Viewer (IGV): high-performance genomics data visualization and exploration. Brief. Bioinf. 14, 178–192. ISSN 1477-4054. https://doi.org/10.1093/bib/bbs017.

Töpfer, A., Höper, D., Blome, S., Beer, M., Beerenwinkel, N., Ruggli, N., Leifer, I., 2013. Sequencing approach to analyze the role of quasispecies for classical swine fever. Virology 438, 14–19. ISSN 1096-0341. https://doi.org/10.1016/j.virol.2012.11.020.

Töpfer, A., Marschall, T., Bull, R.A., Luciani, F., Schönhuth, A., Beerenwinkel, N., 2014. Viral quasispecies assembly via maximal clique enumeration. PLoS Comput. Biol. 10, e1003515. ISSN 1553-7358. https://doi.org/10.1371/journal.pcbi.1003515.

Weiss, R., 2003. Cross-species infections. In: Salomon, D.R., Wilson, C. (Eds.), In: Xeno-Transplantation Springer, Berlin Heidelberg, Germany, pp. 47–71.

Will, S., Reiche, K., Hofacker, I.L., Stadler, P.F., Backofen, R., 2007. Inferring noncoding RNA families and classes by means of genome-scale structure-based clustering. PLoS Comput. Biol. 3 (4), e65.

Yang, X., Charlebois, P., Gnerre, S., Coole, M.G., Lennon, N.J., Levin, J.Z., Qu, J., Ryan, E.M., Zody, M.C., Henn, M.R., 2012. *De novo* assembly of highly diverse viral populations. BMC Genomics 13, 475. https://doi.org/10.1186/1471-2164-13-475.

Zagordi, O., Däumer, M., Beisel, C., Beerenwinkel, N., 2012. Read length versus depth of coverage for viral quasispecies reconstruction. PLoS One. 7, e47046. ISSN 1932-6203. https://doi.org/10.1371/journal.pone.0047046.

Zerbino, D.R., Birney, E., 2008. Velvet: algorithms for *de novo* short read assembly using *de Bruijn* graphs. Genome Res. 18, 821–829. ISSN 1088-9051. https://doi.org/10.1101/gr.074492.107.

Zuker, M., 2003. Mfold web server for nucleic acid folding and hybridization prediction. Nucleic Acids Res. 31, 3406–3415.

FSC
www.fsc.org
MIX
Papier | Fördert
gute Waldnutzung
FSC® C083411

Kolibri 360 GmbH
Ferdinand-Jühlke-Straße 7
99095 Erfurt, Deutschland
produktsicherheit@kolibri360.de